Food and Feed from Legumes and Oilseeds

Food and Feed from Legumes and Oilseeds

Edited by

E. Nwokolo

Lifeline Associates, Maryland, USA

and

J. Smartt

Department of Biology, University of Southampton, UK

CHAPMAN & HALL

London · Glasgow · Weinheim · New York · Tokyo · Melbourne · Madras

Published by Chapman & Hall, 2–6 Boundary Row, London SE1 8HN

Chapman & Hall, 2–6 Boundary Row, London SE1 8HN, UK

Blackie Academic & Professional, Wester Cleddens Road, Bishopbriggs, Glasgow G64 2NZ, UK

Chapman & Hall GmbH, Pappelallee 3, 69469 Weinheim, Germany

Chapman & Hall USA, 115 Fifth Avenue, New York, NY 10003, USA

Chapman & Hall Japan, ITP-Japan, Kyowa Building, 3F, 2–2–1 Hirakawacho, Chiyoda-ku, Tokyo 102, Japan

Chapman & Hall Australia, 102 Dodds Street, South Melbourne, Victoria 3205, Australia

Chapman & Hall India, R. Seshadri, 32 Second Main Road, CIT East, Madras 600 035, India

---

First edition 1996

Typeset in 10/12 Palatino by Words & Graphics Ltd, Anstey, Leicester

ISBN-13:978-1-4613-8050-4 e-ISBN-13:978-1-4613-0433-3
DOI:10.1007/978-1-4613-0433-3

A catalogue record for this book is available from the British Library.

Library of Congress Catalog Card Number: 95–74640

∞ Printed on permanent acid-free text paper, manufactured in accordance with ANSI/NISO Z39.48–1992 and ANSI/NISO Z39.48–1984 (Permanence of Paper).

Contents

Contributors

S.C. Achinewhu, Department of Food Science and Technology, Rivers State University of Science and Technology, Port Harcourt, Nigeria

R.N. Adsule, Department of Biochemistry, Mahatma Phule Agricultural University, Rahari, Maharashtra, India

M. Akpapunam, Department of Food Science and Technology, University of Agriculture, Makurdi, Nigeria

N.M. de Carvalho, Departmento de Fitotechnica, FCA/UNESP 14870, Jabotical, SP, Brazil

B.O. de Lumen, Department of Nutritional Sciences, University of California, Berkeley, CA 94720, USA

S.N. Ilechukwu, Nestlé Foods Nigeria plc, 19/21 Industrial Avenue, Ilupeju, Lagos, Nigeria

K.D. Kostov, Agricultural University, Plovdiv, Bulgaria

L. Lennerts, Hahnenfeldstrasse 9a, 8939 Bad Worishofen, Germany

W. Lennerts, Hahnenfeldstrasse 9a, 8939 Bad Worishofen, Germany

S.S. Negi, Indian Veterinary Research Institute, Regional Station, Palampur (H.P.), 176061, India

E. Nwokolo, Lifeline Associates, 9664 Canterbury Riding, Laurel, MD 20723, USA

O.C. Onwudike, Department of Animal Production, Federal University of Technology, Owerri, Nigeria

D.C. Pavlov, University of Zootechnics and Veterinary Medicine, Stara Zagora, Bulgaria

S.K. Sathe, Department of Nutrition, Food and Movement Sciences, College of Human Sciences, Florida State University, Tallahassee, FL 32306–2033, USA

A. Sotelo, Departamento de Farmacia, Division de Estudios de Posgraduados, Facultad de Quimica, Ciudad Universitaria, 04510 Mexico, D.F. Mexico

N.A. Todorov, University of Zootechnics and Veterinary Medicine, Stara Zagora, Bulgaria

A.C. Uwaegbute, Department of Home Science and Nutrition, University of Nigeria, Nsukka, Nigeria

R.D. Vieira, Departmento de Fitotechnica, FCA/UNESP 14870, Jabotical, SP, Brazil

PART ONE

Legumes

CHAPTER ONE

The need to increase consumption of pulses in the developing world

E. Nwokolo

1.1 INTRODUCTION

For most people in the developing countries of Africa, Asia and Latin America, the 1980s have been a very tumultuous period, a decade in which these people have seen their standards of living dangerously eroded, even devastated, by a steep decline in both real income and buying power. In country after country, devaluation of the local currency in order to meet the International Monetary Fund (IMF) conditions for restructuring loans, to spur foreign investments or to encourage exports, has led to tremendous loss of buying power and left many families reeling under immense economic pressure. For example, a kilogram of beef costs about one tenth of the monthly salary of a janitor in Nigeria, about one fifth of the monthly salary of a similar janitor in Sierra Leone, but costs less than the hourly wage of a janitor in the United States. It has been calculated that a home-cooked breakfast for four children and two adults comprising six large eggs, six medium sausages, toast, milk and coffee would cost about $10 in the US but over N200 in Nigeria, more than half the entire monthly wages of this hypothetical Nigerian janitor. As a consequence of this outrageous increase in local prices of food (especially the price of meat, milk, fish, cheese and eggs), cowpeas, common beans, pigeon pea and other pulses are enjoying a resurgence in interest and an enhanced level of consumption. This is no temporary or seasonal resurgence because food shortages and high prices in Africa seem to be as endemic as malaria and mosquitoes, as common as drought and famine.

Legumes and Oilseeds in Nutrition
Edited by E. Nwokolo and J. Smartt
Published in 1996 by Chapman & Hall. ISBN 0 412 45930 2

Lloyd Timberlake (1985) notes that in a 'normal year', up to 100 million Africans are malnourished and severely hungry. Thus the superimposition of a drought in the 1980s triggered disaster in a continent already on the brink. He notes that almost all African countries were affected by the drought of the early 1980s, 19 countries being adversely affected and nine very adversely. According to Timberlake (1985), the drought affected 4.5 million people in the Sudan, almost 8 million in Ethiopia, about 1.5 million in Chad with another 0.7 million Chadians moving to other countries as refugees, nearly 3 million in Niger, about 1.2 million in Mali, almost 2.5 million in Mozambique, 1.1 million in Mauritania and about 0.5 million in both Burkina Faso and Angola.

1.2 PULSES AS FOOD

Pulses, as distinct from oilseeds, are legumes which are planted and harvested primarily for their mature or immature seeds, a significant source of dietary protein and carbohydrates, but not primarily oil. Whereas oilseed cakes are used to a large extent in feeding of livestock, legume pulses are consumed primarily by humans. Excellent reviews of the effects of legumes in the human diet have been written by Bressani (1975), Walker (1982), Norton *et al.* (1985) and Sgarbieri (1989). Walker (1982) observes that legumes are a valuable source of protein, calcium, iron, thiamine and riboflavin in poorer areas of the world. For some reason, in many parts of the world especially in the tropics, consumption of pulses is associated with poverty, although it is accepted in India where religion or local customs prevent consumption of meat or dairy products. Since most legume seeds contain less than optimal amounts of sulphur amino acids, there is a tendency to consider them as nutritionally inferior to meat and fish, in spite of the very many health-promoting properties of properly cooked legumes.

1.3 PULSES AS SOURCES OF PROTEIN

Protein content of dry legume seeds varies depending on whether the legume is an oilseed or a pulse. In comparison with oilseeds, most pulses contain very little oil, relatively similar amounts of protein and larger amounts of carbohydrates. Winged bean and lupins are, however, pulses which contain large amounts of lipids (16–17%). The United States Department of Agriculture (USDA) publication titled *Nutrient Composition of Foods: Legumes* (USDA, 1986) provides in-depth information on the nutrient composition of over 30 pulses and oilseeds. For most pulses the protein content is in the range 19–25% with an average value of about 22%, whereas for most oilseeds protein content ranges from 20 to 35%, with an average value of about 30%.

1.3.1 Legume proteins

Legume proteins may be classified into three basic groups: (1) globulins representing about 70% of total proteins; (2) albumins which account for 10–20% of total proteins; and (3) glutelins which also account for 10–20% of total proteins of seeds. One of two types of globulins usually predominates in most legume seeds: vicilin which sediments at 7S or legumin at 11S. Another globulin, identified from *Phaseolus vulgaris* is phaseolin (or GI globulin or glycoprotein II). There are three types of phaseolin, namely: phaseolin T, phaseolin S and phaseolin C. Associated with the vicilin fraction of globulin is the protein convicilin isolated from faba beans. Other proteins found in legume seeds include various enzymes, protease and carbohydrase inhibitors, and lectins (haemagglutinins). Most of these proteins are water-soluble, functional proteins as distinct from storage protein and may comprise a significant proportion of the albumin fraction of the seed protein. The various protein fractions are composed of essential and non-essential amino acids, some fractions containing more of certain individual essential amino acids than others. In general, various isolated protein fractions contain higher concentrations of essential amino acids than the entire seed. Amino acid analyses of grain legumes have clearly shown a less than optimal content of the sulphur amino acids cystine and methionine in all food legumes and a deficiency of tryptophan in some. Three approaches to optimizing the nutritional qualities of grain legume proteins have been suggested (Young *et al.*, 1979; Bressani, 1975). Amino acid requirements can be met by consumption of large amounts of these legumes (Young *et al.*, 1979). By increasing intake of the legume, the minimum requirement of a deficient amino acid may still be met. Another approach to enhance utilization of legumes is to feed a mixture of legumes including those with fairly large amounts of sulphur amino acids and those which contain less sulphur amino acids. Bressani (1975) reports nutritional programmes in Brazil and elsewhere in Latin America where mixtures of soybean and the common bean have improved nutrient intake in vulnerable populations. A third approach is to employ the complementarity which exists between high sulphur amino acid cereals and legumes, to supplement the deficiency in the sulphur amino acids.

1.3.2 Protein digestibility

Digestibility of raw legume seeds and products is usually very poor, improving with cooking, autoclaving, toasting or other forms of heat treatment. In general even well-cooked legumes still have poorer protein digestibility than digestibility of animal protein. This is partly due to the presence of protease inhibitors (Tobin and Carpenter, 1978) some residual and others heat-stable. Poor digestibility may partly be due to a deficiency of the sulphur amino acids, thus hindering protein utilization (Bressani 1975; Tobin and Carpenter, 1978). It may also be due to the presence of polyphenols and other antimetabolites or due to the tertiary structure of native proteins making them refractory to enzymatic proteolysis (Tobin and Carpenter, 1978). Whatever the mechanisms, it is undeniable that most pulses have low-to-moderate digestibility. For example Jaffé (1950) reported true digestibility coefficients of 76.8, 79.5 and 84.1% respectively for the protein of black, pink and white *Phaseolus* beans. Various less-than-optimal digestibility coefficients reported for often-consumed grain legumes include the following data reported by Achinewhu (1983) for climbing melon (*Colocynthis vulgaris*) 83%, creeping melon (*Citrullus vulgaris*) 82% and African breadfruit (69%). This was in comparison with casein which had a digestibility of 92%. De-Romana *et al.* (1983) reported nitrogen digestibility coefficients determined with children fed lupin flour or a casein control diet. Digestibility was 81.8% for lupin flour and 87.2% for the casein control. It is important that this less-than-optimal digestibility of legumes be taken into consideration when one is attempting to meet nutritional requirements of humans with diets which are essentially legume-based.

1.4 PULSES AS SOURCES OF CARBOHYDRATE

In most pulses, the major component of the seed is the carbohydrate fraction which usually comprises about 70% of the seed by weight. Starch is the principal carbohydrate, with minor amounts of the smaller molecular weight carbohydrates such as sucrose and sucrosyl oligosaccharides. The oligosaccharides include raffinose, stachyose and verbascose which are present in varying amounts in most legume seeds and are associated with flatulence. Flatus formation is prevalent following consumption of pulses and is a major hindrance to large-scale acceptance of legumes as food. Digestibility of energy in legumes is measured as digestible energy or as metabolizable energy. Metabolizable energy which takes into account metabolic energy losses is easily determined with experimental animals and rather difficult to determine with human subjects. Digestible energy is therefore a more common measure of utilizable energy in humans. Digestible energy coefficients for most legumes as well as mixed diets containing legumes are generally between 85 and 90% of the gross energy of the dry legume seed whereas metabolizable energy values are usually 75–85% of the gross energy of the dry legume seed.

1.5 PULSE LEGUMES AS A SOURCE OF DIETARY FIBRE

Various worldwide health organizations including the American National

Cancer Institute recommend increased consumption of fibre to prevent the onset of various diverticular and degenerative diseases. Suggested daily levels in hospitals, clinics, weight-loss centres and the popular press range from 25 to 50 g of fibre. Whereas most legumes are an excellent source of dietary fibre, most cereals are not, although processed breakfast cereals which contain enhanced levels of bran are excellent sources of dietary fibre. Dietary fibre content of dietary legumes may range from as low as 6.1% in peanuts to as high as 25.5% in kidney beans and mung beans (Paul and Southgate, 1978; Kamath and Belavady, 1980). Dietary fibre has been implicated in many desirable physiological roles in the human body. Based on their study of the relative frequencies of occurrences of the so-called 'diseases of affluence' in Africa as compared with industrialized societies Burkitt and Trowell (1975) hypothesized that the high content of dietary fibre in African diets was responsible for the absence of these diseases among Africans. Thus Burkitt and Trowell (1975) linked low dietary fibre intake in 'western diets' to increased incidence of cancer of the colon and rectum, diverticular disease, coronary heart disease, diabetes and gallstones in these societies. Over the next 10 years researchers established strong relationships between an increase in fibre consumption and a decrease in the incidence of these and other degenerative diseases. It is these results which prompted the American National Cancer Institute and the American Medical Association to recommend increased consumption of fibre as part of the American diet. Whole legumes are good sources of dietary fibre. Most legumes can also be processed to obtain an even more concentrated form of dietary fibre. A concentrated source of dietary fibre from soybean, which is neither hull nor bran, is obtained by processing dehulled, defatted soybean flakes. Slavin (1988) estimates the dietary fibre content of this product at 66–75% as determined by using various methods and observes that whereas it takes only 13 g of soy fibre to provide 10 g of dietary fibre in food, it takes 58 g of oat bran, 23 g of wheat bran, 735 g of lettuce or 502 g of apple to provide 10 g of fibre. Slavin (1988) summarizes the physiological benefits of soy fibre and other dietary fibre sources as:

1. increasing faecal bulk and faecal moisture;
2. reducing plasma cholesterol in subjects with mildly elevated cholesterol and in Type IIA hypercholesterolaemic patients;
3. being a positive influence on blood glucose and insulin concentration in both normal and diabetic patients.

1.6 HYPOCHOLESTEROLAEMIC EFFECTS OF GRAIN LEGUMES

A considerable body of epidemiological data has accumulated over the last 30 years linking the type of diet consumed with the incidence of coronary heart disease. Serum cholesterol, which is directly influenced by diet has been directly implicated as a major factor in coronary heart disease (Kritchevsky, 1977). There is incontrovertible evidence that the type of diet consumed has a considerable effect on serum cholesterol. Interest in this relationship was aroused by the initial observation 30 years ago (Groen *et al.*, 1962) that the serum cholesterol levels of Trappist monks consuming large quantities of legumes (100–150 g/day) was significantly lower than the cholesterol levels in Benedictine monks consuming 'western-type' diets. Since then, several researchers have shown that consumption of various legume diets was associated with a decrease in serum cholesterol. These legumes include black gram (Devi and Kurup, 1972), bengal gram (Mathur *et al.*, 1964, 1968) as well as various leguminous seeds (Grande *et al.*, 1965). This hypocholesterolaemic effect was attributable to the dietary fibre fraction of legumes (Cummings, 1978; Hellendoorn, 1979) because of its high content of pectins, gums and galactans. Dietary fibre has also been shown to absorb bile salts (Eastwood and Hamilton, 1968). Oakenfull and co-workers (Oakenfull and Fenwick, 1978; Oakenfull *et al.*, 1979) showed that saponins are essential for the absorption of bile salts by plant fibres. Since legumes tend to be higher in saponins than other vegetable foods, this may partly explain their pronounced hypocholesterolaemic effects.

1.7 ANTINUTRITIONAL FACTORS IN SOME PULSES

Most raw legume products contain some antinutritional factors. These are chemical substances which, although non-toxic, generate adverse physiological responses in animals consuming these legumes and, in many cases, interfere with the utilization of nutrients in these products. Antinutritional factors which have been identified in soybean include protease inhibitors, haemagglutinins (lectins), goitrogens, antivitamins and phytates. Others include saponins, oestrogens, flatulence factors, allergens and lysinoalanine (Liener, 1981). In addition to those mentioned above, other antinutritional factors have been identified in other legumes. These include cyanogens, favism factors, lathyrism factors, amylase inhibitors, tannins, aflatoxin and pressor amines. Although only a few legume seeds may contain all these antinutritionals, many contain a few of them. Walker (1982) notes that haemagglutinins, aflatoxin, cyanogens, favism factors and lathyrism factors pose the greatest threat to the health and well-being of humans and animals. Some antinutritional factors are heat labile, including the protease inhibitors, haemagglutinins, goitrogens, antivitamins and phytates. These are inactivated or destroyed by heat; the degree of destruction usually being related to the intensity and duration of heat application (Jaffé and Vega Lette, 1968). Complete thermal destruction of antinutritionals, however, increases the danger of heat-damage to dietary proteins. The chemistry, types, significance and mechanisms of action of protease inhibitors have been extensively researched (Liener, 1966, 1973, 1979; Birk 1976a, b). Haemagglutinins (lectins) bind to the carbohydrate moiety of some compounds; for example, the glycoprotein moiety located on the surface of red blood cells, causing the red cells to clump together. Whereas most lectins are relatively harmless and can be destroyed by mild heat treatment, others like ricin in castor bean are extremely toxic to animals and humans. Indeed, Liener (1981) suggests that the lectin of soybean plays only a minor role in determining the nutritional quality of soybean protein. This seems to be the case with most edible legumes which also contain very low lectin contents. Goitrogenic and rachitogenic effects have been reported following the consumption of raw soybeans by experimental animals, effects which can be eliminated by autoclaving or application of moist heat as in cooking (Liener, 1981). Since most of these antinutritional factors are heat labile, and since humans only consume legume seeds after cooking, it would appear that they do not constitute any major health hazard as long as the legume seeds or legume products are properly cooked.

Given the scientific information that is currently available on the health-promoting effects as well as the excellent physiological and nutritional qualities of pulses, it is only prudent to counsel the increased production and consumption of these grain legumes worldwide. It is, however, of extreme importance that governments in developing countries pay special attention to encouraging programmes which advocate increased production and consumption of pulses. As is common usage everywhere, it is the right thing to do.

REFERENCES

Achinewhu, S.C. (1983) The nutritive qualities of plant foods. 2. Protein quality of breadfruit (*Artocarpus altilis*), climbing melon (*Colocynthis vulgaris*) and creeping melon (*Citrullus vulgaris*). *Nutr. Rep. Int.*, **27**, 541–5.

Birk, Y. (1976a) Proteinase inhibitors from plant sources. *Methods Enzymol.* **45**, 695–7.

Birk, Y. (1976b) Proteinase inhibitors from legume seeds. *Methods Enzymol.*, **45**, 697–700.

Bressani, R. (1975) Legumes in human diets and how they might be improved, in *Nutritional Improvement of Food Legumes by Breeding*, (ed. M. Milner), Wiley, New York.

Bressani, R., Elias, L.G. and Molina, M. R. (1977) Protein digestability of some legume foods. *Arch. Latinoam. Nutr.*, **27**, 215–31.

Burkitt, D.P. and Trowell, H.C. (1975) *Refined Carbohydrate Foods and Disease*, Academic Press, London.

Cummings, J.H. (1978) Nutritional implications of dietary fiber. *Am. J. Clin. Nutr.*, **31**, 521–7.

De-Romana, G.L., Graham, G.G., Morales, E. *et al.* (1983) Protein quality and oil digestibility of *Lupinus mutabilis* Metabolic studies in children. *J. Nutr.*, **113**, 773–8.

Devi, K.S. and Kurup, P.A. (1972) Hypolipidemic activity of *Phaseolus mungo* L. (black gram) in rats fed a high-fat, high-cholesterol diet. *Atherosclerosis*, **15**, 223–30.

Eastwood, M.A. and Hamilton, D. (1968) Studies on the adsorbtion of bile salts to non-absorbed components of diet. *Biochim. Biophys. Acta*, **152**, 165–73.

Grande, F., Anderson, J.T. and Keys, A. (1965) Effects of carbohydrates of leguminous seeds, wheat and potato on serum cholesterol concentration in man. *J. Nutr.*, **86**, 313–17.

Groen, J.J., Tijong, K.B., Kostev, M. *et al.* (1962) The influence of nutrition and ways of life on blood cholesterol and the prevalence of hypertension and coronary heart disease among Trappist and Benedictine monks. *Am. J. Clin. Nutr.*, **10**, 456–70.

Hellendoorn, E.W. (1979) Beneficial physiological activity of leguminous seeds. *Qual. Plant/Plant Foods Human Nutr.*, **29**, 227–35.

Jaffe, W.G. (1950) Protein digestibility and trypsin inhibitor activity of legume seeds. *Proc. Soc. Exp. Biol. Med.*, **75**, 219–20.

Jaffe, W.G. and Vega Lette, C.L. (1968) Heat-labile growth inhibitory factors in beans (*Phaseolus vulgaris*). *J. Nutr.*, **94**, 203–10.

Kamath, M.V. and Belavady, B. (1980) Unavailable carbohydrates of commonly consumed Indian foods. *J. Sci. Food Agric.*, **31**, 194–202.

Kritchevsky, D. (1977) Diet and cholesterolemia. *Lipids*, **12**, 49–52.

Liener, I.E. (1966) Toxic substances associated with seed proteins, in *World Protein Resources*. Advanced Chem. Series no. 57, pp. 178–94.

Liener, I.E. (1973) Antitryptic and other antinutritional factors in legumes, in *Nutritional Improvement of Food Legumes by Breeding*, (ed. M. Milner), Protein Advisory Group, New York.

Liener, I.E. (1979) Significance for humans of biologically active factors in soybeans and other food legumes. *J. Am. Oil Chem. Soc.*, **56**, 121–9.

Liener, I.E. (1981) Factors affecting the nutritional quality of soya products. *J. Am. Oil Chem. Soc.*, **58**, 406–15.

Mathur, K.S., Singhal, S.S. and Sharma, R.D. (1964) Effect of bengal gram on experimentally induced high levels of cholesterol in tissues and serum of albino rats. *J. Nutr.*, **84**, 201–4.

Mathur, K.S., Singhal, S.S. and Sharma, R.D. (1968) Hypocholesterolemic effect of bengal gram: a long term study in man. *Br. Med. J.*, **1**, 30–1.

Norton G., Bliss F.A. and Bressani R. (1985) Biochemical and nutritional attributes of grain legumes, in *Grain Legume Crops*. (eds R.J. Summerfield and E.H. Roberts), Collins Professional and Technical Books, London.

Oakenfull, D.G. and Fenwick, D.E. (1978) Adsorption of bile salts from aqueous solution by plant fiber and cholestyramine. *Br. J. Nutr.*, **40**, 299–309.

Oakenfull, D.G., Fenwick, D.E., Good, R.L. *et al.* (1979) Effects of saponins on bile acids and plasma lipids in rats. *Br. J. Nutr.*, **42**, 209–316.

Paul, A.A. and Southgate, D.A.T. (1978) *McCance and Widdowson's The Composition of Foods*, 4th Revised and Extended Edition of MRC Special Report no. 297. HMSO, London.

Sgarbieri, V.C. (1989) Composition and nutritive value of beans (*Phaseolus vulgaris* L.). *World Rev. Nutr. Diet.*, **60**, 132–98.

Slavin, J.L. (1988) Nutritional value of soybean fiber, in *Soybean Utilization Alternatives*. Proceedings of a conference by The Center for Alternative Crops and Products, (ed. L. McCann). University of Minnesota, St Paul, Feb. 16–19.

Timberlake, L. (1985) *Africa in Crisis: the Causes, the Cures of Environmental Bankruptcy*. An Earthscan Book, International Institute for Environment and Development, London, Washington, DC.

Tobin, G. and Carpenter, K.J. (1978) The nutritional value of the dry bean (*Phaseolus vulgaris*): a literature review. *Nutr. Abstr. Rev.*, **48**, 919–36.

USDA (1986) *Composition of Foods: Legumes and Legume Products*. Human Nutrition Information Service. Agriculture Handbook No. 8–16 United States Department of Agriculture, Washington, DC.

Walker, A.F. (1982) Physiological effects of legumes in the human diet: a review. *J. Plant Foods*, **4**, 5–14.

Young, V.R., Scrimshaw, N.S., Torun, B. and Vitieri, F. (1979) Soybean protein in human nutrition: an overview. *J. Am. Oil Chem. Soc.*, **56**(3), 110–20.

CHAPTER TWO

The nutritional value of selected Asiatic pulses: chickpea, black gram, mung bean and pigeon pea

S.K. Sathe

2.1 INTRODUCTION

Plants are an important source of nutrients in the human diet. Over 3000 plant species have been used as human food throughout history (Borlaug, 1981). Of these, at least 150 different species are grown in sufficient quantity and are thus important in world food trade. Cereals constitute the single most important group of food plants in the human diet (excluding the large number of species of fruits and vegetables). Legumes are next to cereals in terms of their economic and nutritional importance as human food resources. They are valued not only for their protein and carbohydrate contents (just as cereals) but also because of the oil content of oilseed legumes such as soybeans and peanuts. The ability of legumes to fix atmospheric nitrogen in the soil–crop ecosystem is one of their unique and beneficial characteristics among all plant species. The amount of nitrogen fixed in the soil equals the amount of nitrogen input from commercial fertilizers (Delwiche, 1978).

In the underdeveloped and developing countries, legumes are an important source of proteins and in some areas they are the major source of dietary proteins. In addition to being an excellent and relatively inexpensive source of dietary proteins, legumes possess several desirable attributes. Among these are included their low sodium content, high

Legumes and Oilseeds in Nutrition
Edited by E. Nwokolo and J. Smartt
Published in 1996 by Chapman & Hall. ISBN 0 412 45930 2

potassium content, abundance of complex carbohydrates, ability to lower serum cholesterol in humans, high fibre content, low fat content (excluding the oilseeds), high concentration of polyunsaturated fatty acids (particularly the essential fatty acids linoleic and linolenic), long shelf-life, and the diversity of foods that can be made from them. In addition, legumes contain several bioactive compounds whose beneficial effects in human health remain largely unexplored. Despite these qualities, legumes remain underexploited as human food partly because of the lower protein quality (compared to animal proteins) due to sulphur amino acid deficiency, presence of several antinutritional factors such as the protease inhibitors, lectins, tannins, phytates, etc. and some undesirable attributes such as the flatulence producing ability, beany/grassy flavours, and long cooking times required in their preparation.

Among legumes, dry beans lag behind oilseeds in both production and consumption. Among dry beans, common beans (*Phaseolus vulgaris* L.) are produced and consumed in highest quantity on a global basis. However, there are several dry bean varieties that are preferred in certain regions of the world and are therefore important in the local economy as well as human health. Four such beans, chickpea (*Cicer arietinum*), black gram (*Phaseolus mungo* or *Vigna mungo*), mung bean (*Vigna radiata*) and pigeon pea (*Cajanus cajan*), are discussed in this chapter.

2.2 GENERAL BOTANICAL, ECOLOGICAL AND PRODUCTION INFORMATION

The botanical and common names for chickpea, black gram, mung bean and pigeon pea are summarized in Table 2.1. In the general classification of legumes, chickpea, black gram, mung bean and pigeon pea belong to the Leguminosae (or Fabaceae). Among flowering plants, the Leguminosae is the third largest (after Compositae and Orchidaceae) in size and in economic importance second only to grasses (Gramineae). The Leguminosae has an estimated 16 000–19 000 species in 750 genera (Allen and Allen, 1981). As with many other legumes, the chickpea, black gram, mung bean and pigeon pea have related wild species. For example, the chickpea is known to have 39 related wild species. *Cicer reticulatum* (from which *Cicer arietinum* is derived), *Cicer pinnatifidum*, *Cicer echinosperum* and *Cicer bijugum*, are all closely related to *Cicer arietinum* in terms of botanical characteristics. Based on geographical distribution and seed colour, the chickpea is divided into two major types, desi (Indian origin) and kabuli (Mediterranean and Middle Eastern origin). The desi varieties are typically wrinkled at the beak and colours range from light brown, fawn, brown, yellow, orange, green to black. Kabuli seeds are often white to cream colour. Black gram has early as well as late

Table 2.1 Botanical and common names for chickpea, black gram, mung bean and pigeon pea

Botanical name(s)	*Common names*
Cicer arietinum L.	Bengal gram, garbanzo bean, chickpea, hommes, pois chiche, boot, chana, chola, chhole, desi chana, kabuli
Vigna mungo (L.) Hepper	Black gram, mungo bean, mung bean, urd, urid, urad
Vigna radiata (L.) R. Wilczek	Green gram, mung bean, golden gram, moog, moong
Cajanus cajan L.	Red gram, pigeon pea, tur, arhar, toovar, toor

maturing varieties. The seeds of black gram are small and typically have an olive-green or black seed coat. The mung bean has several botanical synonyms including *Azukia radiata* (L.) Ohwi, *Phaseolus aureus* Roxb., *Phaseolus radiatus* L., *Rudua aurea* (Roxb.) Maekawa, and *Vigna aureus* (Roxb.) Hepper. The seeds are usually green or brown coloured and have a flat hilum. Pigeon pea (*Cajanus cajan*) is a single polymorphic species. The genus *Atylosia* (containing many species) is closely related to *Cajanus* with which it is now considered to be synonymous. Chickpea seeds have the largest size (especially the kabuli type). Black and mung bean seeds are comparable in size but generally slightly smaller than those of the pigeon pea.

Production data for dry beans are summarized in Table 2.2, which shows that Asia ranks first both in area harvested and production. The yield (kg/ha) is however, highest in the USSR. Among the dry beans, chickpea and pigeon pea are produced in highest quantities and are ranked fifth in the world after soybeans, peanuts, beans and peas (Milner, 1972). India accounts for 75% and 96% of the total global production of the chickpea and pigeon pea, respectively. The major production areas for chickpea are tropical and subtropical areas (India, Middle East, Mediterranean region). Black gram is primarily grown in India, Greece, Southeast Asia and Eastern Africa. Mung bean is grown throughout Southeast Asia, Central America, and the warmer parts of China and the USA. Pigeon pea is mainly grown in India and to some extent, in East Africa, the Caribbean, and Latin America.

2.3 NUTRIENT COMPOSITION AND NUTRITIVE VALUE

The proximate composition of chickpea, black gram, mung bean and pigeon pea (Table 2.3) indicates that they all are high in protein and

Table 2.2 Production data for dry beans[a]

	Area harvested (ha)x10^{-3}	*Yield (kg/ha)*	*Production (MT)x10^{-3}*
World	26 207	563	14 750
Africa	2592	737	1911
North Central America	3138	837	2627
South America	5218	457	2839
Asia	12894	494	6366
Europe	1293	642	830

[a] Source: FAO Production Yearbook (1986).

carbohydrates, and contain significant amounts of minerals and fibre, and are low in fat. The majority of these nutrients are primarily located in the cotyledons with the exception of fibre (which is mostly in the seed coat) and certain minerals such as the calcium which is found in significant quantities in the seed coat (for example, in mung bean, most of the calcium is in the seed coat). Generally dry beans are isocaloric with cereals and provide about 300–350 kcal/100 g dry seeds. However, because dry beans contain more protein than cereals, they are more nutrient dense than cereals.

2.3.1 Proteins

The total protein content in legumes is generally determined by the Kjeldahl procedure which measures total nitrogen. The protein content is

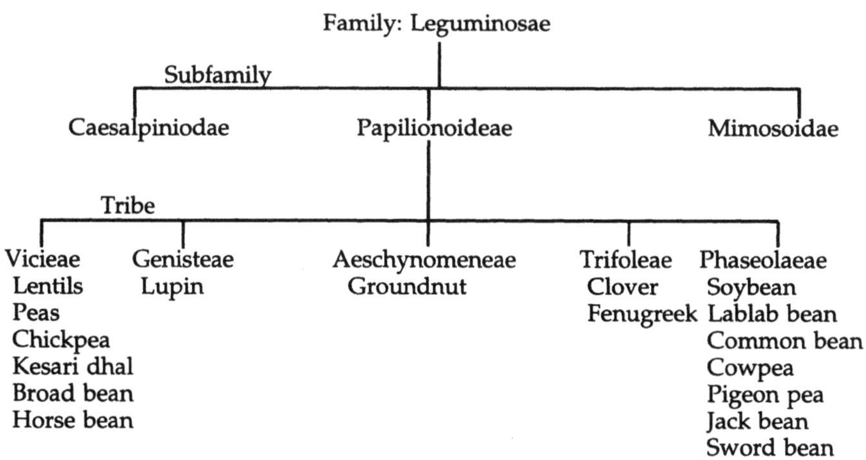

Fig. 2.1 Botanical classification of food legumes.

Table 2.3 Proximate composition of chickpea, black gram, mung bean and pigeon pea[a]

	Content (g/100 g)			
Component	*Chickpea*	*Black gram*	*Mung bean*	*Pigeon pea*
Protein	25.5	21.57	22.9	22.9
Fat	6.0	1.43	1.2	3.8
Total carbohydrates	64.0	60.00	61.8	58.0
Crude fibre	1.2	5.40	4.4	5.0
Ash	2.9	3.33	3.5	4.15

[a] Compiled from Salunkhe and Kadam (1989).

calculated by multiplying Kjeldahl nitrogen by a factor of 6.25. Because many dry beans contain 8–15% nitrogen as non-protein nitrogen which includes free amino acids, pyridines, pyrimidines, nucleic acids, alkaloids, amines and complex lipids; the protein content may be overestimated by 1–2% on a seed weight basis (Earle and Jones, 1962; Deshpande and Nielsen, 1987).

Legume proteins consist of storage and metabolic proteins. Storage proteins usually account for the largest portion of the total proteins and may constitute up to 80–90% of the total proteins. The other proteins in legumes, which include enzyme inhibitors, lectins, enzymes, structural, transport, and recognition proteins, account for less than 20% of the total proteins. Consequently, storage proteins are important from the human nutritional view point. Based on solubility classification, legume storage proteins are primarily globulins (soluble in dilute salt solutions) followed by the albumins (water-soluble proteins). In most seeds, globulins typically constitute up to 70% of the total proteins whereas albumins account for up to 30% of the total proteins (Sathe *et al.* 1984a). Storage proteins are synthesized at the endoplasmic reticulum and deposited in the protein bodies. The size of protein bodies varies but is usually in the range 40–100 μm (Sathe *et al.*, 1984a). The total protein content is affected by several parameters including soil type, climatic conditions, region, use of fertilizers, and genetic factors (Mosse and Pernollet, 1982; Deshpande and Damodaran, 1990).

The major storage proteins in dry beans are further divided into two major categories vicilin and legumin, based on their sedimentation velocity (S). In this system of classification, vicilin storage proteins are termed 7S and legumins 11S proteins (Derbyshire and Boulter, 1976; Derbyshire *et al.*, 1976). Typically, the 7S globulins are glycosylated whereas the 11S are not. The relative proportion of 7S and 11S globulins depends on the type and cultivar of the beans. The legumins are usually hexameric proteins and consist of six subunits each with an estimated molecular weight (MW) of 60 000. Each of these six subunits contains two polypeptides, with estimated molecular weights of 40 000 and 20 000, linked together via disulphide bond(s). The molecular weight of the native legumin is therefore in the range 350 000–400 000. Vicilins are somewhat less well defined proteins which have molecular weights in the range 150 000–180 000. Typically, vicilins mainly consist of subunit polypeptide of MW 50 000 and non-equimolar amounts of several subunits of MW 33 000, 29 000, 13 000 and 12 500. Both legumin and vicilin are co- and post-translationally modified *in vivo* prior to deposition in the protein bodies. The vicilin and legumin subunits also exhibit size and charge heterogeneity. This heterogeneity has been used to identify the type of subunits and to classify storage proteins into distinct types (Derbyshire *et al.*, 1976; Boulter, 1983; Osborn, 1988).

Chickpea globulins have been reported to be similar to those of peas (*Pisum sativum* or *Vicia faba*) in terms of distribution and polypeptide composition and contain both vicilin and legumin (Jackson *et al.*, 1969; Kumar and Venkataraman, 1978). Schroeder *et al.* (1988) have shown that the albumin to globulin ratio in chickpea is 0.21–0.23. The major polypeptides in globulins had MW 75 000 (one polypeptide), 50 000 (two polypeptides), 30 000–40 000 (three polypeptides), 18 000–20 000 (three polypeptides), and 6000 (one polypeptide), whereas those in albumins had estimated MWs of 94 000, 20 000 and 9000. Black gram proteins have been well characterized with respect to their polypeptide composition and amino acid content (Padhye and Salunkhe, 1979). Typically, the major storage protein in black gram is similar to phaseolin (the major storage protein of *Phaseolus* beans) with respect to its polypeptide composition (Sathe, unpublished work). The total protein profile of black gram proteins is also similar to that of *Phaseolus vulgaris* seed proteins. Padhye and Salunkhe (1979) reported at least six different protein subunits with estimated MWs of 21 000, 25 000, 34 000, 55 000, 140 000 and 200 000. The major subunit (MW 55 000) accounted for 65% of the total seed globulins (based on densitometric analysis of gels). Mung bean contains both legumin and vicilin of which the latter is the more abundant. Legumin is composed of three polypeptides of MWs 37 000, 34 000 and 20 000 (Derbyshire and Boulter, 1976), whereas vicilin (8S) contains four subunits with MWs 63 000, 50 000, 29 500 and 24 000. Pigeon pea contains 78% salt-soluble proteins of which 61% are globulins that have been partially characterized (Krishna *et al.*, 1977). Sucrose density gradient fractionation of globulins resolved them into three major fractions with sedimentation values of 9.3S (α), 10.6S (β), and 12.6S (γ). Sodium dodecyl sulphate polyacrylamide gel electrophoresis indicated that α, β and γ fractions contained seven, three and three polypeptides, respectively. The molecular weights of these polypeptides ranged from 20 000 to 50 000. The γ fraction had four times more S-containing amino acids than the α fraction. Data on water-soluble proteins of chickpea, mung bean and pigeon pea are lacking but black gram albumins have been characterized to some extent (Padhye and Salunkhe, 1979). The

Table 2.4 Essential amino acid composition of chickpea, black gram, mung bean and pigeon pea[a]

	Content (g/100g protein)					
Amino acid	*Chickpea*	*Black gram*	*Mung bean*	*Pigeon pea*	*FAO reference protein*	*Egg protein*
Lysine	6.3	6.5	7.3	6.8	4.2	7.2
Threonine	3.4	3.9	3.4	3.8	2.8	5.2
Valine	5.5	5.6	6.9	4.8	4.2	7.4
Leucine	8.2	7.2	7.7	6.8	4.8	7.8
Isoleucine	6.0	5.8	6.3	5.7	4.2	7.4
Methionine	1.2	1.1	1.5	1.1	2.2	3.4
Tryptophan	0.8	0.5	0.4	0.8	1.4	1.5
Phenylalanine	4.9	5.5	5.3	9.0	2.8	5.8
Arginine	6.9	5.7	6.3	5.4	–	6.7
Histidine	2.3	2.7	2.7	3.4	–	2.4

[a] Compiled from Gupta (1982).

major polypeptides in black gram albumins had MWs of 140 000, 84 000, 56 000 and 39 000 and a total of eight polypeptides in the MW range 27 000–140 000 were reported (Padhye and Salunkhe, 1979).

The amino acid composition of chickpea, black gram, mung bean and pigeon pea proteins has been determined by several researchers. Depending on the variety and seed maturity, total amino acid composition varies. The essential amino acid composition (Table 2.4) of these bean proteins indicates that the sulphur amino acid methionine is the limiting amino acid. Based on literature data, the acidic amino acids (Asx and Glx) dominate amino acid composition and usually account for 30–40% of the total amino acids. These beans contain large amounts of lysine, an essential amino acid. Cereals typically lack lysine but contain adequate amounts of methionine. For this reason, when cereals are consumed with legumes, they complement each other thus improving the nutritive value of both. The biological value, digestibility coefficient, and the protein efficiency ratio for chickpea, black gram, mung bean and pigeon pea proteins are summarized in Table 2.5. These protein quality indices show that the nutritional quality of these bean proteins is low when compared with animal proteins such as milk casein and egg proteins.

2.3.2 Carbohydrates

The total carbohydrates in chickpea, black gram, mung bean and pigeon pea range from 50.6 to 70.9% of which starch is the major component (Table 2.6). The total crude fibre content ranges from 1.2 to 13.5% of which lignin (a non-carbohydrate), cellulose and hemicellulose are the primary

Table 2.5 Nutritional quality of chickpea, black gram, mung bean and pigeon pea proteins[a]

Legume	*Biological value (BV) %*	*Digestibility coefficient (DC) %*	*Protein efficiency ratio (PER)*
Chickpea	52–78	80–89	0.7–1.8
Black gram	60–64	78–85	1.0–1.9
Mung bean	39–70	62–91	0.8–2.12
Pigeon pea	61–74	67–90	0.7–1.7

[a] Compiled from Salunkhe and Kadam (1989) and Deshpande and Damodaran (1990).

Table 2.6 Carbohydrate composition of chickpea, black gram, mung bean and pigeon pea[a,b]

	% of seed weight			
	Chickpea	*Black gram*	*Mung bean*	*Pigeon pea*
Total carbohydrates	50.6–70.9	56.6–63.4	53.3–61.2	57.3–58.7
Starch	37.2–50.8	32.3–47.9	37.0–53.2	39.0–58.9
Amylose[b]	31.8–45.8	26.7–44.0	13.8–35.0	26.5–38.6
Total soluble sugars	4.8–9.3	3.0–7.1	3.0–7.2	2.4–6.3
Sucrose	0.7–2.9	1.46	0.3–2.0	1.6–2.7
Raffinose	Trace-3.0	0.0–0.9	0.3–2.6	1.0–1.1
Stachyose	0.5–6.48	0.9–2.3	1.2–2.8	2.7–3.0
Verbascose	Trace-6.48	3.4–3.5	1.7–3.8	4.0–4.1
Total crude fibre	1.2–13.5	1.2–7.1	1.2–12.8	1.2–8.1
Lignin	2.2–5.9	0.2–3.8	2.2–7.2	2.9
Cellulose	7.1–9.7	0.3–5.0	2.5–4.6	7.3
Hemicellulose	3.5–8.7	10.7	0.3–9.1	10.1

[a] Compiled from Salunkhe and Kadam (1989), Reddy *et al.* (1984), and Sathe *et al.* (1984b).
[b] Data expressed as % of the total starch.

components. The soluble sugars are usually about 10% of the total seed weight. Raffinose, stachyose, verbascose and sucrose form the bulk of the total soluble sugars. Raffinose, stachyose and verbascose are known incitors of human flatulence (Reddy *et al.*, 1984; Sathe and Salunkhe, 1984; Deshpande *et al.*, 1984). Verbascose is the major soluble sugar in chickpea, black gram, mung bean and pigeon pea. The flatulence production in humans is summarized in Table 2.7.

The variation in carbohydrate composition is partly due to variety and partly due to environmental conditions. None the less, starch is the major constituent of total carbohydrates in these beans. The starch granules vary in size and usually range from 7 to 40 μm (for both length and width of the granule). The appearance of starch granules normally reflects their

Table 2.7 Flatulence activity of chickpea, mung bean and pigeon pea in preschool children[a]

Bean preparation	*Weight in basal diet (g)*	*Flatus (ml/h)*
Basal diet	–	19
Chickpea cotyledons (boiled)	40	44
Chickpea cotyledons (roasted)	40	58
Chickpea whole (boiled)	40	52
Mung bean cotyledons (boiled)	40	30
Mung bean whole (germinated and boiled)	40	29
Pigeon pea (boiled)	40	48

[a] Compiled from Geervani and Theophilus (1979).

origin. The starch granules in these beans vary from round to oval to elliptical to irregular. The main components of starch are amylose and amylopectin with the latter being in higher proportion (up to 68% of the total starch). Based on X-ray diffraction spectra, starches are divided into three major types. The A type refers to cereal starches, the B type to potato and tuber starches (and also high amylose cereal starches), and the C type which is a mixture of the A and B types. With the exception of mung bean (A type), chickpea, black gram and pigeon pea starches are all C type. Typically, these bean starches gelatinize at 60–80°C (judged on the loss of birefringence). They also exhibit restricted swelling characteristics which makes them thermally more stable than many cereal and tuber starches. The legume carbohydrates have several desirable attributes which include slow digestion and therefore steady release of glucose, ability of the fibre to reduce serum cholesterol, ability to remove bile and bile salts and decreasing the food transit time through the digestive tract. These beneficial effects have been demonstrated *in vivo* in humans (Rao, 1969, 1976; Shurpalekar *et al.*, 1979). Generally, the carbohydrate digestibility improves with heat processing (Table 2.8). However, in the case of chickpea and mung bean, starch digestibility decreased with heating. The reason for such a decrease in whole beans but not in cotyledons has not yet been elucidated. These researchers also noted that digestibility of bean starches *in vivo* and *in vitro* did not correlate.

2.3.3 Vitamins and minerals

Chickpea black gram, mung bean and pigeon pea contain appreciable amounts of several vitamins and minerals (Table 2.9). B-group vitamins are present in these beans in particularly high concentrations compared to many foods. Vitamins C and D do not occur in mature seeds to a

Table 2.8 *In vitro* digestibility of chickpea, black gram, mung bean and pigeon carbohydrates[a]

	Digestibility coefficient (%)			
Process	*Chickpea*	*Black gram*	*Mung bean*	*Pigeon pea*
Bean cotyledons				
Raw (unprocessed)	87.2	89.9	92.2	84.7
Boiled	92.3	89.2	92.6	91.6
Pressure-cooked	92.8	90.3	93.2	89.0
Roasted	88.4	90.3	92.3	86.0
Fermented	–	93.7	–	–
Whole beans				
Raw (unprocessed)	93.8	95.6	96.7	93.9
Boiled	87.5	–	86.1	–
Germinated (24 h)	87.5	–	85.1	–

[a] Compiled from Shurpalekar *et al.* (1979) and Geervani and Theophilus (1981).

Table 2.9 Vitamin and mineral content of chickpea, black gram, mung bean and pigeon pea[a]

	Chickpea	*Black gram*	*Mung bean*	*Pigeon pea*
Carotene (μg/100 g)	120	107[b]	94	99
Folic acid (μg/100 g)	150	140	140	100
Niacin (mg/100 g)	2.9	2.0	2.1	2.1
Riboflavin (mg/100 g)	0.3	0.37	0.27	0.16
Thiamin (mg/100 g)	0.4	0.42	0.47	0.63
Calcium (mg/100 g)	150	154	124	124
Copper (mg/100 g)	2.3	0.72	0.97	1.25
Iron (mg/100 g)	6.9	9.1	7.3	5.8
Magnesium (mg/100)	168	185	171	133
Phosphorus (mg/100 g)	331	385	326	304
Potassium (mg/100 g)	797	–	843	1009
Sodium (mg/100 g)	26	40	28	28
Zinc (mg/100 g)	2.85	3.5	–	–

[a] Data compiled from Salunkhe and Kadam (1989), Deshpande and Damodaran (1990) and Gupta (1982).
[b] Recalculated assuming 10 IU= 1 RE and 1 RE= 6 μg β-carotene.

significant extent. However, germinated beans do contain small amounts of vitamin C. Consequently, the mature dry seeds of these beans are a good source of thiamin, niacin, riboflavin and folic acid. Data on bioavailability of bean vitamins are lacking (Sathe *et al.*, 1984b). The mineral content of chickpea, black gram, mung bean and pigeon pea (Table 2.9) indicates that they are an excellent source of calcium, potassium, copper, iron, phosphorus, magnesium and zinc and are low in sodium. The low sodium content of these beans coupled with high

Table 2.10 Lipid composition of chickpea, black gram, mung bean and pigeon pea[a]

	% of total lipids			
	Chick pea	*Black gram*	*Mung bean*	*Pigeon pea*
Total saturated fatty acids	10.42	23.70	27.70	28.20
Palmitic acid	9.22	17.80	14.10	20.50
Stearic acid	1.20	5.90	4.30	6.90
Arachidic acid	–	–	–	0.80
Behenic acid	–	–	9.30	–
Total unsaturated fatty acids	67.13	76.30	72.80	71.80
Oleic acid	21.84	17.25	20.80	10.50
Linoleic acid	43.29	11.60	16.30	56.30
Linolenic acid	2.00	47.45	35.70	5.00

[a] Adapted from Salunkhe *et al.* (1982).

protein and high complex carbohydrate content makes them an excellent food source of both proteins and energy for hypertensive people. The bioavailability of minerals may be affected due to the presence of high concentrations of phytic acid, a known chelator of many divalent metal ions (Reddy *et al.*, 1989).

2.3.4 Lipids

The total lipids in chickpea, black gram, mung bean and pigeon pea are quite low (Table 2.3) and therefore these beans are not considered as a major source of dietary lipids. The low lipid content coupled with high amounts of other nutrients makes these beans nutrient dense. The detailed lipid composition of these beans (Table 2.10) indicates that the polyunsaturated essential fatty acids (linoleic and linolenic) account for half the total lipids. Because of the presence of these polyunsaturated fatty acids, many products prepared from these beans are susceptible to oxidative damage. Of particular significance to consumer acceptability is the development of rancidity (Salunkhe *et al.*, 1982).

2.3.5 Antinutritional factors

Like other beans, chickpea, black gram, mung bean and pigeon pea also contain several antinutritional factors (Table 2.11) in appreciable quantities. Since protease inhibitors can be adequately inactivated by simple heat treatments such as cooking or autoclaving (pressure cooking) and HCN producing glycosides can easily be removed by discarding the soaking and/or cooking water, they do not pose serious threat to human health. However, heat-stable antinutrients such as the phytates and tannins are of concern. Chickpea and black gram do not contain lectins

Table 2.11 Antinutritional factors in chickpea, black gram, mung bean and pigeon pea[a]

Antinutrient	*Chickpea*	*Black gram*	*Mung bean*	*Pigeon pea*
Trypsin inhibitor (U/g)[b]	0.76	0.80	0.64	0.74
Chymotrypsin inhibitor (U/g)[b]	0.68	0.71	0.00	0.65
Subtilisin inhibitor (U/g)[b]	0.11	0.33	0.35	0.16
Tannins (mg/100 g)	78–272	540–1197	437–799	380–1710
Phytates (g/100 g)	0.28–1.26	0.72–1.46	0.67–1.10	0.71–7.0
HCN (mg/100 g)	0.8	–	–	0.5

[a] Compiled from Salunkhe and Kadam (1989), Reddy *et al.* (1985, 1989) and Liener (1982). Data are for whole beans.
[b] U= Units. One unit of inhibitory activity was defined as the amount of inhibitor that could inhibit 1 mg enzyme.

(Salunkhe and Kadam, 1989; Reddy *et al.*, 1982, 1986). Amylase inhibitor is shown to be absent in black gram (Reddy *et al.*, 1982). This absence of certain antinutritional factors in some of these beans should be carefully evaluated in an effort to breed varieties that lack them.

2.4 REMOVAL OF ANTINUTRIENTS AND UNWANTED COMPONENTS

In order to utilize beans effectively as human food, it is essential to inactivate/remove antinutritional and unwanted components. Such inactivation or removal of these compounds depends on the type of bean, the amount of these compounds present, and the final product to be prepared. There are several methods of processing beans which include dehusking, milling, soaking, cooking, germination, fermentation, autoclaving, roasting, frying, parching and protein extract extrusion depending on the type of bean. Generally, adequate heat processing inactivates the protease inhibitors (Liener, 1982; Sathe and Salunkhe, 1984). Water-soluble components such as the raffinose oligosaccharides can be removed to a significant extent by discarding the soak water which helps reduce the flatulence potential of beans which contain them. (Sathe and Salunkhe, 1984). Heat-stable compounds such as the polyphenols (including tannins) and phytates are, however, not easily removed by simple soaking. In such instances, other methods that can hydrolyse these components may be more useful. For example, in chickpea, 91% of verbascose + stachyose and 88% of raffinose are removed 4 days after germination. Selected data for removal of some of the antinutrients/unwanted components from chickpea black gram, mung bean and pigeon pea by various processing techniques are summarized in Tables 2.12–2.15. As can be seen from these tables, the extent of removal of undesirable attributes depends on the

Table 2.12 Influence of dehulling, cooking and germination on the tannin content of chickpea, black gram, mung bean and pigeon pea[a] (mg/g)

Processing method	*Chickpea*	*Black gram*	*Mung bean*	*Pigeon pea*
Dehulling	74.5–92.6	96.1–98.2	92.6–97.4	94.2–98.0
Cooking	77.0	69.4	71.9	58.4
Germination				
0 h	53[b]	22[b]	28[b]	48[b]
24 h	59	36	46	52
48 h	64	53	52	59

[a] Data compiled from Reddy *et al.* (1985).
[b] Beans were soaked in water for 15 hours at room temperature prior to germination. The soaking water was discarded.
Results expressed as equivalent of mg of tannic acid per gram of sample.

Table 2.13 Influence of germination on raffinose oligosaccharides in chickpea, black gram, mung bean and pigeon pea[a]

Legume	*Germination (h)*	*Oligosaccharides hydrolysed (%)*
Chickpea	72	88.57
Black gram	48	100
Black gram (cv. $M_{1\text{-}1}$)	96	93.1
Mungbean (cv. T_1)	96	93.62
Pigeon pea	72	87.88

[a] Compiled from Sathe and Salunkhe (1984).

Table 2.14 Phytic acid hydrolysis during germination of chickpea, black gram, mung bean and pigeon pea[a]

Legume	*Germination (h)*	*Phytic acid hydrolysed (%)*
Chickpea	120	41.14
Black gram	120	60.42
	240	57.24
Mung bean	120	30.24–58.97
Pigeon pea	120	20.16

[a] Compiled from Sathe and Salunkhe (1984).

Table 2.15 Influence of fermentation on phytate content of black gram[a]

Fermentation time (h)	*Phytate content (mg/100 g)*	*Phytate hydrolysed (%)*
Raw	17.04	–
0[b]	17.04	–
4	16.90	0.8
8	16.54	2.9
12	16.54	2.9
16	16.54	2.9
20	16.40	3.8
24	14.98	12.1
45	14.77	13.3

[a] Compiled from Reddy *et al.* (1989).
[b] Samples were soaked (2 h) and blended. Data are expressed on a dry weight basis.

component, the bean and the method of processing.

2.5 CURRENT AND POTENTIAL USAGE AS ANIMAL AND HUMAN FOOD

The green leaves, stalks, and the seed coats of chickpea black gram, mung bean and pigeon pea are used as animal feed in many underdeveloped countries. These parts of the plants are not used as human food. On a household level, chickpea seeds are sometimes used as animal feed, especially for horses. Usually the whole plant (except the roots) is fed to horses. Seeds of chickpea, black gram, mung bean and pigeon pea are, however, primarily used for human food purposes in a variety of ways. One of the most common ways that these seeds are consumed is in the form of a dhal. Typically, the seed coats are removed (either manually or mechanically), cotyledons split, and the resulting dehusked split cotyledons (dhal) are soaked in water and cooked (or directly cooked in sufficient water). The cooked dhal is then seasoned appropriately and usually consumed with cereal-based products such as wheat bread (including chapati, nan, paratha, etc.) or cooked rice. Sometimes the dhal and cereal (especially rice) are cooked together and seasoned (khichadi). The type of seasonings used in such preparations varies considerably. Onions, garlic, fresh cilantro (coriander leaves or Chinese parsely), green chillies, ginger root, black mustard seeds, cumin seeds, anise seeds, coriander (whole seeds or powder) and turmeric are the most frequently used spices in such preparations. Typically, chopped raw onions, chopped cilantro, crushed/chopped garlic, and small pieces of green chillies and ginger roots are lightly fried in a small amount of cooking oil in which mustard, coriander and cumin seeds have been popped first. The whole mixture is then mixed with the cooked dhal and turmeric, cumin and coriander powders are added. The preparation is brought to a boil and adequate amount of salt is added. Boiling is continued for a few more minutes and if desired, butter is added to improve flavour. This preparation is usually soup-like in consistency. Other spices alone or in combination are also used for such preparations depending on regional preferences. In addition to this common mode of consumption, a variety of products are prepared from these beans. Such preparations are made from bean alone or after combining them with cereals.

2.5.1 Chickpea

Chickpea is consumed as whole seeds or dhal. Cooking, germination, composite flours and preparation of a variety of snack products are some of the ways of processing this bean for human consumption. High protein foods using low fat groundnut flour, chickpea flour and fish flour have been developed (Salunkhe and Kadam, 1989). Chickpea flour, also known as besan in India, has been added to wheat flour (up to 20% replacement) to produce acceptable quality bread (Naikare and Kadam, 1985). Fermented foods prepared from chickpea flour include dhokla, khaman and dosa and these products are popular in the Indian subcontinent. Typically, dhokla is prepared by soaking the chickpea flour in water containing 25% buttermilk (12 h at 32°C). Seasonings (ginger, garlic, green chillies, salt, sugar, and baking soda) are added to the batter which is then steamed for 10–15 min. The resulting puffed product is topped with mustard seeds cracked in hot cooking oil plus shredded coconut and chopped cilantro leaves. It is served as cut pieces, when hot. Khaman is prepared from a 1:1 (w/w) mixture of chickpea and black gram flour soaked in buttermilk for 8 h. The general procedure for the preparation is similar to that of dhokla. Dosa is prepared by soaking chickpea dhal (5 h at room temperature), grinding, seasoning the slurry and then frying the thin batter in paper-thin circular discs in a flat pan. These are consumed as such or after rolling in the cooked and spiced vegetables.

Deep fat fried products such as vada, pakoda (also known as bhajia), chakli, and sev are also prepared from chickpea. Chakali and sev are extruded products whereas vada and pakoda use chickpea flour–water slurry as the casing material. Due to its crisp texture and mellow flavour, chickpea flour is the material of choice for such preparations. Sev is typically prepared from seasoned chickpea flour paste which is extruded (hand-operated extruders) directly into the hot cooking oil and deep fat fried. The seasonings added to the paste include ajwan, asafoetida, salt, baking powder, red chilli powder and turmeric. Chakli is prepared from a composite flour of chickpea dhal, rice and black gram dhal (1:4:1, by weight). The dhals and polished rice are dry roasted for 2–3 minutes, mixed with spices, coarsely ground and a dough is prepared by adding sufficient water. The dough is kneaded and processed in a similar way to sev. Sev is similar in appearance to spaghetti and chakli is like a circular pretzel with a spiked surface. The thickness of these products is varied depending on local preferences. Vada is prepared from cut boiled potatoes mixed with seasonings (chopped onion, garlic, ginger root, green chillies, cilantro leaves, and lemon juice) and chickpea flour in a thin batter containing salt and baking soda. The potatoes with seasonings are made into balls which are dipped into the thin batter and deep-fat fried in cooking oil for a few minutes. Pakoda is also a deep-fat fried product where the chickpea flour (eight parts) and rice flour (one part) is used to make a thin batter. The fillings are numerous and include raw onion rings, thin slices of potatoes, several vegetables such as cauliflower, cabbage, green chillies, sliced tomatoes, spinach, radishes, etc. either singly or in combination. Among these products, sev and chakli have an excellent shelf-life if properly stored (2–3 weeks). Pakoda and vada are usually consumed fresh. Other deep-fat fried products similar to sev in composition are called papdi and ganthia. Papdi is usually a flat stick 1–2 inches (2.5–5 cm) wide and 6–12 inches (15–30 cm) long whereas ganthia is about 0.25 inch (0.6 cm) in diameter and 2–6 inches (5–15 cm) long. All these products enjoy great popularity in the Indian subcontinent as snack foods and are consumed year round. Bundi is another deep-fat fried product made from chickpea flour batter. The flour is mixed with water to prepare a thin batter which is then dropped into the hot cooking oil in the form of small droplets that are fried for 2–5 minutes. There are two types of bundi. One is a salted version (salt added to the batter) and the other is a sweetened variety (the fried product is dipped into a sugar syrup for a few minutes). The sweetened variety is usually moulded into balls by hand and is known as bundi laddu. These moulded balls vary in diameter (1–6 inches (2.5–15 cm) is the normal range) and are used in many religious functions. Another type of laddu prepared from chickpea flour involves mixing the flour alone or in combination with wheat flour (1:1 w/w) with hot sugar syrup, cooling the mass, and moulding it into suitable size (usually 1–2 inches (2.5–5 cm) diameter) balls. Fat added to such preparations can either be butter or partially hydrogenated vegetable oils. Laddu has an excellent shelf-life and can last from a few days to a few months. Mysore pak is another sweet product made from chickpea flour in which the flour is mixed with hot hydrogenated fat and then mixed with a hot thin sugar syrup. The mixture is heated and boiling hydrogenated fat added to the mixture intermittently. The product is then cooked until the excess fat starts separating. Upon cooling, the set mixture is cut into suitably sized pieces. Chickpea dhal cooked with sugar or jaggery (1:1 w/w) is used as a filling in stuffed chapati and is called puran poli. The cooked flour–sugar mixture is rolled into a wheat flour dough (two parts filling to one part wheat flour dough) in thin flat discs. These discs are then baked on flat open pans (2 min each side at 200°C).

The baked product is consumed after seasoning with butter or oil. When butter or oil is not used, hot milk is usually used.

Whole or dehusked chickpea seeds, or separated cotyledons are also consumed in dry roasted (200–300°C) form (with or without seasonings) as crunchy snacks. The dehusked seeds or cotyledons are also deep-fat fried in cooking oil and seasoned for consumption as snacks. Green whole seeds are cooked in boiling water until tender and, after seasoning with salt, red pepper and lemon juice, are also used as snacks in certain parts of India. The kabuli variety is usually preferred for dry roasting.

In the Western countries, the chickpea is used canned in brine, canned with meats, or as a salad.

2.5.2 Black gram

Black gram is used in the production of fermented foods such as idli, dosa, hopper, waries (spicy hollow balls) and papad. Among these products, idli and dosa are the most popular breakfast and snack foods in India. Papad and waries are generally used with other vegetables, rice and wheat preparations.

Idli is a fermented steamed product prepared by fermenting black gram dhal and rice (black gram to rice proportion may vary from 1:1 to 1:4 w/w) slurry for 12–20 h (until volume is doubled) and steaming the batter for 5–10 min. The resulting product has a soft, spongy texture. Dosa (described under chickpea) is also prepared from this same batter after adding more water or buttermilk to thin the batter. The thin batter is then fried in thin circular discs on a lightly greased pan (cooking vegetable oil such as peanut oil is commonly used for this purpose) for a few seconds on each side. The product is consumed as such or after rolling in a vegetable dressing (cooked, seasoned vegetables such as potatoes are most commonly used).

Papad is prepared from black gram flour dough by rolling small dough portions into thin circular discs and drying them (sun drying or mechanical drying). Typically, the dough is made from 100 parts flour (85 mesh), 45 parts water, 7–8 parts common salt, and 1–1.5 parts sodium bicarbonate to yield the optimum product. Use of spices in the dough preparation is optional. The dough is mixed for a few minutes (until it becomes non-sticky), rested for 30 min or longer, and rolled into 2 cm diameter (or larger) balls, rolled into thin discs (1 mm thick), and dried (12–17% final moisture content). Traditional home preparations use papad khar or sajji khar in place of sodium bicarbonate. Fermentation occurs during the resting period. The diameter of the final product varies greatly (6–24 cm) depending on the regional and personal preferences (Shurpalekar, 1986). Shelf-life of this product can be up to one year if properly stored (dry, air-tight containers). Papad is either roasted or fried just before consumption. Roasting or frying is quick and does not take more than a minute or two.

Black gram dhal is soaked and coarsely ground into a thick dough. The dough is then divided into small pieces and dried. Usually some spices, such as turmeric and red chilli powder, and salt are added to the dough. The resulting product is known as sandage which is consumed either roasted or fried with other foods primarily as a condiment.

Waries are spicy hollow balls 3–8 cm in diameter and are friable, brittle condiments prepared from black gram paste inoculated with a portion of the previous batch. Spices are mixed with the paste which is then allowed to ferment for 1–5 days at room temperature. Spices used in such preparations typically include asafoetida, caraway, cardamom, cloves, fenugreek, ginger, red pepper and salt. The fermented paste is moulded into balls by hand and dried for 6–8 days in the sun. During drying, a mucilaginous seal develops on the outside which helps retain flavours and moisture. The microorganisms responsible for fermentation are *Candida krusei* and *Saccharomyces cerevisiae*. The role of bacteria in this fermentation is not well defined.

2.5.3 Mung beans

The mung bean is used primarily for cooking curry in India. On a worldwide scale, sprouted mung beans as a part of salads is the most popular method of consumption. Chinese cooking also uses sprouted mung beans in soup. Mung bean noodles are also prepared in China, Taiwan, Korea and the Philippines. Roasting and frying the whole seeds or dhal is a popular way in India to prepare snack products which are usually spiced.

In addition to these uses, the mung beans are also used in the preparation of papads and composite flours, porridge preparations, and in the manufacture of confectionaries and bean cakes.

2.5.4 Pigeon pea

Pigeon pea is primarily used in the preparation of curry in India. The two major types of such preparations which are popular in India are sambar and rasam. Both are soup-like in consistency and appearance and are prepared from the dehusked split cotyledons that are cooked until tender and seasoned with spices. Both contain added vegetables (optional) such as egg plant, drum sticks, carrots, tomatoes, potatoes, green peas, etc. Sambar and rasam are usually consumed with cooked rice or snack foods such as idli, dosa, and vada.

The foregoing discussions have shown that the food uses of chickpea, black gram, mung bean and pigeon pea are primarily indigenous and commercial processing is limited to very few methods such as the milling and dehusking operations.

2.6 RESEARCH NEEDS

Although used extensively in Asia, the Middle East and Africa; chick pea, black gram, mung bean and pigeon pea remain underused in other parts of the world. One of the major factors which hinders their full utilization is the inherent beany/grassy flavour. In the indigenous uses, this is partially overcome by use of spices and processing techniques. In regions where the use of spices is limited, removal of beany/grassy flavours from these beans would increase their utilization. Controlling pre- and post harvest losses is of economic importance where these beans already enjoy acceptance as human food. Reducing such losses would therefore help to improve availability of these beans at an affordable price. The development of new and novel products would be useful in helping to popularize food uses of these beans in Western countries. One of the products that would be of help in this regard is the protein isolate/ concentrate. Preparation of starch would be another such product that can be used in food preparations. The study of fundamental properties of proteins and carbohydrates (particularly starch) would help to develop novel products. Of particular interest in this regard are the adhesive properties of chickpea and black gram. These two beans are used as binders in more food preparations than any other dry beans. The beneficial health effects of dry beans are now well recognized; however, beneficial effects of several bioactive compounds present in these beans remain largely unexplored. Efforts to discern the usefulness of these bioactive compounds in humans therefore warrant intensive research.

REFERENCES

Allen, O.N. and Allen, E.K. (1981) *The Leguminosae*. The University of Wisconsin Press, Madison, WI.

Borlaug, N.E. (1981) Using plants to meet World food needs in *Future Dimensions of World Food and Population* (ed. R.G. Woods) Westview Press, Boulder, Co, pp. 101–82.

Boulter, D. (1983) Protein composition of grains of the Leguminosae. *Qual. Plant. Plant Foods Hum. Nutr.*, **32**, 247–52.

Delwiche, C.C. (1978) Legumes: past, present, and future. *Bioscience*, **28**, 565–70.

Derbyshire, E. and Boulter, D. (1976) Isolation of legumin-like protein from *Phaseolus aureus* and *Phaseolus vulgaris*. *Phytochemistry*, **15**, 411–14.

Derbyshire, E., Wright, D.J. and Boulter, D. (1976) Legumin and vicilin, storage proteins of legume seeds. *Phytochemistry*, **15**, 3–24.

Deshpande, S.S. and Damodaran, S. (1990) Food legumes: chemistry and technology. *Adv. Cereal Sci. Technol.*, **10**, 147–241.

Deshpande, S.S. and Nielsen, S.S. (1987) Nitrogenous constituents of selected grain legumes. *J. Food Sci.*, **52**, 1321–5.

Deshpande, S.S., Sathe, S.K. and Salunkhe, D.K. (1984) Dry beans of *Phaseolus*: a review. Part 3. *CRC Crit. Rev. Food Sci. Nutr.*, **21**, 137–95.

Earle, F.R. and Jones, O. (1962) Analysis of seed samples from 113 plant families. *Econ. Bot.*, **16** 221–50.

FAO (1986) Food and Agriculture Organization of the United Nations, *Production Yearbook*, **40**, 102–3.

Geervani, P. and Theophilus, F. (1979) Flatus inducing effect of processed legumes in pre-school children. *Indian J. Med. Res.*, **70**, 750–5.

Geervani, P. and Theophilus, F. (1981) Studies on digestibility of selected legume carbohydrates and its impact on the pH of the gastrointestinal tract in rats. *J. Food Sci.*, **32**, 71–8.

Gupta, Y.P. (1982) Nutritive value of food legumes, in *Chemistry and Biochemistry of Legumes*, (ed. S.K. Arora), Oxford and IBH Publishing Co., New Delhi, pp. 287–327.

Jackson, P., Boulter, D. and Thurman, D.A. (1969) A comparison of some properties of vicilin and legumin isolated from seeds of *Pisum sativum, Vicia faba* and *Cicer arietinum. New Phytol.*, **68**, 25–33.

Krishna, T.G., Mitra, R.K. and Bhatia, C.R. (1977) Seed globulins of *Cajanus cajan. Qual. Plant. Plant Foods Hum. Nutr.*, **27**, 313–25.

Kumar, K.G. and Venkataraman, L.V. (1978) Chickpea seed proteins: modification during germination. *Phytochemistry*, **17**, 605–9.

Liener, I.E. (1982) Toxic constituents in legumes, in *Chemistry and Biochemistry of Legumes*, Oxford and IBH Publishing Co., New Delhi, pp. 217–57.

Milner, M. (1972) *Nutritional Improvement of Food Legumes by Breeding*, Proc. Symp., Protein Advisory Group, Food and Agriculture Organization, Rome, Italy.

Mosse, J. and Pernollet, J.C. (1982) Storage proteins of legume seeds, in *Chemistry and Biochemistry of Legumes*, (ed. S.K. Arora), Oxford and IBH Publishing Co., New Delhi, pp. 111–94.

Naikare, S.M. and Kadam, S.S. (1985) Bread additives: their role and importance in making quality bread, in *Quality of Wheat and Wheat Products*, (eds D.K. Salunkhe, S.S. Kadam and A. Austin), Metropolitan Press, New Delhi.

Osborn, T.C. (1988) Genetic control of bean seed protein. *CRC Crit. Rev. Plant. Sci.*, **7**, 93–116.

Padhye, V.W. and Salunkhe, D.K. (1979) Biochemical studies on black gram (*Phaseolus mungo* L.) seeds: amino acid composition and subunit constitution of fractions of the proteins. *J. Food Sci.*, **44**, 606–10.

Rao, P.S. (1969) Studies on digestibility of carbohydrates in pulses. *Indian J. Med. Res.*, **57**, 2151–7.

Rao, P.S. (1976) Nature of carbohydrates in pulses. *J. Agric. Food Chem.*, **24**, 958–61.

Reddy, N.R., Salunkhe, D.K. and Sathe, S.K. (1982) Biochemistry of black gram (*Phaseolus mungo* L.): a review. *CRC Crit Rev. Food Sci. Nutr.*, **16**, 49–114.

Reddy, N.R., Pierson, M.D., Sathe, S.K. and Salunkhe, D.K. (1984) Chemical, nutritional, and physiological aspects of dry bean carbohydrates: a review. *Food Chem.*, **13**, 25–68.

Reddy, N.R., Pierson, M.D., Sathe, S.K. and Salunkhe, D.K. (1985) Dry bean tannins: a review of nutritional implications. *J. Am. Oil Chem. Soc.*, **62**, 541–9.

Reddy, N.R., Pierson, M.D. and Salunkhe, D.K. (1986) *Legume-based Fermented Foods*, CRC Press, Boca Raton, FL, pp. 254.

Reddy, N.R., Pierson, M.D., Sathe, S.K. and Salunkhe, D.K. (1989) *Phytates in Cereals and Legumes*, CRC Press, Boca Raton, FL, pp. 152.

Salunkhe, D.K. and Kadam, S.S. (eds) (1989) *Handbook of World Food Legumes: Nutritional Chemistry, Processing Technology, and Utilization*, Vols. I, II, and III, CRC Press, Boca Raton, FL.

Salunke, D.K., Sathe, S.K. and Reddy, N.R. (1982). Legume lipids, in *Chemistry and Biochemistry of Legumes*, (ed. S.K. Arora), Oxford and IBH Publishing Co., New Delhi, pp. 51–109.

Sathe, S.K. and Salunkhe, D.K. (1984) Technology of removal of unwanted components of dry beans. *CRC Crit. Rev. Food Sci. Nutr.*, **21**, 263–87.

Sathe, S.K., Deshpande, S.S. and Salunkhe, D.K. (1984a) Dry beans of *Phaseolus*: a review. Part 1. Proteins. *CRC Crit. Rev. Food Sci. Nutr.*, **20**, 1–46.

Sathe, S.K., Deshpande, S.S. and Salunkhe, D.K. (1984b) Dry beans of *Phaseolus*: a review. Part 2. Chemical composition: carbohydrates, fiber, minerals, vitamins, and lipids. *CRC Crit. Rev. Food Sci. Nutr.*, **21**, 41–93.

Schroeder, H., Gibson, A.H., Oram, R.N. and Shaikh, M.A.Q. (1988) Seed protein characterisation and nitrogen fixation rates in the chickpea mutant Hyprosola and its parent. *J. Sci. Food Agric.*, **44**, 31–41.

Shurpalekar, K.S., Sundaravalli, O.E. and Rao, M.N. (1979) *In vitro* and *in vivo* digestibility of legume carbohydrates. *Nutr. Rep. Int.*, **19**, 111–17.

CHAPTER THREE

The nutritive value of wild Mexican legumes with a potential for nutritional use

A. Sotelo

3.1 GENERAL BOTANICAL, ECOLOGICAL AND PRODUCTION INFORMATION

Several thousands of years ago, humans focused attention on legumes and cereals because of their great importance in satisfying human dietary needs. The family Leguminosae has a great number of species only surpassed by the Compositae and Orchidaceae families; in fact there are 560–690 genera and 12 000–17 000 species recognized worldwide (Heywood, 1971). Furthermore, it is very important from an economic point of view, not only because of the great variety of uses (e.g. as food, drugs used in medicine, dyes, etc.) but also because of its botanical importance in fixing nitrogen (in symbiosis with *Rhizobium*).

As is common in other developing countries, Mexico has many natural resources, which are not sufficiently studied or exploited as yet. It possesses one of the most varied floras in America, since it is located between a moderately northern climate and a southern tropical zone, with an extensive subtropical area. This variety in the flora is the result of the incredible diversity of climates and soil as a consequence of its rolling topography. The vegetation of the country varies from high jungles in the humid southeast region and the Atlantic shores, to the dry desert areas in Chihuahua, Sonora and Baja California, and to the high and cold regions in the transversal volcanic chain (Miranda and Hernandez, 1985).

Mexico has about 134 genera of legumes, with 1707 species of which 884 are endemic to the country (Sousa and Delgado, 1991). In the

Legumes and Oilseeds in Nutrition
Edited by E. Nwokolo and J. Smartt
Published in 1996 by Chapman & Hall. ISBN 0 412 45930 2

catalogue of Mexican plants, Martinez (1987) has recorded 576 species including both wild and edible legumes.

3.2 CURRENT AND POTENTIAL USE AS HUMAN FOOD

The importance of the Leguminosae is based on the fact that it includes numerous useful species and it is distributed worldwide. Legumes have been recognized for their nutritionally important protein content, their high calorific value and the important vitamins and minerals they contain. The part most utilized as food in legumes is the seed; although, the green pods, sprouts, roots, leaves and flowers can also be used as food. The majority of the wild or semicultivated legumes which in Mexico are considered edible, are only for local consumption. In other words, they are mainly consumed in the places they grow. However, there are a few wild legumes which grow in one location and are transported to other towns and cities where they are utilized. One such example is the green *Leucaena leucocephala*, which grows in the States of Puebla, Guerrero and Morelos and is sent to the markets of important cities as Cuernavaca, Puebla and Mexico City. Similarly some species of the genus *Inga*, which grow in the tropics, can be purchased in the above-mentioned markets. One legume with edible leaves is *Crotalaria pumila* which is cultivated and consumed all year round in the warm humid area of Chiapas and although it is considered a representative dish of Tapachula, a city in that state, it is also consumed in Oaxaca and Tabasco. *Canavalia ensiformis* or jack bean is endemic in India and is considered an edible in many countries where it is cultivated for human and animal feeding. Many studies have been conducted on this legume which is cultivated in the Yucatan and Central America (Bressani and Sosa, 1990; de la Vega and Sotelo, 1982). In all respects, the area most representative of the conditions for cultivation of legumes is the southeast part of the country, where members of the genera *Inga, Hymenea, Enterolobium, Leucaena* and other wild species are consumed. Along the Pacific south seashore, some of the previously mentioned species are consumed, together with the huizache (*Acacia farnesia*) and mesquite (*Prosopis juliflora*). The latter is one of the few that grow in the desert zone of the Mezquital valley in central Mexico. Another legume of commercial value which has an edible root is the jicama (*Pachyrhizus erosus*), cultivated universally and available in any market in the country. Although its nutritional content is low, with a high water content (80–90%), and an insipid flavour, it can be eaten raw, spiced with salt, lemon and hot pepper. Another interesting legume is the velvet bean, *Stizolobium cinerium*, which is nicknamed 'nescafe' in the Indian markets in the highlands of Chiapas State and is used to prepare a coffee-like beverage; it is also used as an adulterant for real coffee, since the toasted seed has a coffee-like odour. The edible flowers with the highest consumption, mainly in southeast Mexico, are the flowers of *Gliricidia sepium* and of *Erythrina mexicana*. The latter is eaten after being boiled in water, the cooking water discarded, and the cooked flowers carefully washed. If eaten otherwise, they are known to cause sleepiness. It is well known that plants in the genus *Erythrina* contain tryptophan-derived alkaloids such as erythroidine, which has a curare-like action. It is important to note that *Erythrina* seeds can not be consumed because of their high alkaloid content and toxicity.

About 30 species of the genus *Lupinus* have been reported to thrive in Mexico. Juarez (1991) investigated the nutritional value of three wild and one cultivated species after detoxification. In all of them, the protein content was very high, 42–46%. In one of the wild samples, *L. campestris*, the sulphur amino acid content was similar to the FAO reference pattern (FAO/WHO, 1973), although the content of alkaloids in the wild species was higher than the cultivated ones.

3.3 CURRENT AND POTENTIAL USE AS ANIMAL FEED

Currently, a certain proportion of basic grains used for human feeding are also used in animal feeding. For example, corn, oats, barley and certain legumes such as soybean, chickpea and others, are used as both human and animal feeds. If some wild legume species which up till now have merely been considered potentially comestible, could be employed as animal feed, this would facilitate the use of corn and oats, etc. exclusively for human purposes. This would halt the huge corn, bean and soybean importations made during the last few years to satisfy human needs (FAO, 1990).

Wild legumes with potential for use as animal feed include *Leucaena leucocephala*, which is native to the Yucatan peninsula, and currently used in the Caribbean as animal feed. Additional research may be needed as some authors have reported some undesirable effects when it is fed (Huang *et al.*, 1988). Other wild legumes which have a potential for use as animal feed include mesquite (*Prosopis juliflora*), huizache (*Acacia farnesiana*), parota (*Enterolobium cyclocarpum*) and others.

3.4 NUTRIENT COMPOSITION AND NUTRITIVE VALUE

Over 40 years ago, extensive analyses of various local Mexican foods was conducted by Massieu *et al.* (1950) and by Cravioto *et al.* (1951). Similar nutritional investigations in Guatemala, aimed at improving the usefulness of the tropical and semitropical flora of Central America have been conducted by Bressani and his colleagues at the Instituto de

Table 3.1 Proximate analysis of wild legumes

	Content (g/100g sample)					
Scientific name	*Total solids*	*Protein*	*Fat*	*Fibre*	*Ash*	*Carbohydrates*
Seeds						
Albizzia lebbek	90.53	33.67	3.13	13.17	3.57	46.44
Cassia floribunda	90.26	18.97	3.98	16.04	3.09	56.92
Acacia pennatula	91.51	29.11	4.18	18.40	4.85	43.46
Canavalia ensiformis	92.73	31.15	2.22	14.12	3.34	49.17
Caesalpinia crista	97.94	24.73	21.43	14.86	3.33	35.65
Gliricidia sepium	93.83	43.29	24.03	21.66	4.09	64.12
Cassia grandis	95.10	28.99	2.68	35.03	3.82	29.48
Mucuna argyrophylla	90.66	22.40	0.37	13.11	3.61	60.51
Erythrina americana	94.00	28.24	19.79	17.33	4.08	30.56
Cassia cornigera	96.08	20.97	10.44	19.77	4.06	44.76
Acacia paniculata	92.65	27.56	5.59	13.19	4.20	49.47
Cassia occidentalis	93.52	26.76	5.13	15.22	5.37	47.53
Swartzia guatemalensis	89.08	15.03	0.75	9.27	1.79	73.16
Leucaena esculenta	89.32	27.70	4.61	12.52	5.11	50.07
Leucaena leucocephala	95.35	31.09	3.24	14.77	0.47	50.44
Caesalpinia velluda	99.56	17.70	16.21	18.07	3.14	44.88
Pithecellobium dulce	89.77	22.47	15.28	9.42	2.70	50.13
Hymenaea courbaril	89.98	10.64	8.59	10.93	1.81	68.03
Phaseolus lunatus (wild)	92.41	21.59	1.32	8.21	4.74	64.14

Table 3.1 Proximate analysis of wild legumes (continued)

	Content (g/100g sample)					
Scientific name	*Total solids*	*Protein*	*Fat*	*Fibre*	*Ash*	*Carbohydrates*
Acacia coelliacantha	94.24	26.78	2.59	19.77	4.37	46.49
Acacia farnesiana	96.18	25.12	3.31	18.29	3.89	49.39
Senna reticulata	97.04	17.62	4.04	12.10	4.34	61.90
Pithecellebium flexicaule	91.71	31.55	14.95	14.67	3.44	35.39
Caesalpinia pulcherrima	95.94	21.94	25.11	7.50	3.93	41.49
Lysiloma divaricata	95.62	31.21	13.10	8.98	4.30	42.43
Caesalpinia sp.	90.30	17.18	6.30	20.53	3.10	52.89
Stizolobium cinerium brown var.	96.87	24.97	4.19	9.35	3.29	58.20
Stizolobium cinerium pale var.	89.99	28.36	3.50	3.71	3.50	55.93
Entada polystachia	94.33	17.95	21.91	11.35	3.30	45.49
Cymbopetalum penduliflorum	92.84	11.26	27.36	47.36	2.57	11.20
Pithecellobium undulatum	97.83	28.58	5.39	15.97	4.06	46.00
Enterolobium cyclocarpum	91.70	21.30	1.71	16.12	3.65	57.22
Bauhinia purpurea	96.71	30.91	17.22	5.04	3.73	43.12
Erythrina breviflora	97.41	34.79	16.26	10.99	3.28	34.68
Senna multiglandulosa	96.58	19.58	4.40	12.43	3.71	59.58
Delonix regia	92.36	17.87	2.27	27.02	3.84	49.00
Lysiloma acapulcensis	94.20	33.94	12.94	13.51	3.86	35.75
Phaseolus lunatus (Wild)	91.00	31.82	1.8	12.06	4.27	50.05

Table 3.1 Proximate analysis of wild legumes (continued)

	Content (g/100g sample)					
Scientific name	*Total solids*	*Protein*	*Fat*	*Fibre*	*Ash*	*Carbohydrates*
Inga radians	89.11	25.55	1.26	3.11	2.89	67.19
Entada scandens	93.64	24.39	13.81	1.99	3.23	56.58
Total legume (fruit)						
Cassia fistula L.	95.00	8.66	1.09	38.46	2.54	49.25
Mimosa tenella	95.74	15.74	4.95	30.65	3.99	44.64
Acacia coelliacantha	92.89	13.70	1.13	31.60	5.26	48.31
Acacia farnesiana	94.95	16.13	3.28	16.26	3.90	60.42
Prosopis juliflora	98.68	13.45	2.08	39.51	3.54	41.42
Pithecellobium undulatum	97.73	22.94	3.13	33.16	3.89	36.88
Delonix regia	93.02	9.07	1.18	58.97	3.01	27.77
Cassia spectabilis	89.93	14.69	1.17	51.44	5.00	7.79
Mimosa albida	95.90	18.20	2.98	23.70	5.42	49.69

Nutricion de Centro America y Panama (INCAP). These results have been compiled in several publications (Bressani, 1972; Bressani, and Elias 1974). Wild Mexican legumes have also been intensively studied, and nutrient analyses carried out (Giral *et al.*, 1978; Sotelo *et al.*, 1980, 1986; de la Vega *et al.*, 1981; de la Vega and Sotelo 1982; Lucas *et al.*, 1988). Wild legumes collected from different parts of Mexico include those listed in Table 3.1.

Table 3.1, shows the proximate analysis of various wild legumes from which it is evident that their protein content varies greatly. In general, only a few of these legumes have a high fat content of 12–30% (Sotelo *et al.*, 1990); in contrast, there are abundant species with a high fibre content of 12–50%. This high fibre content is a general characteristic of wild vegetable materials many of which can only be used as food for ruminants. This group of wild legumes with a high protein content includes *Gliricidia sepium*, whose seed has a composition similar to that of soybean (43% protein and 24% fat). However, it also contains certain toxic constituents which affect rodents, to such as extent that in some places it is known as 'mouse-killer'. Nevertheless, because of its fat composition and lipid characteristics it is considered as an edible fat. The same finding can be applied to other wild legumes (Sotelo *et al.*, 1986, 1990). The Yucatan is an area with an abundance of legume species which have been studied by different authors (Standley, 1930; Miranda, 1978; Duke, 1981; Lundell and Lundell, 1983; Hernandez, 1987). Nutrient analysis of 25 legumes from Yucatan conducted by Flores *et al.* (1988) found that *Sesbania emercus* and *S. grandiflora* had relatively higher protein content than others (35% and 44%, respectively). Table 3.2 shows the essential amino acids present in some wild legumes (Giral *et al.*, 1978; Sotelo *et al.*, 1980, 1986; de la Vega *et al.*, 1981). In all the legume samples sulphur amino acids were limiting.

3.5 PRESENCE, CHEMICAL NATURE AND DETOXIFICATION OF ANTINUTRITIONAL SUBSTANCES IN WILD LEGUMES

Members of the legume family contain, perhaps, the greatest variation in number and concentration of toxic or antinutritional substances. The concentration of these toxic substances is usually higher in wild than in edible types (Liener, 1969; Committee on Food Protection, 1973). Table 3.3 lists the antinutritional factors present in some wild Mexican legumes. These substances include trypsin inhibitors, phytohaemagglutinins, cyanogenic glucosides, alkaloids, etc. (Giral *et al.*, 1978; Sotelo *et al.*, 1980). The presence of saponins and tannins was investigated in 20 wild Mexican legumes. It was found that 40% of the species contained tannins in low concentration and three (*Entada scandens, Mucuna argyrophylla* and *Stizolobium cinerium*) showed high concentrations (Muñoz, 1979).

Table 3.2 Essential amino acids in Mexican wild legume seeds

Scientific name	Amino acid content (mg/g/N)								
	Isol	Leu	Lys	Met+Cys	Phen+Tyr	Thre	Try	Val	Tot EAA
Albizzia lebbek	186	386	358	88	395	285	61	221	1981
Cassia floribunda	226	397	396	140	470	228	92	329	2278
Acacia pennatula	125	349	236	110	302	173	16	171	1482
Canavalia ensiformis	207	415	271	86	337	309	72	226	1963
Caesalpinia crista	181	393	340	136	491	167	67	236	2011
Gliricidia sepium	182	386	265	136	364	142	46	183	1704
Cassia grandis	206	388	320	77	437	196	66	216	1906
Mucuna argyrophylla	190	382	296	61	255	349	89	223	1845
Erythrina americana	199	390	321	136	379	201	53	236	1915
Phaseolus lunatus	240	389	311	97	464	232	90	255	2078
Cassia cornigera	158	349	204	121	431	178	51	217	1709
Cassia occidentalis	210	363	322	173	386	266	98	273	2091
Swartzia guatemalensis	137	225	647	27	408	120	0	187	1751
Phaseolus lunatus	251	440	351	84	528	279	71	256	2260
Leucaena esculenta	235	461	328	166	432	238	68	226	2154
Leucaena leucocephala	141	448	207	91	277	112	39	159	1474
Caesalpinia velluda	190	365	297	124	467	242	51	233	1969
Pithecellobium dulce	219	414	491	400	422	200	62	226	2434
Hymenaea courbaril	237	442	282	147	241	249	0	265	1863
Phaseolus lunatus (wild)	397	656	453	191	693	343	0	592	3325
Acacia coelliacantha	137	327	215	127	376	144	47	191	1564
Acacia farnesiana	168	579	277	138	577	167	68	252	2226

Table 3.2 Essential amino acids in Mexican wild legume seeds (continued)

	Amino acid content (mg/g/N)								
Scientific name	*Isol*	*Leu*	*Lys*	*Met+Cys*	*Phen+Tyr*	*Thre*	*Try*	*Val*	*Tot EAA*
Senna reticulata	274	458	356	173	510	302	106	368	2547
Lysiloma divaricata	233	443	363	149	451	201	0	257	2097
Caesalpinia sp.	324	524	426	180	577	350	112	366	2859
Stizolobium cinerium brown var.	211	477	287	153	521	266	0	294	2209
Stizolobium cinerium pale var.	227	397	364	85	559	193	29	239	2093
Entada polystachia	273	452	365	133	485	552	99	312	2671

Table 3.3 Antinutritional factors present in some wild Mexican legume seeds

Scientific name	*Trypsin inhibitors TUI[a]/mg sample*	*Haemagglutinins[b]*				
		C	R	H	HA	HO
Canavalia ensiformis	22	–	10	13	–	–
Erythrina americana[d]	68	–	1	6	–	7
Stizolobium cinerium	16	–	–	5	–	–
Bauhinia purpurea[c]	16	–	–	5	–	–
Delonix regia	2	–	–	2	–	–
Phaseolus lunatus (wild)[c]	94	–	–	–	9	–
Gliricidia sepium[c]	66	–	1	–	–	–
Leucaena esculenta	25	–	–	–	–	–
Leucaena leucocephala	14	–	–	–	–	–
Acacia pennatula	11	–	–	–	–	–
Lysiloma acapulcensis	43	–	–	2	–	–
Entada polystachia	121	–	–	5	–	–
Phaseolus lunatus cv. guinze[c]	67	–	–	8	–	–
Mucuna argyrophylla	17	–	–	3	–	–
Cassia occidentalis	0	–	1	2	–	–
Senna reticulata	7	–	1	4	–	–
Lysiloma divaricata	49	–	–	4	–	–
Erythina breviflora[d]	64	–	–	7	–	7
Enterolobium cyclocarpum	20	–	–	5	–	–
Prosopis juliflora	38	ND[e]	ND	ND	ND	ND

[a] Trypsin unit inhibited (Kakade *et al.*, 1974).
[b] The highest dilution that produces agglutination. C=cow; R=rabbit; H=hamster; HA=human type A; HO=human type O erythrocytes (Jaffé *et al.*, 1974).
[c] Cyanogenic glucosides present.
[d] Alkaloid present.
[e] ND=Not determined.

3.5.1 Destruction of antinutritional substances

The most common antinutritional substances are protease inhibitors and phytohaemagglutinins, both of which are protein in nature and thus inactivated by heat. For this reason, cooking increases the nutritive quality of legumes, although the degree of destruction especially of protease inhibitors is variable, depending on the heating process and on the condition of the grain, such as whether or not it was previously soaked (Gomez *et al.*, 1973). In general, it is considered that thermal treatment in an autoclave at 120°C and at a pressure of 15 psi is enough to destroy most of the trypsin inhibitors. However, the various types of inhibitors in each legume may behave differently when subject to different destruction methods (Belitz and Weder, 1990).

Table 3.4 Main components of wild Mexican legumes for ruminant feeding

Wild legumes	*Content (g/100 g)*						
	Protein	*Fat*	*Carbohydrate*	*Crude fibre*	*ADF*[a]	*Lignin*	*Digestibility*[b] *(ruminal liquid) (% t ± SD)*
Delonix regia	9.0	1.2	27.8	59.0	60.7	15.3	32.0±4.8
Canavalia ensiformis	31.1	2.2	49.2	14.1	11.6	0.8	94.0±0.6
Wild *Phaseolus lunatus*	29.0	1.6	45.3	11.0	14.4	2.3	32.0±6.4
Enterolobium cyclocarpum (total legume)	13.5	1.7	64.8	15.8	22.5	7.2	75.0±1.1
Enterolobium cyclocarpum (seed)	21.3	1.7	57.2	16.1	18.8	3.5	86.3±1.7
Pithecellobium undulatum	22.2	3.1	36.9	33.2	32.3	8.3	61.0±1.3
Bauhinia purpurea	29.5	15.0	39.1	12.5	6.6	1.3	62.6±0.5
Acacia farnesiana	16.1	3.3	60.4	16.3	12.6	5.8	77.6±1.2
Prosopis juliflora	16.13	1.5	37.6	41.9	45.4	10.4	51.9±2.5
Tamarindus indica	14.6	2.7	67.2	12.6	26.9	9.4	21.0±4.8

[a] Acid detergent fibre.
[b] From Goering and van Soest (1975).

Table 3.5 True protein and non-protein amino acids present in wild Mexican legumes[a]

Sample	Nitrogen distribution (g/100 g sample)			Non-protein amino acids (mg/g N)				
	Total N	True protein N	True protein[b] (g/100 g sample)	Homoserine	Djenkolic	Mimosine	DOPA	Canavanine
C. ensiformis	4.09	3.05	19.06	34.9	3.0	–	–	429.4
G. sepium	7.06	5.56	34.75	3.7	–	–	–	566.9
M. argyrophylla	3.09	2.32	14.50	ND[d]	–	–	342.8	–
E. americana	4.91	4.55	28.43	3.7	–	–	–	1.0
L. esculenta	4.42	3.50	21.87	13.2	49.7	363.6	–	–
L. leucocephala	4.40	2.99	18.69	ND	16.9	864.4	–	–
S. cinerium	4.16	3.58	22.37	2.7	–	–	108.0	–
P. undulatum	4.40	2.93	18.31	11.0	64.0	192.3	–	–
E. cyclocarpum	4.18	2.93	18.31	3.1	–	–	–	–
B. purpurea	4.73	4.18	26.12	2.8	–	–	–	2.8
P. flexicaule	3.09	2.22	13.87	ND	–	–	–	–

[a] From Lucas *et al.* (1988).
[b] True protein = True protein N × 6.25.
[c] DOPA = Dehydroxyphenylalanine.
[d] ND = Not determined.

3.6 ANIMAL AND HUMAN STUDIES WITH WILD LEGUMES

It is difficult to obtain solid scientific information on the long-term effects in humans of consumption of some of these wild legumes. Most often there is only information on temporary effects which some legumes may have when used as food for local consumption. Also, most of these wild legumes are consumed as food only when there is a serious shortage of other grains. In some of the legumes only a part of the fruit is used, whereas in others, only the flowers or the fresh seeds are consumed, and then only over short periods of time and not on a daily basis. The same situation holds true for cattle and other farm animals.

In order to decide which animals can be fed with the legumes under study, it is important to determine their proximate contents, especially the crude fibre content of the seeds. Table 3.4 shows those legumes with a high fibre content and an acceptable digestibility which could be considered as forage (Aguilar, 1978). In general, animal experiments in which wild legume seeds have been fed have usually been designed (a) to estimate the nutritive value of the selected legumes, (b) to investigate any acute or subacute toxic effects of the wild materials, and (c) to determine the degree of effectiveness of destruction of certain toxicants by various processing methods. In many of the biological tests performed with seeds of wild legumes, the PER (protein efficiency ratio) test gave very low or negative values in raw seeds, and much higher values in cooked seeds. Generally speaking, when the PER value of cooked seeds is not significantly increased it may be due to the quality of the protein or to the presence of such thermostable toxicants as alkaloids, saponins, non-protein amino acids, cyanogenic glycosides, etc. The nutritive value of those wild legumes with low fibre content and low protein quality, was considerably enhanced by heat treatment and supplementation with cereals or methionine (de la Vega *et al.*, 1981; de la Vega and Sotelo, 1986). In some raw legumes, the toxicity is associated with the presence of lectins which cause agglutination of cow's red blood cells, as demonstrated by Jaffé and Gomez (1975) and confirmed by others (Gonzalez-Garza *et al.*, 1982; Sotelo *et al.*, 1985). Although dry heat is not adequate to destroy the haemagglutinins, they are almost entirely eliminated by autoclaving. This increases the nutritive value of the legumes, as not only haemagglutinins are destroyed, but also other growth-inhibiting substances (Jaffé, 1969). Additionally, some non-protein amino acids, such as mimosine, are partially destroyed by heat and cyanide present in the seed in the form of HCN is released during the soaking and cooking processes (Liener, 1975). The non-protein amino acid content of selected wild Mexican legumes is shown in Table 3.5.

From the above discussion, it is clear that the great majority of Mexican legumes in the wild flora constitutes a great reservoir of protein with an enormous potential use for human or animal feeding. Undoubtedly, the same holds true for many other countries, especially in the developing world. What is needed is for modern processing technology to be applied to transform most of the wild legumes into edible vegetable protein, through elimination of the toxic factors and antinutritional substances which currently limit the use of these wild legumes in human and animal nutrition.

REFERENCES

Aguilar, R.M. (1978) Evaluacion quimica y biologica de nuevas fuentes de alimentos para rumiantes. Tesis Facultad de Quimica UNAM. Mexico.

Belitz, H.D. and Weder, J.K.P. (1990) Protein inhibitors of hydrolases in plant food stuffs. *Food Rev Int.*, **6**, 151–211.

Bressani, R. (1972) Legumes in human diets and how they might be improved, in *Nutritional Improvement of Food Legumes By Breeding*, (ed. M. Milner), Wiley, New York.

Bressani, R. and Elias, E.G. (1974) Legumes foods, in *New Protein Foods*, Vol I, *Technology*, Part A, (ed. A.M. Altschul), Academic Press, New York, pp. 230–97.

Bressani, R. and Sosa, L.J. (1990) Effect of processing on the nutritive value of Canavalia jackbeans (*Canavalia ensiformis* L.). *Plant Foods Hum. Nutr.*, **40**, 207–14.

Chel, G.L. and Castellanos, R.A. (1984) El valor nutritivo de la harina de hojas de *Leucaena* en dietas para ratas. *Prod. Animal Trop.*, **9**, 307–13.

Committee of Food Protection. Food and Nutritional Board of the NRC (1973) *Toxicants Occurring Naturally in Foods*. National Academy of Sciences. Washington, DC.

Cravioto, O.R., Massieu, H.G., Guzman, G.J. and Calvo de la Torre, J. (1951) Composicion de alimentos Mexicanos. *Ciencia*, **11**, 129–55.

de la Vega, A. and Sotelo, A. (1982) Valor nutritivo de la semilla de *Canavalia ensiformis* sola y suplementada con metionina y cereales. *Rev. Soc. Quim. Mex.*, **26**, 92–6.

de la Vega, A. and Sotelo, A. (1986) The nutritional quality and toxin content of wild and cultivated lima beans (*Phaseolus lunatus*). *Qual. Plant. Plant Foods Hum. Nutr.*, **36**, 75–83.

de la Vega, A., Giral, F. and Sotelo, A. (1981) Nutritional evaluation of the velvet bean (*Stizolobium cinerium*) alone and supplemented with methionine or wheat flour. *Nutr. Rep. Int.*, **24**, 817–23.

Duke, J.A. (1981) *Handbook of Legumes of World Economic Importance*. Plenum Press, New York, London, pp. 345.

FAO / WHO (1973) Energy and Protein Requirements. Report of a Joint FAO / WHO Committee. Nutrition Report Series no. 52. FAO, Rome.

FAO (1990) *Boletin Trimestral de Estadistica*, **3**, 75.

Flores, J.S., Martinez, C.A., Olvera, M.A. *et al.* (1988) Potencial de algunas leguminosas de la flora yucatense como alimento humano o animal. *Turrialba*, **38**, 159–62.

Giral, F., Sotelo, A., Lucas, B. and De la Vega, A. (1978) Chemical composition and toxic factors content in fifteen leguminous seeds. *Q. J. Crude Drug Res.*, **16**, 143–9.

Goering, H. and van Soest, P. (1975) *Forage Fiber Analysis*. Agriculture Handbook, Agriculture Research Service, USDA.

Gomez, B.R., Elias, L.G., Molina, M.R. *et al.* (1973) Changes in chemical composition and nutritive value of common beans and other legumes during house cooking, in *Nutritional Aspects of Common Beans and Other Legume Seeds as Animal and Human Foods*. Proceedings of a meeting held in Ribeirao Preto. (ed. W.G. Jaffé). Arch. Latinoam. Nutr. Venezuela, pp. 93–107.

Gonzalez -Garza, M.T., Sousa, V. and Sotelo, A. (1982) Differential cytotoxicity of the isolated protein fraction of escumite bean (*Phaseolus acutifolius*). *Qual. Plant. Plant Foods Hum. Nutr.*, **31**, 319–25.

Hernandez, X.E. (1987) Los frijoles y otras leguminosas cultivadas en otras Chiapas, in *Xolocotzia, Geografia Agricola*. II Univ. Autom. de Chapingo, pp. 769–75.

Heywood, V.H. (1971) The leguminosae – A systematic purview, in *Chemotaxonomy of the Leguminosae* (eds J.B. Harborne, D. Boulter and B.L. Turner), Academic Press, London.

Huang, J.C., Wen, S.H. and Li, P.H. (1988) Serum thyroid hormone levels of heifers grazing *Leucaena leucocephala* in Southern Taiwan. *Chin. J. Physiol*, **31**, 9–23.

Jaffé, W.G. (1969) Hemagglutinins. in *Toxic Constituents of Plant Foods*, (ed. I.E. Liener), Academic Press, New York.

Jaffé, W.G. and Gomez, M. (1975) Beans of high or low toxicity. *Qual. Plant. Plant Food Hum. Nutr.*, **24**, 359–65.

Jaffé, W.G., Levy, A. and Gonzalez, D.I. (1974) Isolation and partial characterizations of bean phytohemagglutinins. *Phytochemistry*, **13**, 2685–93.

Juarez, C.H. (1991) Destoxificacion comparativa de tres especies de Lupinus silvestres y de *Lupinus mutabilis* cultivada en Mexico. Tesis Instituto Politecnico Nacional Mexico.

Kakade, M.L., Rackis, J.J., McGhee, J.E. and Puski, G. (1974) Determination of trypsin inhibitor activity of soy products: a collaborative analysis of an improved procedure. *Cereal Chem.*, **51**, 376–82.

Liener, I.E. (1969) *Toxic Constituent of Plant Foodstuffs*. Academic Press, New York.

Liener, I.E. (1975) Effects of anti-nutritional and toxic factors on the quality and utilization of legume proteins, in *Protein Nutritional Quality of Foods and Feeds*, Part II, (ed. M. Friedman), Marcel Dekker, New York, pp. 523–50.

Lucas, B., Guerrero, A., Cigales, L. and Sotelo, A. (1988) True protein content and non-protein amino acids present in legume seeds. *Nutr. Rep. Int.*, **37**, 545–53.

Lundell, C.L. and Lundell, A.A. (1983) The flora of the northern Yucatan and the Coba area of Quintana Roo. Mexico: Collections and Observations in 1938. *Wrightia*, **7**, (3) The University of Texas at Dallas, USA.

Martinez, M. (1987) *Catalogo de nombres vulgares y cientificos de Plantas mexicanas*. Fondo de Cultura Economica Mexico.

Massieu, H.G., Guzman, G.J., Cravioto, R.O. and Calvo, T.J. (1950) Contenido en aminoacidos indispensables en algunas semillas Mexicanas. *Ciencia*, **10**, 142–4.

Miranda, F. (1978) Vegetacion de la Peninsula de Yucatan. Colegio de Postgraduados. Chapingo, SARH. Mex. 537.

Miranda, F. and Hernandez, X.E. (1985) The types of vegetation of Mexico and their classification. I: Xolocotzia. *Rev. Agricola*, **I**, 41–55.

Muñoz, R.M. (1979) Determinacion de saponinas, taninos, y accion antibiotica de algunas plantas silvestres Mexicanas. Tesis. Facultad de Quimica, UNAM, Mexico.

Sotelo, A., Lucas, B., Uvalle, A. and Giral, F. (1980) Chemical composition and toxic factors content of sixteen leguminous seeds (II). *Q. J. Crude Drug Res.*, **18**, 9–16.

Sotelo, A., Artega, M.E., Frias, M.I. and Gonzalez -G. M.T. (1980) Cytotoxic effect of two legumes in epithelial cells of the small intestine. *Qual. Plant. Plant Food Hum. Nutr.*, **30**, 79–85.

Sotelo, A., Uvalle, A. and De la Vega, A. (1985) Composicion quimica y aspectos toxicologicos de siete variedades de frijoles cultivados en el Estado de Chiapas. *Tecnol. Aliment Mex.*, **20**, 18–25.

Sotelo, A., Lucas, B., Blanc, F. and Giral, F. (1986) Chemical composition of seeds of *Gliricidia sepium*. *Nutr. Rep. Int.*, **34**, 315–21.

Sotelo, A., Lucas, B., Garza, L. and Giral, F. (1990) Characteristics and fatty acid content of the fat of seeds of nine wild Mexican plants. *J. Agric. Food Chem.*, **38**, 1503–5.

Sousa, M. and Delgado, A. (1991) Mexican Leguminosae: Phytogeography, endemism and origins, in *Biological Diversity of Mexico: Origins and Distribution*. Oxford University Press. New York.

Standley, P.C. (1930) *Flora of Yucatan*. Field Museum of Natural History, Chicago, USA, pp. 492.

CHAPTER FOUR

Peanut (*Arachis hypogaea* L.)

E. Nwokolo

4.1 INTRODUCTION

The peanut (*Arachis hypogaea* L.) is better known worldwide as groundnut and to a much lesser extent as earthnut, monkeynut and goobers. It is one of nature's most nutritious seeds and one of the world's most popular and universal crops, cultivated in nearly 100 countries in all six continents. Because it is so flavourful while containing only a few antinutritional factors, it is consumed as a snack in a wide variety of ways, the only limitation being one's imagination and inventiveness. The peanut is also a major source of edible oil and protein meals considered to be highly valuable in human and animal nutrition.

The peanut is not a true nut but rather a legume much like the bean or a pea. It is an annual herb, a member of the family Leguminosae (Fabaceae), whose origins are in South America probably Brazil (Higgins, 1951) or Bolivia (Rosengarten, 1984). It was introduced by early European explorers to Europe and the old world tropics where it became an important food and export crop. India is the world's largest producer of peanuts, accounting for nearly 33% of total world production and together with China accounting for nearly 60% of world production (FAO, 1990). Other major producers USA, Nigeria, Indonesia and Senegal (Table 4.1). Peanuts, peanut oil and peanut protein meal constitute an important segment of world trade in oilseeds and products, peanut being the fifth most important oilseed in the world (Table 4.2).

Legumes and Oilseeds in Nutrition
Edited by E. Nwokolo and J. Smartt
Published in 1996 by Chapman & Hall. ISBN 0 412 45930 2

Table 4.1 Major peanut producers: area planted, yield and production (1990)

Country	*Area (1000 ha)*	*Yield (kg/ha)*	*Production (1000 MT)*
India	8000	900	7200
China	3085	2127	6563
USA	729	2242	1634
Nigeria	1000	1166	1166
Indonesia	628	1463	919
Senegal	886	787	698
Myanmar	524	851	459
Zaire	575	748	430
Argentine	180	2056	370
Vietnam	220	955	2100
Total Africa	5877	811	4765
Total North and Central America	873	2048	1789
Total Asia	12826	1238	15879
Total South America	346	1785	618
Grand Total World	19968	1157	23109

Source: FAO, (1990).

Table 4.2 World production of major oilseed crops, oil and meal in 1990

	Production (million metric tonnes)		
Oilseed	*Seed*	*Edible oil*	*Protein meal*
Soybean	103.94	15.92	69.67
Cottonseed	33.43	3.82	9.98
Rapeseed	25.12	8.67	10.23
Sunflower seed	22.57	7.94	8.49
Peanut	22.27	3.41	5.45
All others	10.29	17.18	2.09

Source: USDA Foreign Agricultural Service Circular Series, 1992.

4.2 BOTANY OF CROP

Arachis hypogaea L. is a member of the genus *Arachis* in the family Leguminosae (Fabaceae). The genus comprises 15 species, of which only *A. hypogaea* is cultivated. Members of this genus are annuals or herbaceous perennials and may be bunch or spreading (runner) types. The leaves are pinnate, the flowers pea-shaped. Following fertilization, a stalk-like structure rapidly elongates carrying the fertilized ovules into the soil where each structure enlarges into a pod, and fertilized ovules develop and mature. The bunch form of the peanut is 45–60 cm high with short branches, whereas the runner type, although shorter (30–45 cm), spreads over longer distances on the soil. The pods are usually 2.5–5 cm long and contain two or three seeds, each seed separated from the other by a partial constriction of the pod. The shape of the pod is roughly cylindrical, with a spongy brittle hull and a characteristic net-like pattern on the outside. The seeds vary in shape from small and almost round, to large and nearly oblong. There is wide variation in seed colour, from cream to dark purple, although red, mahogany and rose are most common. In the United States and elsewhere, there has been a trend toward increased cultivation of the runner type and away from the bunch type because of the higher yield typical of the runner varieties.

4.3 FOOD USES OF PEANUTS

Glowing reports about the peanut have been written in both the scientific and lay press over the past one hundred years. Rosengarten (1984) has traced the history of the peanut noting that it was an important food to the Indians of South America before the arrival of Columbus and other Spanish explorers. It was taken by the Portuguese to West Africa where it very quickly adapted to the ecology and became a major dietary supplement for slaves in slave ships. During the American civil war the peanut although initially despised, proved to be a source of badly needed nutrients.

The history of the uses of the peanut as food would be incomplete without mention of the great African–American scientist George Washington Carver who at Tuskegee Institute in Alabama, USA, conducted the most extensive research on record on the peanut. He is credited with having made entire meals and desserts from peanut as well as having obtained patents for the utilization of peanut and peanut by-products in the manufacture of nearly 300 industrial and food products (Elliot, 1965). Food products made by George Washington Carver from peanuts included mayonnaise, cheese, chili sauce, various peanut candies and cookies and various peanut soups, cream vegetables and salads. In the United States, peanuts are produced primarily for manufacture of peanut butter but also for consumption as blanched, dry or salt-roasted peanuts, as honey-roasted peanuts, as 'ball-park' peanuts (roasted in-shell), as peanut cookies and candies and as toppings to various desserts.

In Africa, peanuts are prepared as fresh boiled peanuts or as roasted peanuts. Roasted peanuts are eaten with a variety of starchy foods including popcorn, dried corn and garri (fried cassava). They may be ground with pepper and other spices to form a paste for the traditional consumption of kolanuts. They may also be milled with shrimp, peppers, spices and other ingredients as a replacement for melon seeds in preparation of traditional sauces. Peanut stews cooked with meat/fish, red peppers, onions and tomatoes are favourite accompaniments for rice and yam dishes. Unfortunately there is very little use of processed peanut products as supplements to traditional diets. There is a need to develop ways of incorporating significant quantities of peanut protein into local food systems without altering the organoleptic properties of the indigenous foods. A fairly recent review of food uses and other nutritional characteristics of peanuts (Rhee, 1985) addresses some of these and other issues. In the United States and elsewhere peanuts are consumed in the following ways.

4.3.1 Fresh cooked peanuts

Following the harvest season, fresh unshelled mature peanuts are cooked in light brine solution and are consumed as snack foods. All varieties of peanuts may be cooked fresh but the larger-seeded cultivars are preferable because they are easier to shell.

4.3.2 Roasted peanuts

Peanuts may be roasted in their shell or decorticated prior to roasting. Roasting in-shell peanuts may be unsalted or salted prior to roasting. Roasting is usually by radiant heat. Decorticated peanuts may be salted or unsalted and may be roasted in oil or dry roasted by radiant heat. There are specific time and temperature requirements for optimal roasting by various procedures.

4.3.3 Peanut butter

Preparation of peanut butter involves sorting and cleaning to obtain the best quality peanuts, following which the peanuts are blanched prior to roasting. This is followed by blending of ingredients, grinding, mixing and packaging. Salt, sugar and oil may be added to a basic formation of 90–95% roasted peanuts. Antioxidants and stabilizers may be needed to prevent deterioration and separation of oil after processing and during storage. Peanut butter is a very nutritious product which is ideal for preparation of sandwiches and capable of meeting the protein requirements of children and adults alike.

4.3.4 Peanut flour, concentrates and isolates

Full-fat peanut flour is produced by grinding decorticated blanched peanuts and contains the same quantity of protein (28–30%) and fat (45–48%) as the whole seed. Partially defatted peanut flour is produced by mechanical or hydraulic extraction of oil from decorticated, blanched peanuts. It contains 45–48% protein and 30–35% oil. Low-fat peanut flour can also be produced by direct solvent extraction of milled peanuts. Fully defatted flours, grits and flakes can be prepared by cooking, followed by mechanical/hydraulic expression, and in a final solvent extraction are cut into thin slices for effective removal of peanut lipids. Peanut protein concentrate containing 60–70% protein may be produced either from peanuts or from already defatted peanut flour. When concentrates are to be produced from blanched peanuts, the seeds are processed in a manner similar to that for production of defatted flour. There are, however, additional steps of alcohol and acid washing, neutralization of acid and flash-drying by superheated air. Washing peanut flour with aqueous alcohol and dilute acid removes bitter compounds, oligosaccharides, free amino acids, colour and flavour components as well as residual lipids (Rayner *et al.*, 1970; Nagaraj and Subramanian, 1974). Aflatoxin may be removed from peanuts by using sodium hypochlorite and hydrogen peroxide as oxidizing agents (Rhee *et al.*, 1977).

Protein isolates usually contain 90% or more protein and are produced by first solubilizing peanut flour protein in alkali, followed by acid precipitation of the protein. The protein curd is washed, flash-dried and packaged.

4.3.5 Confectionery uses of peanut products

Hundreds of culinary uses have been demonstrated or patented for processed peanuts or peanut derivatives.

Peanut flour has supplemented wheat flour, replacing up to 20% of the flour in acceptable bread formulations (Rooney *et al.*, 1972; Khan *et al.*, 1975). Khan *et al.* (1975) observed that bread baked with peanut flour is superior to bread baked with full-fat soy flour in respect of loaf volume, flavour and texture, although heating of peanut flour is necessary for production of breads with acceptable flavour and aroma. Peanut protein concentrate produced by an aqueous extraction process yields the best bread in terms of crumb structure, colour and taste. Peanut flour may be used to fortify breakfast cereals and snack foods, and may replace milk or milk plus eggs in baking. When peanut flour replaces milk and eggs in baking, bread with light resilient texture and excellent loaf volume, and doughnuts with good crumb structure and excellent surface colour are produced (Ayers and Davenport, 1977). In the 1970s and 1980s research into the supplementation of cookies with peanut products confirmed the high nutritional quality of these peanut-cookies, without any deleterious effects on flavour and acceptability (Ranhotra, 1980; Ranhotra *et al.*, 1980; Conkerton and Ory, 1985). Results from the large number of studies on peanut flour supplementation of cookies, indicate that the flour can replace 10–30% of wheat flour in cookies without any significant effect on sensory and physical attributes (Mathews *et al.*, 1970; Rooney *et al.*, 1972).

4.3.6 Use as meat extenders and in meat formulations

Rhee (1985) notes that various peanut products including flour, grits, concentrates and isolates serve as potential extenders in many meat formulations. Thus any of those products may be used in the formulation of meat patties and meat balls, luncheon meat, salisbury steaks, frankfurters, bolognas and meat loaves etc. He notes that in experiments where peanut flakes replaced meat in manufactured meat products at levels of 5–10%, use of peanut flakes improved tenderness and cohesiveness in ground beef patties. Levels of peanut grits as high as 30% in frankfurters and levels of flakes as high as 50% in turkey products have been used successfully (Joseph *et al.*, 1978).

4.3.7 Uses as dairy extenders

Peanut protein isolates have for very many years been used in the manufacture if imitation milk or as an extender to cow or buffalo milk. According to Rhee (1985), this synthetic milk consists of protein isolates, sodium caseinate, vegetable fat, sugar, minerals and vitamins, stabilizer, emulsifier, preservatives, flavour, colour and water.

Chandrasekhara *et al.* (1971) found that a proprietary product, Miltone, in India is manufactured by mixing buffalo milk with peanut protein isolate, glucose syrup, minerals and vitamins. The mixture is diluted with water, homogenized in much the same way as fresh milk, and is subsequently pasteurized.

4.4 NUTRITIONAL QUALITY OF THE PEANUT

Worldwide, peanuts are utilized as snack foods by young and old alike, and in many countries they are a significant component of national school lunch programmes. In countries of Africa, Asia and Latin America, where malnutrition is endemic, increased production and use of peanuts holds out the best prospect for meeting the nutritional needs of huge segments of these populations.

4.4.1 Proximate composition

It is estimated that the shell represents about 25% of the dry weight of in-shell peanut, whereas the kernel comprises the other 75%. The two fleshy cotyledons comprise 93% by weight of the kernel, and the germ and the testa comprise 4 and 3% of kernel weight, respectively. The chemical composition of peanut components shown in Table 4.3 is from the review by Rhee (1985) based on the data of Cobb and Johnson (1973). The

Table 4.3 Chemical composition of peanut components

Components	*Cotyledons (%)*	*Shells (%)*	*Testa (%)*	*Germ (%)*
Moisture	5–8	–	9.0	–
Protein	25.4–33.8	4.8–7.2	11.0–13.4	26.5–27.8
Oil	44.5–56.3	1.2–2.8	0.5–1.9	39.4–43.0
Total carbohydrates	6.0–24.9	10.6–21.2	48.3–52.2	–
Reducing sugars	0.1–0.4	0.3–1.8	1.0–1.2	7.9
Disaccharides	2.9–6.4	1.7–2.5	–	12.0
Pentosans	2.2–2.7	16.1–17.8	–	–
Starch	0.9–5.3	0.7	–	–
Hemicellulose	–	10.1	–	–
Crude fibre	1.6–1.9	65.7–78.3	21.4–34.9	1.6–1.8
Ash	1.8–2.9	1.9–4.6	2.1	2.9–3.2
Calories (100/g)	564	–	–	–

Source: Cobb and Johnson (1973).

Table 4.4 Proximate composition of various peanut products

Nutrients	*Raw peanuts*	*Low-fat flour*	*Defatted flour*	*Peanut butter*
Water (g/100 g)	6.5	7.8	7.8	1.4
Protein (g/100 g)	25.8	33.8	55.2	24.6
Fat (g/100 g)	49.2	21.9	0.6	50.0
Carbohydrate (g/100 g)	16.1	31.3	34.7	20.7
Crude fibre (g/100 g)	4.9	3.0	4.1	2.41
Ash (g/100 g)	2.3	5.2	4.8	3.3
Food energy (Cal/100 g)	567	428	327	588

Source: USDA (1986).

cotyledons are the main nutrient storage tissues and are a concentrated source of protein, lipids and dietary energy, whereas the shell is high in crude fibre, a structural carbohydrate. Table 4.4 gives the proximate composition of raw peanuts and various processed peanut products. As expected, protein and carbohydrate content increase with a more thorough fat extraction from peanuts. Peanut butter is exceptionally well balanced in nutrient content.

4.4.2 Amino acids

The profiles of amino acids in raw peanut, low fat peanut flour, defatted peanut flour and peanut butter are shown in Table 4.5. Amino acid profile of raw peanut is in many respects inferior to the profile of raw soybean.

Table 4.5 Amino acid content of peanut products

Amino acids	Content (g/100 g product)			
	Raw peanuts	*Low-fat flour*	*Defatted flour*	*Peanut butter*
Tryptophan	0.25	0.33	0.51	0.24
Threonine	0.88	1.16	1.79	0.84
Isoleucine	0.91	1.19	1.84	0.87
Leucine	1.67	2.19	3.38	1.59
Lysine	0.93	1.21	1.87	0.88
Methionine	0.32	0.41	0.64	0.30
Cystine	0.33	0.43	0.67	0.32
Phenylalanine	1.34	1.75	2.71	1.28
Tyrosine	1.05	1.37	2.12	1.00
Valine	1.08	1.42	2.19	1.03
Arginine	3.09	4.04	6.24	2.94
Histidine	0.65	0.85	1.32	0.62
Alanine	1.03	1.34	2.07	0.98
Aspartic acid	3.15	4.12	6.37	3.00
Glutamic acid	5.39	7.06	10.91	5.14
Glycine	1.55	2.04	3.14	1.48
Proline	1.14	1.49	2.30	1.09
Serine	1.27	1.67	2.57	1.21

Source: USDA (1986).

This is probably as expected since the protein content of raw peanuts (25–26%) is only about 70% of that of raw soybean (35–36%). However, a comparison of defatted flours of peanut and soybean indicates a measurable superiority of soybean protein over peanut protein. In general for each gram of milled raw peanut or raw soybean, there is twice as much amino acid in soybean as in peanut. For low-fat peanut flour and defatted peanut flours, the concentration of most amino acids is 70–85% of the level in a similar soybean product. These differences in amino acid content are important when it is desired to replace peanut flour with soyflour in formulated foods especially for children. As in most oilseeds and legumes, including soybean, peanuts, peanut flours and peanut protein products are low in sulphur amino acids and would therefore need to be supplemented with foods high in these essential amino acids.

4.4.3 Fatty acids

The fatty acid composition of peanut oil is about 1% palmitic acid, over 80% oleic and linoleic acids. The fatty acid profiles of raw peanuts, low-fat peanut flour and peanut butter (Table 4.6) confirm that peanut oil is an

Table 4.6 Fatty acid content of peanut products

Fatty acids	*Content (g/100 g product)* Raw peanuts	Low-fat flour	Peanut butter
Saturated total	6.83	3.04	9.59
14:0	0.03	0.01	0.05
16:0	5.15	2.29	5.50
18:0	1.10	0.49	2.14
Mono-unsaturated total	24.43	10.87	23.58
16:1	0.0	0.0	–
18:1	23.76	10.57	22.90
20:1	0.66	0.30	0.62
Polyunsaturated total	15.56	6.92	14.36
18:2	15.56	6.92	14.10
18:3	0.0	0.0	0.08

Source: USDA (1986).

Table 4.7 Mineral content of various products

Minerals	*Content (mg/100 g product)* Raw peanut	Low-fat flour	Defatted flour	Peanut butter
Calcium	92	130	140	34
Iron	4.6	4.7	2.1	1.7
Magnesium	168	48	370	157
Phosphorus	376	N/D	7600	323
Potassium	705	1358	1290	721
Sodium	18	1	180	17
Zinc	3.3	6	5.1	2.5
Copper	1.1	2.0	1.8	0.6
Manganese	1.9	4.2	4.9	1.5

Source: USDA (1986).
N/D, not determined.

excellent source of mono- and polyunsaturated fatty acids, exceeding the levels of these fatty acids in soybean and corn oil, but significantly lower than in sunflower seed oil or safflower oil.

Mineral content of peanut and other processed products is shown in Table 4.7 and the concentration of vitamins as reported by both the USDA (1986) and Cobb and Johnson (1973) is shown in Table 4.8. Peanuts are a reasonable source of dietary minerals especially potassium, phosphorus

Table 4.8 Vitamin content of peanut products

	(Content per 100 g product)			
Vitamins	*Raw peanut*	*Low-fat flour*	*Defatted flour*	*Peanut butter*
Ascorbic acid (mg)[a]	–	N/D	–	–
Thiamin (mg)[a]	0.6	N/D	0.7	0.1
Riboflavin (mg)[a]	0.1	N/D	0.5	0.1
Niacin (mg)[a]	12.1	N/D	27.0	13.1
Pantothenic acid (mg)[a]	1.8	N/D	2.7	0.9
Vitamin B6 (mg)[a]	0.3	N/D	0.5	0.4
Folacin (mg)[a]	239.8	N/D	248.2	78.2
Vitamin A (IU)[b]	26	N/D	–	–
Vitamin E (mg)[b]	41.6	N/D	N/D	N/D
L-Tocopherol (mg)[b]	17.1	N/D	N/D	N/D
R-Tocopherol (mg)[b]	22.9	N/D	N/D	N/D
O-Tocopherol (mg)[b]	1.6	N/D	N/D	N/D
Inositol (mg)[b]	180	N/D	N/D	N/D
Choline (mg)[b]	165–174	N/D	N/D	N/D

[a] Source: USDA (1986).
[b] Source Cobb and Johnson (1973).
ND, not determined.

and magnesium, although a significant proportion of the phosphorus may exist, as in most legumes, as salts of phytic acid. Peanuts are a very poor source of fat-soluble vitamins A, D and K, and only a moderate source of vitamin E.

4.5 ANTINUTRITIONAL FACTORS OF PEANUTS

Raw peanuts have very low concentrations of most of the antinutritional factors found in raw soybean. For example, Mir and Hill (1979) reported trypsin inhibitor content of 5.8–9.4 TIU (trypsin inhibitor units) in four varieties of peanut meal, compared to 66.8 TIU for soybean. Anantharaman and Carpenter (1969a) found that trypsin inhibitor content of raw peanut was only about 20% of the level in raw soybean. Other antinutritional factors which have been isolated, purified and identified in raw peanut include various lectins (Lotan *et al.*, 1975; Ahmed, 1986; Ahmed *et al.*, 1988) and trypsin inhibitors (Ahmed and Applewhite, 1988). A goitrogenic factor has also been isolated and identified in the testa of the peanut and some saponin-like compounds which are bitter tasting have been identified in the germ (Rhee, 1985). Peanut skins have also been reported to possess trypsin-inhibitors and growth-retarding principles (Anatharaman and Carpenter, 1969b). It is likely that the few reported cases of headaches and stomach cramps resulting from the consumption of raw peanuts may be due to ingestion of these antinutritional factors present in the skin. Removal of peanut skins is usually one of the initial steps in further processing of peanuts.

It has been suggested (Walker, 1982) that peanuts are atherogenic and this property should be considered an antinutritional effect. Walker argues that it was not the high fat diet that precipitated the atherogenesis but the type of fat, as other lipids have not been implicated in this type of atherogenicity.

4.6 AFLATOXINS IN PEANUTS

In warm wet climates, grains are easily infected with toxigenic microorganisms. Undoubtedly, the best known toxic compounds of peanut are the aflatoxins, metabolic by-products of the moulds *Aspergillus flavus* and *Aspergillus parasiticus*. There are four distinct types of aflatoxin, B_1, B_2, G_1 and G_2, the chemical and physiological aspects of which have been well studied. Of these four, aflatoxin B_1 is the most toxic and best known because it is a very potent hepatocarcinogen (Heathcote and Hibbert, 1977). Aflatoxin B_1 is reported to be mitogenic, causing chemical modifications of the DNA, inhibiting RNA synthesis and interfering with protein synthesis (Heathcote and Hibbert, 1977; Yu, 1977; Clifford and Rees, 1967).

Toxigenic microorganisms are ubiquitous in the environment, and mould infection of badly harvested and/or poorly stored peanuts, occurs around 20–25°C. When these mouldy peanuts are eaten or processed into food or feed, aflatoxin poisoning occurs. Since aflatoxin is adventitious, extreme care in harvesting and storage of the seeds, and uncompromising quality control practices, which ensure that infected peanut batches are promptly rejected, are absolutely necessary. Unfortunately, although aflatoxin infection is no longer a problem in peanuts produced and used in North America, it is still a major problem in peanuts produced in may tropical countries. As a rule therefore, peanuts originating from tropical countries are rather held suspect in North America, until they are shown to be free of aflatoxin. Whether the reputation is deserved or not, there is no doubt that as a result of poor handling and storage conditions, peanuts from Africa, Asia and South America are likely to be infected with aflatoxins. Very stringent inspection and quality control practices are recommended to all peanut processors in the tropics and elsewhere. Since end-users of peanuts produced in the tropics can not effectively control the quality of the peanuts coming in from the field, methods for eliminating aflatoxin from contaminated peanuts have been suggested (Rhee, 1985). These include removal of the toxins from the peanuts by extraction using polar solvents or solvent systems to which have been added 0.5% hydrogen peroxide or 0.2% sodium hypochlorite. The alternative method for elimination of aflatoxicosis has not been as successful as extraction. This method involves the detoxification of aflatoxin with such chemicals as ammonia, chlorine gas and sodium hydroxide.

4.7 NUTRITIONAL EVALUATION OF PEANUT PROTEIN

It is generally known that the concentration and availability of individual essential amino acids are the major determinants of the nutritive value of a protein source. Protein quality determinations of peanut and other protein sources are routinely conducted with rat assays. However, it is important to remember that such protein efficiency ratio (PER) bioassays underestimate protein quality because the rat has a higher requirement for sulphur amino acids than human beings. The level at which peanut protein is fed to experimental animals is critical. For example, Miller *et al.* (1978) fed rats with diets containing 12, 16 or 20% protein supplied by peanut meal. Rats fed 12% protein from peanuts grew at a very slow rate in comparison with those fed 12% protein from casein. At 16% dietary protein (from either peanuts or casein), growth was still slower on peanut protein although it was closer to growth on casein protein. In other words, the higher the level of dietary peanut protein, the closer the utilization of peanut protein approached the utilization of casein. Miller *et al.* (1978) had earlier observed that lysine and methionine, the most limiting amino acids, were not provided at recommended levels even by the high peanut protein diets. Threonine, isoleucine and valine, in that order, were reported to be the next most limiting amino acids at low levels of peanut protein. Results similar to these have also been reported by Miller and Young (1977) who observed that Florunner peanut meal providing a dietary protein level of 16.7%, supported rat growth at a rate comparable with growth on a 12% casein diet.

The scientific literature containing numerous protein quality assays employing protein efficiency ratios (PER) and net protein utilization (NPU) conducted on peanut protein. Alid *et al.* (1981) determined the PER of extrusion-textured peanut protein and peanut flour as 1.54 and 1.57, respectively. Addition of 0.3% DL-threonine, 0.2% lysine and 0.2% DL-methionine to the peanut flour or the texturized peanut protein raised the PER value to 2.18. This was in comparison with casein which had a PER value of 2.50.

From a nutritional quality evaluation of various cultivars of Canadian peanuts Mir and Hill (1979) using a rat bioassay noted that methionine was the first limiting amino acid. Supplementation with lysine alone was not effective but lysine and methionine supplementation was as effective as methionine supplementation alone, indicating that lysine was not a limiting amino acid. A combination of lysine and methionine with threonine increased both PER and NPU beyond previous limits, indicating that threonine was probably the second limiting amino acid. These workers showed that moist heat treatment of 20 min was enough to enhance the nutritional quality of raw peanut meal; heat treatment beyond 20 min depressed nutritive value in proportion to the duration of heat treatment. The effect of superimposition of moist heat on roasting seems to further increase the PER and NPU of roasted peanuts.

Khalil and Chughtai (1983) found, in an investigation of American and Chinese cultivars of peanuts grown in Pakistan, that with the exception of leucine, phenylalanine and tyrosine, all the essential amino acids were in lower amounts than the internationally recommended reference pattern (FAO/WHO, 1973). They observed that lysine, threonine and the sulphur amino acids were the first, second and third limiting amino acids, respectively. Although these reports on the limiting amino acids are similar to some other published data (Chopra and Sidhu, 1967; Daghir *et al.*, 1969), they are at variance with those which find that the sulphur-amino acids are the first limiting acids in some cultivars (Miller and Young, 1977; Miller *et al.*, 1978). It is clear therefore, that the peanut cultivars common in each locality should be studied to determine which are the limiting amino acids.

Begum *et al.* (1991) have reported that supplemental feeding of pregnant women in rural Pakistan with peanut–chickpea mixtures can have a significant positive effect on birth weight and neonatal growth of babies. Devi *et al.* (1990) have shown that the nutritive quality of an indigenous weaning food in Maharashtra, India can be improved by supplementation with peanuts and soybeans, to provide a highly nutritious food for all ages of children and adults. Similar results using peanut, blackeyed peas and melon seed have been reported in a West African diet (Sanchez *et al.*, 1972). Nutritional supplementation of diets of pregnant rural Gambian women has been shown to have a real positive impact by Prentice *et al.* (1983). These workers provided supplemental high energy/protein peanut-based biscuits to pregnant women in negative energy balance. This enhanced their energy/protein intake and resulted in increased birth weights and a reduction in the incidence of low birth weights. In Nigeria a chemical and clinical evaluation of peanut–corn gruel without supplemental milk found that children consuming the peanut–corn gruel gained more weight than controls fed regular village diets (Afolabi *et al.*, 1988).

REFERENCES

Afolabi, O.A., Ojofeitimi, E.O. and Oke, O.L. (1988) Chemical and clinical evaluation of ground nut–maize gruel (Epa-Ogi) in the amelioration of

protein-energy malnutrition in developing countries. *Nutr. Rep. Int.*, **36**, 621–8.
Ahmed, E.M. (1986) Lectin quantification in peanut and soybean seeds. *Peanut Sci.*, **13**(1), 4–7.
Ahmed, E.M. and Applewhite, J.A. (1988) Characterization of trypsin inhibitor in Florunner peanut seeds (*Arachis hypogaea* L). *Peanut Sci.*, **15**(2), 81–4.
Ahmed, E.M., Ali, T. and Sitren, H.S. (1988) Extraction and purification of lectins from Florunner peanut seeds. *Peanut Sci.*, **15**(2), 44–7.
Alid, G., Yamez, E., Aguilera, J.M. *et al.* (1981) Nutritive value of an extrusion-texturized peanut protein. *J. Food Sci.*, **46**, 948–9.
Anantharaman, K. and Carpenter, K.J. (1969a) Effects of heat processing on the nutritional value of groundnut products. 1. Protein quality of groundnut cotyledons for rats. *J. Sci. Food Agric.*, **20**, 703–8.
Anantharaman, K. and Carpenter, K.J. (1969b) Trypsin-inhibitory and growth depressing activities of groundnut skins. *Proc. Nutr. Soc.*, **28**, 46A.
Ayers, J.L. and Davenport, B.L. (1977) Peanut protein: a versatile food ingredient. *J. Am Oil Chem. Soc.*, **54**, 109A–111A.
Begum, N., Hussain, T., Afridi, B. and Hamid, A. (1991) Effect of supplemental feeding on pregnant women on birth weight of the new born. *Plant Foods Human Nutr.*, **41**, 329–36.
Chandrasekhara, M.R., Ramanna, B.R., Jagannath, K.S. and Ramanathau, P.K. (1971) Miltone vegetable toned milk: use of peanut expands supply of milk. *Food Technol.*, **25**, 32–4.
Chopra, A.I.K. and Sidhu, G.S. (1967) Nutritive value of groundnut (*Arachis hypogaea*). 2. Biological evaluation of different varieties of groundnuts grown in the Punjab. *Br. J. Nutr.*, **21**, 583–6.
Clifford, J.I. and Rees, K.R. (1967) The action of aflatoxins B_1 on the rat liver. *Biochem J.*, **102**, 65–75.
Cobb, W.Y. and Johnson, B.R. (1973) *Peanuts: Culture and Uses*. American Peanut Research Education Association Publication. Stone Printing Press, Roanoke, VA.
Conkerton, E.J. and Ory, R.L. (1985) Peanuts as food proteins. In *Developments in Food Proteins* – 5. (ed. B.J.F. Hudson), Elsevier Applied Science Publishers, London.
Daghir, N.J., Ayyash, B. and Pellet, P.L. (1969) Evaluation of groundnut meal protein for poultry. *J. Sci. Food Agric.*, **20**, 349–54.
Devi, R., Boralkar, M.A. and Hamdapukar, V.R. (1990) Nutritional improvement of a traditional weaning food mix (Sattu). *Food Nutr. Bull.*, **12**, 323–4.
Elliot, L. (1965) Beyond fame and fortune. *Readers Digest*, **86**, 261–85, 288–310.
FAO (1990) Production Yearbook. Food and Agriculture Organization of the United Nations, Rome.
FAO/WHO (1973) Energy and Protein Requirements. FAO Nutritional Meeting Series no. 52. WHO Technical Report Series no. 522. Food and Agriculture Organization of the United Nations, Rome.
Heathcote, J.G. and Hibbert, J. (1977) *Aflatoxins: Chemical and Biological Aspects*. Elsevier Science Publishers, Amsterdam.
Higgins, B.B. (1951) Origin and early history of peanuts, and economic importance of peanuts, in *The Peanut*, National Fertilizer Association, Washington, DC.
Joseph, A.L., Berry, B.W., Wells, L.H. *et al.* (1978) Utilization of textured peanut grits in frankfurters. *Peanut Sci.*, **5**, 61–4.
Khalil, J.K. and Chughtai, M.I.D. (1983) Chemical composition and nutritional quality of five peanut cultivars grown in Pakistan. *Qual. Plant. Plant Foods Hum. Nutr.*, **33**, 63–70.
Khan, M.N., Rhee, K.C., Rooney, L.W. and Cater, C.M. (1975) Bread making properties of aqueous processed peanut protein concentrates *J. Food Sci.*, **40**,

580–3.

Lotan, R., Skutelsky, E., Danon, D. and Sharon, N. (1975) The purification, composition and specificity of the anti T Lectin from peanut (*Arachis hypogaea*). *J. Biol. Chem.*, **250**, 8518–23.

Mathews, R.H., Sharpe, E.J. and Clarke, W.M. (1970) The use of some oilseed flours in bread. *Cereal Chem.*, **47**, 181–9.

Miller, J. and Young, C.T. (1977) Protein nutritional quality of Florunner peanut meal as measured by rat bioassy *J. Agric. Food Chem.*, **25**, 653–7.

Miller, J., Dixon, R. and Young, C.T. (1978) Protein nutritional quality of meals made from several cultivars of peanuts as measured by rat bioassay. *Peanut Sci.*, **5**, 19–22.

Mir, Z. and Hill, D.C. (1979) Nutritional value of peanut meals from Ontario grown peanuts compared with a meal from US grown peanuts and a soybean meal. *J. Can. Inst. Food Sci. Technol.*, **12**(2), 56–60.

Nagaraj, H.K. and Subramanian, N. (1974) Studies on groundnut protein concentrates prepared by alcohol and acid washing of the defatted flour. *J. Food Sci. Technol.*, **11**(2), 54–7.

Prentice, A.M., Whitehead, R.G., Watkinson, M. *et al.* (1983) Prenatal dietary supplementation of African women and birth weight. *Lancet*, **i**, 489–92.

Ranhotra, G.S. (1980) Nutritional profile of high protein cookies. *Cereal Foods World*, June, 308–9.

Ranhotra, G.S., Lee, C. and Gelroth, J.A. (1980) Nutritional characteristics of high protein cookies. *J. Agric. Food Chem.*, **28**, 507–9.

Rayner, E.T., Dollear, F.G. and Codifer, L.P. (1970) Extraction of aflatoxins from cottonseed and peanut meals with ethanol. *J. Am. Oil Chem. Soc.*, **47**, 26.

Rhee, K.C. (1985) Peanuts (groundnuts). in *New Protein Foods*, (eds A.M. Altschul and H.L. Wilcke), Food Science and Technology Series, Academic Press, New York.

Rhee, K.C., Natarajan, K.R., Cater, C.M. and Mattil, K.F. (1977) Processing edible peanut concentrates and isolates to inactivate aflatoxins *J. Am. Oil Chem. Soc.*, **54**, 245A–249A.

Rooney, L.W., Gustafson, C.B., Clark, S.P. and Cater, C.M. (1972) Comparison of the baking properties of several oilseed flours. *J. Food Sci.*, **37**, 14–18.

Rosengarten, F. (1984) *The Book of Edible Nuts*. Walker Press, New York.

Sanchez, A., Fuller, A.B., Yahiku, P.Y. and Baldwin, M.V. (1972) Supplementary value of black-eye peas, peanuts and egusi seed on the typical West African diet of plant origin. *Nutr. Rep. Int.*, **6**, 171–9.

USDA (United States Department of Agriculture) (1986) *Composition of Foods: Legumes and Legume Products*. United States Department of Agriculture, Human Nutrition Information Service. Agriculture Handbook no. 8–16. Washington, DC.

USDA (1992) Foreign Agricultural Service Circular. United States Department of Agriculture, Washington, DC.

Walker, A. (1982) Physiological effects of legumes in the human diet. *J. Plant Foods*, **4**, 5–14.

Yu, F.L. (1977) Mechanisms of aflatoxin B_1 inhibition of rat hepatic nuclear RNA systems. *J. Biol Chem.*, **252**, 3245–51.

CHAPTER FIVE

Pigeon pea
(*Cajanus cajan* (L.) Millsp.)

E. Nwokolo

5.1 INTRODUCTION

The pigeon pea, *Cajanus cajan* (L.) Millsp., also known as red gram, is found throughout tropical and subtropical areas of the world, from about 30°N to about 30°S of the equator. There is some uncertainty about the origins of the pigeon pea, and although it is believed to have originated in India, it is found very extensively in Africa (Duke, 1981). Greater use of the pigeon pea as food, is made in India than in most other parts of the world. In India, it is split and skinned into dhal which is relished as excellent food. According to Duke (1981), preparation of dhal involves sun-drying, followed by partial splitting of the seeds by a stone mill. The seeds are then treated with a vegetable oil and stored. During storage, oil is absorbed by the seed coat, which facilitates its final removal. The resulting product is sieved and winnowed to remove any residual seed coats, has an excellent half-globular shape, cooks soft and commands a high price. It may also be cooked and eaten in the green form, or as dried red gram. In West Africa, the mature seeds are soaked in water, cooked and eaten alone or with rice, yam or a variety of vegetables. In the Caribbean, pigeon peas may be harvested green, cut, steamed and canned or as dry seeds, cooked and canned for export. For domestic use, the peas are harvested green or dried and used in a variety of local dishes. Estimates of dry seed yield are in the range 400–600 kg/ha in Africa, 700 kg/ha in Asia and about 1400 kg/ha in North America, whereas green pod yields are estimated at between 1000 and 9000 kg/ha (Duke, 1981).

Pigeon pea and chickpea are the most important pulses of the

Legumes and Oilseeds in Nutrition
Edited by E. Nwokolo and J. Smartt
Published in 1996 by Chapman & Hall. ISBN 0 412 45930 2

developing countries. The Indian subcontinent produces about 90% of the world's supply of pigeon pea, the remaining 10% being produced in Africa and the Caribbean (Kanwar, 1987). Appropriately therefore, a great majority of the research into pigeon pea production and utilization occurs in the Indian subcontinent in universities and institutes, such as The International Crops Research Institute for the Semi-Arid Tropics (ICRISAT).

In Africa, pigeon pea is an important field crop in Kenya, Malawi and Uganda and to a much smaller extent in other African countries. Pigeon pea is considered to have great potential as an important grain legume for human nutrition, in many protein-deficient tropical countries. Because of its potential as a protein supplement to cereal-based foods, the Protein Advisory Group (PAG) of the United Nations recommended that urgent attention be paid to research into the production and nutritional evaluation of pigeon pea and seven other grain legumes (PAG, 1972). Although cowpeas and other beans are the legumes of choice, pigeon pea has the potential to become an important dietary legume because of its relatively cheaper price in comparison with cowpeas. This is indeed happening in Nigeria, Ghana and other West African countries, where the rural and urban poor have turned to pigeon pea for part of their dietary protein requirements. In parts of Nigeria, pigeon pea is referred to a the 'poor man's cowpeas'.

Considerably less research attention has, however, been paid to pigeon pea than most of the other grain legumes notably, soybean, cowpea, peanut, dry bean, broad bean and chickpea. The nutritional value of pigeon pea has been much less studied than the agronomy of the legume. In fact, most research on pigeon pea has concentrated on breeding and other aspects of legume production, to the almost total exclusion of nutritional evaluation.

5.2 NUTRIENT COMPOSITION AND NUTRITIVE VALUE

Studies on the chemical composition of pigeon pea (Mtanga and Sugiyama, 1974; Singh *et al.*, 1975; Deosthale and Rao, 1981) indicate that variety has a significant influence on chemical composition. Deosthale and Rao (1981) found that among varieties, protein content varies from 19.8% to 23.6%, methionine from 1.2% to 1.9% and tryptophan from 0.43% to 0.62%. From a study of amino acid composition of East African legumes Mtanga and Sugiyama (1974) found that pigeon pea is limiting in methionine and tryptophan. Based on calculation of protein content and protein scores, these workers rank pigeon pea ahead of cowpea, but behind soybean in nutritive value. Nutrient profiles of dehusked, mature seed and immature seed vary quite distinctly, apparently depending on the maturity of the seed. Moisture content is

Table 5.1 Proximate composition of pigeon pea seeds

Nutrient	*Dehusked seed*	*Mature seed*	*Immature seed*
Moisture (%)	15.2	9.9	69.5
Protein (%)	22.3	19.5	7.2
Fat (%)	1.7	1.3	0.6
Ash (%)	3.6	3.8	1.4
Fibre (%)	–	1.3	3.3
Carbohydrate (%)	57.2	65.5	21.3
Calcium (%)	0.09	0.16	0.03
Phosphorus (%)	0.26	0.28	0.14
Iron (mg/100 g)	ND	15.0	1.3
Calories (Cal/100 g)	ND	345	117
Thiamine (mg/100 g)	ND	0.72	0.4
Riboflavin (mg/100 g)	ND	0.14	0.25
Niacin (mg/100 g)	ND	2.9	2.4
Carotene (μg/100 g)	45	55	145

From Duke (1981).

low in the mature seed (10%) and is very high (69.5%) in the immature seed (Table 5.1).

Consequently, protein is much higher in dehusked seed (22.5%) and mature seed (19.5%) than in the immature seed (7.5%). Pigeon pea has a low content of lipids (1.3%), a little higher than in many cereals but similar to the fat content in many beans and peas. Pigeon pea, like these other peas and beans, is not a major source of lipid in human diets or diets of animals. As a result, its lipid profile, although interesting, is not that of an important source of unsaturated fatty acids.

Mature pigeon pea has a low fibre content (1.3%) and a high carbohydrate content (65.5%), which is significantly higher than in immature seed (21.8%). Starch is the major component of the carbohydrate fraction in pigeon pea, accounting for 70–80% of all carbohydrates. The oligosaccharides are soluble sugars which are reputed to be responsible for some of the flatulence of beans and peas. Table 5.2 shows the carbohydrate profile of pigeon pea. According to Reddy *et al.* (1984), oligosaccharides account for 6–17% of carbohydrates in pigeon pea. Of the fibre components, hemicellulose and cellulose content are high, and lignin content is low. Fractionation of pigeon pea (Singh *et al.*, 1968) shows that the cotyledons comprise 83% of the whole seed, whereas seed coat comprises 15.5%. The embryo is only a negligible component (1.5%). The seed coat is very high in carbohydrate and fibre, moderate in protein and ash and extremely low in fat content (Table 5.3).

Pigeon pea cotyledons contain very little crude fibre (0.4%), low amounts of ash and fat (4.2% and 4.4%, respectively), moderate amounts of protein (24.3%) and high levels of carbohydrates (66.7%). Pigeon pea

Table 5.2 Carbohydrate profile of pigeon pea

	%
Total carbohydrates	57.3–58.7
Starch	40.4–48.2
Total oligosaccharides	3.5–10.2
Sucrose	2.7
Raffinose	1.0–1.1
Stachyose	2.7–3.0
Verbascose	4.0–4.1
Total crude fibre	1.2–8.1
Lignin	2.9
Cellulose	7.3
Hemicellulose	10.1

From Reddy *et al.*, (1984)

Table 5.3 Nutrient composition of pigeon pea fractions

Constituent	*Seed coat*	*Cotyledon*	*Embryo*
% of whole seed	15.50	83.00	1.50
Ash (%)	3.50	4.17	5.98
Protein (%)	5.60	24.30	48.10
Ether extract (%)	0.30	4.40	13.50
Crude fibre (%)	31.90	0.40	1.40
N-free extract (%)	58.70	66.70	31.00
Phosphorus (mg/100 g)	314	238	90
Calcium (mg/100 g)	917	176	400
Iron (mg/100 g)	9.50	6.10	13.00

From Singh *et al.*, (1968).

embryos contain low levels of fibre (1.4%), and moderate levels of mineral elements (6.0%) and lipids (13.5%). In the seed embryo, carbohydrate content is much lower and the protein content very much higher, than in the seed coat or the cotyledon. Raw pigeon pea has a metabolizable energy value (2.680 kcal/g), that is much lower than the metabolizable energy of toasted pigeon pea (2.909 kcal/g). The metabolizable energy of pigeon pea is similar to values for cowpeas and other edible peas and beans (Nwokolo and Oji, 1985), and slightly lower than that for full-fat soybean. This is because full-fat soybean, although containing less carbohydrate than most beans, contains 15–20 times the fat content of the average bean.

Pigeon pea is therefore not just a protein source but also a contributor of a considerable amount of carbohydrate in the diets of people who

Table 5.4 Mineral content and mineral availability in pigeon pea

Mineral	*Mineral content*	*Mineral availability (%)*
Phosphorus (%)	0.25	59.18
Calcium (%)	0.15	78.45
Magnesium (%)	0.14	50.78
Potassium (%)	1.25	44.54
Zinc (mg/kg)	24	54.10
Iron (mg/kg)	39	55.19
Manganese (mg/kg)	13	54.02
Copper (mg/kg)	18	69.43

From Nwokolo (1987).

Table 5.5 Distribution of lipid fractions

Legumes	*Neutral lipids (%)*	*Phospholipids (%)*	*Glycolipids (%)*
Pigeon pea	51.05	34.37	8.84
Cowpea	46.88	36.82	8.98
Soybean	88.10	9.80	1.60

From Salunke *et al.* (1985).

consume it. As in many oilseeds and legumes, potassium content is very high (1.25%), and phosphorus, calcium and magnesium content moderately high. Iron, zinc, manganese and copper content are very low (Table 5.4). Average availability of minerals in pigeon pea is moderate (58.2%). Of the macro minerals (phosphorus, calcium, magnesium and potassium), availability is high in calcium (78.5%), moderate in phosphorus (59.2%), low in magnesium (50.8%) and very low in potassium (44.5%). Mineral availability is also moderate (58.2%) among the micro elements (zinc, iron, manganese and copper). Table 5.5 compares the distribution of lipids in pigeon pea to that in cowpeas and soybeans. The lipid profiles of pigeon pea and cowpea are similar but quite dissimilar to that of soybean. There is a high content of neutral lipids, a moderate content of phospholipids and a low content of glycolipids in pigeon pea and cowpea. However, in soybean lipids, there is an extremely high content of neutral lipids (88.1%), a low content of phospholipids (9.8%) and an extremely low content of glycolipids (1.6%). The fatty acid profile of pigeon pea lipids indicates a preponderance of saturated fatty acids, accounting for 59% of total fatty acids in the oil (Table 5.6) compared with soybean lipid in which only 14% of the fatty

Table 5.6 Fatty acid profile of pigeon pea and soybean lipids

Fatty acids	*Pigeon pea lipids*	*Soybean lipids*
10:0	2.45	–
12:0	18.56	–
14:0	23.73	0.12
16:0	10.98	10.45
18:0	3.32	3.50
18:1	8.44	22.53
18:2	22.54	54.50
18:3	–	7.90

From Nwokolo (1987).

Table 5.7 Distribution of protein and non-protein nitrogen in seed fractions (dry weight basis)

Seed fraction	*% of seed*	*Protein %*	*Non-protein nitrogen %*
Whole seed	–	20.5	12.8
Embryo	0.7	49.6	6.2
Cotyledon	85.1	22.2	9.5
Seed coat	14.2	4.9	27.4

From Singh and Jambunathan (1982).

Table 5.8 Distribution of protein fractions in pigeon pea components

	Protein fractions					
Seed fraction	*Albumin*	*Globulin*	*Glutelin*	*Prolamin*	*Residue*	*Total*
Whole seed	10.2	59.9	17.4	3.0	5.3	95.8
Embryo	17.0	52.7	21.3	2.7	2.1	95.8
Cotyledon	11.4	64.5	18.2	3.5	1.8	99.4
Seed coat	2.6	26.3	32.8	4.2	23.0	88.9

From Singh and Jambunathan (1982).

acids are saturated. Linoleic acid comprises only 22.5% of total fatty acid content in pigeon pea, compared with a linoleic acid content of soybean lipid of about 54.5%. Table 5.7 shows the percentage of the seed in the embryo, cotyledon and seed coat. Cotyledon (85.2%) and seed coat (14.2%), comprise about 99% of the seed, with the embryo comprising less

Table 5.9 Amino acid content and availability in toasted pigeon pea

Amino acids	*% of total amino acids*	*Content (mg/g seed)*	*% availability*
Aspartic acid	9.9	17.4	82.6
Threonine	3.7	6.4	84.7
Serine	4.9	8.7	84.9
Glutamic acid	19.7	34.7	87.4
Proline	4.8	8.4	51.5
Glycine	4.2	7.3	57.6
Alanine	4.6	8.1	83.0
Cystine	0.8	1.8	81.4
Valine	4.4	7.7	83.6
Methionine	1.1	1.9	77.7
Isoleucine	3.7	6.6	84.8
Leucine	7.8	13.8	85.6
Tyrosine	2.9	5.1	82.9
Phenylalanine	8.9	15.6	90.8
Histidine	5.0	8.8	78.3
Lysine	7.4	13.0	85.4
Arginine	6.9	12.2	87.8
Total amino acids	100.0	175.6	82.3

From Nwokolo (1987).

than 1% of the seed. The embryo, though very small as a proportion of the seed, is extremely high in protein content (about 50%). The cotyledon has a protein content of 22%, whereas the seed coat contains only about 5% protein. Only the seed coat has a large amount of non-protein nitrogen (27.4%), compared with 6.2% and 9.5% in the embryo and cotyledon, respectively. Fractionation of the protein of pigeon pea (Singh *et al.*, 1975) showed a high content of globulins in the whole seed, the embryo and the cotyledon, but not in the seed coat (Table 5.8). There is a moderate content of glutelins and albumins in the whole seed, embryo and cotyledons.

The seed coat, on the other hand, has a very low content of albumins and a high content of glutelins. The amino acid profile of pigeon pea is presented in Table 5.9. When expressed as a percentage of total amino acids, the amino acid profile is similar to that in most beans, peas and edible pulses. However, concentration of amino acids (mg/g) is inferior to that of most edible pulses including cowpea, kidney bean and field pea. As is the case with many oilseeds and legumes, there is a deficiency of the sulphur amino acids, methionine and cystine. Other essential amino acids like histidine and threonine are marginally deficient and may need to be supplemented, through use of other protein sources which are high in these amino acids. The protein of pigeon pea may not be considered as being deficient in lysine, though it is not a particularly high source of this amino acid. The distribution of amino acids in the embryo, cotyledon and seed coat of pigeon pea is shown in Table 5.10. There is an approximately

Table 5.10 Amino acid composition of components of pigeon pea

	Content (g/16 g N)		
Amino acid	*Embryo*	*Cotyledon*	*Seed coat*
Lysine	7.0	7.1	3.9
Histidine	3.3	3.9	1.0
Arginine	8.4	7.1	3.2
Aspartic acid	10.1	11.0	5.7
Threonine	4.7	4.3	2.5
Serine	5.3	5.5	3.5
Glumatic acid	16.2	20.6	6.9
Proline	4.7	4.3	3.1
Glycine	4.5	4.0	4.5
Alanine	6.0	4.5	3.0
Cystine	1.7	1.3	–
Valine	6.6	5.6	3.2
Methionine	1.4	1.2	0.7
Isoleucine	4.4	4.3	2.9
Leucine	7.1	7.8	4.0
Tyrosine	3.8	3.2	1.7
Phenylalanine	6.8	10.7	2.3

From Singh and Jambunathan (1982).

equal distribution of most amino acids between the embryo and cotyledon fractions of pigeon pea, with very little in the seed coat. Cystine is absent from the pigeon pea seed coat, methionine and histidine are in such low proportions as to be practically lacking, and threonine, phenylalanine, valine, arginine and isoleucine are marginally deficient. This is indicative of the fact that the seed coat protein is of very poor quality, and cannot meet the amino acid requirements of human subjects or monogastric animals.

Availability of amino acids in pigeon pea (Table 5.9) ranges from 51.5% (proline) to 90.8% (phenylalanine), with a mean availability of 82.3%. Most essential amino acids in pigeon pea are highly available, indicating that they will be well utilized in meeting the body's requirement for amino acids for maintenance, growth or production. Average availability of pigeon pea amino acids is lower than in soybean, melon seed or pumpkin seed, higher than in African yam bean and African breadfruit seed, and similar to bambara groundnuts, all products consumed by human subjects in West Africa and many other tropical countries of Africa. Toasting seems to improve the digestibility of protein in pigeon pea. A comparison of protein digestibility in toasted and raw pigeon pea (Nwokolo, 1987) has shown that in raw pigeon pea, true protein digestibility was 69.4%, whereas in toasted pigeon pea, the value was 80.6%, similar to a digestibility coefficient of 84.1% in cooked soybean.

5.3 TOXICOLOGICAL COMPONENTS

Most legumes contain various antinutritional factors. These may range from the relatively innocuous phytates, oligosaccharides, polyphenols and antivitamins, to the harmful enzyme inhibitors, haemagglutinins and saponins. Those in soybean and other more popular legumes have been extensively documented (Dijkhof and Poort, 1978; Sotelo *et al.*, 1980; Rackis, 1981). There is, however, little information on the toxicological components of pigeon pea. Elias *et al.* (1976), report the existence of trypsin inhibitors, estimated at 10.1 TUT, in contrast to a much higher content of 25.5 TUT in soybean. Duke (1981) found both trypsin and chymotrypsin inhibitors, although the level at which these inhibitors are contained in pigeon pea is not indicated. Whatever the nature and level of antinutritional factors in pigeon pea, they seem to be completely destroyed or inactivated by heat treatment. Cooking, toasting and other heat treatment procedures seem enough to enhance the nutritional qualities of pigeon pea, well above the nutritive value of raw pigeon pea.

5.4 ANIMAL GROWTH EXPERIMENTS

Only a few experiments have been reported, in which animal growth or productivity was measured, to evaluate pigeon pea as a feed source. In growth trials with rats fed pigeon pea, cowpeas or soybeans (Elias *et al.*, 1976), observed protein efficiency ratios (PER) were 0.89, 1.40 and 2.15, respectively. Growth trials of pigeon pea-based diets with rats (Sotelo *et al.*, 1980), showed a net protein utilization value of 46.0, a biological value of 53.7 and true protein digestibility of 85.8%. Other studies with rats and pigs, compared pigeon pea with chickpea and soybean. These workers observed that for both animals, the quality of pigeon pea protein was significantly less than that of chickpea or soybean.

REFERENCES

Deosthale, K.G. and Rao, O.S.S. (1981) Mineral and trace element composition of red gram (*Cajanus cajan, indicus*). *Ind. J. Nutr. Diet.*, **18**, 130–5.

Dijkhof, J. and Poort, C. (1978) Changes in rat pancreatic synthesis after a single feeding with diets containing raw or heated soybeans. *J. Nutr.*, **108**, 1222–8.

Duke, J.A. (1981) *Handbook of Legumes of World Economic Importance*. Plenum Press, New York.

Elias, L.G., Cristales, F.R., Bressani, R. and Miranda, H. (1976) Chemical composition and nutritive value of some grain legumes. *Nutr. Abs Rev.* (Series B 1977), **47**, 864.

Kanwar, J.S. (1987) Relevance of ICRISAT research to Africa, in *Research on Grain Legumes in Eastern and Central Africa*. International Crops Research Institute for the Semi-Arid Tropics (ICRISAT), Andhra Pradesh, India.

Mtanga, L.A. and Sugiyama, T. (1974) A note on the amino acid composition of some legume seeds grown in Tanzania. *East African Agric. For. J.*, **39**, 307–10.

Nwokolo, E. and Oji, U.I. (1985) Variation in metabolizable energy of raw or autoclaved white and brown varieties of three tropical legumes. *Anim. Feed Sci. Technol.*, **13**, 141–6.

Nwokolo, E. (1987) Nutritional evaluation of pigeon pea meal. *Plant Foods Hum. Nutr.*, **37**, 283–90.

PAG (1972) *Nutritional Improvement of Food Legumes by Breeding*. Proceedings of Protein Advisory Group Symposium. Food and Agriculture Organization of the United Nations, Rome, July 3–5.

Rackis, J.J. (1981) The significance of soya trypsin inhibitors in nutrition: effect on digestibility. *J. Am. Oil Chem. Soc.*, **58**, 495–501.

Reddy, N.R., Pierson, M.D., Sathe, S.K. and Salunkhe, D.K. (1984) Chemical, nutritional and physiological aspects of dry bean carbohydrates: a review: *Food Chem.*, **13**, 25–68.

Salunkhe, D.K., Kadam, S.S. and Chavan, J.K. (1985) *Postharvest Biotechnology of Food Legumes*, pp. 29–52. CRC Press, Boca Raton, FL.

Singh, U. and Jambunathan, R. (1982) Distribution of seed protein fractions and amino acids in different anatomical parts of chickpea and pigeon pea. *Qual. Plant. Plant Foods Hum. Nutr.*, **31**, 347–54.

Singh, L., Sharman, D., Deodhar, A.D. and Sharman, Y.K. (1975) Variation in protein, methionine, tryptophan and period in pigeon pea (*Cajanus cajan* L.). *Indian J. Agric. Sci.*, **43**, 795–8.

Singh, S., Singh, H.D. and Sikka, K.C. (1968) Distribution of nutrients in anatomical parts of common Indian pulses. *Cereal Chem.*, **45**, 13–17.

Sotelo, A., Arteaga, M.E., Frias, M.I. and Gonzales-Garza, M.I. (1980) Cytotoxic effects of two legumes in epithelial cells of the small intestine. *Qual. Plant. Plant Foods Hum. Nutr.*, **30**, 79–85.

CHAPTER SIX

Jack bean (*Canavalia ensiformis* (L.) D.C.)

E. Nwokolo

6.1 INTRODUCTION

Jack bean *Canavalia ensiformis* (L.)D.C. belongs to an immense family, the Leguminosae (Fabaceae), which consists of 500 genera and 15 000 species of both cultivated and wild plants. Members of the *Leguminosae* are spread worldwide and are cultivated in all climates. The jack bean consists of both bush and climbing types. Kessler (1990) describes the bush types as compact, erect and self-supporting, forming a canopy at a height of about 100 cm, and developing stiff lateral branches at the lower nodes. The climbing types have been described (Kessler, 1990) as lax, with a tendency to wind their stems around supports, and possessing very long internodes.

Germination of sown seed occurs within a week, foliage and stems are green, the stems turning to a reddish colour with onset of maturity. Flowering in the jack bean may be classified into three categories: (1) early season flowering cultivars (58 days after sowing); (2) mid-season flowering cultivars (84 days after sowing); and (3) late-season flowering cultivars (90–110 days after sowing). Kessler observes that early season and mid-season flowering jack bean cultivars have a much shorter flowering period than late-season flowering cultivars. Most of the bush types are early flowering whereas, most of the climbing types are late-season flowering plants. Early flowering, bush type jack bean cultivars are relatively small plants growing to an average height of about 77 cm, mid-season flowering cultivars grow to an average height of about 140 cm,

Legumes and Oilseeds in Nutrition
Edited by E. Nwokolo and J. Smartt
Published in 1996 by Chapman & Hall. ISBN 0 412 45930 2

and late-flowering bush types attain an average height of about 130 cm. Late-flowering climbing types grow to an average height of about 200 cm. One, two or three pods may be borne on a peduncle, and there are four to seven pods per plant. Early flowering bush types have an average of 9–10 seeds per pod, whereas mid-season and late-flowering bush types have 11–12 seeds per pod. Pods are ready for harvest in about 160 days after planting. Most common jack bean seeds are white in colour and weigh 1.5–2 g each. In a study of 18 accessions of jack beans from different countries of the world, Kessler (1990) reported that average seed yield was 2150 kg/ha. with a range of 1733–2655 kg/ha. Bressani *et al.* (1987) report the yield of grain from 11 experimental plots as 2400–3400 kg/ha, although in general, yields may be as low as 800–1000 kg/ha and as high as 4600 kg/ha. The yields of jack bean in Latin America compare well with yields of common bean (*Phaseolus vulgaris*) and soybean (*Glycine max*), estimated at 500–700 kg/ha and 600–1000 kg/ha, respectively (Bressani *et al.*, 1987).

6.2 CURRENT AND POTENTIAL USAGE AS FOOD

Even though jack bean has a very high protein content, it is considered a minor food legume because its present use is very highly limited as food for humans. It is, however, used to a slightly greater extent in livestock feeding. There is considerable potential for enhanced use of jack bean both as human food and as a major source of nutrients in livestock diets. In the 1970s the National Academy of Sciences of the USA recommended the jack bean as a crop whose cultivation should be encouraged because it is capable of producing dependable yields under conditions where other common food legumes would not thrive. It therefore has the potential to increase human nutrient intake and enhance animal productivity in hot and arid climates. Gopalan *et al.* (1982) describe the jack bean as a rich source of proteins, essential amino acids, minerals, thiamine and riboflavin. Barbar *et al.* (1988) note that in spite of its nutrient content, jack bean is not accepted or utilized as human food due to the presence of fairly high concentrations of various antinutritionals like lectins, polyphenols and trypsin inhibitors. There is also the presence of oligosaccharides which cause flatulence manifested as intestinal discomfort, rectal gas discharge, stomach rumbling, diarrhoea and nausea. Bressani and Sosa (1990) from a study of the effects of processing on the nutritive value of jack bean conclude that both extrusion and pressure cooking with lime are superior to pressure cooking without lime, or roasting, in improving the nutritive value of the jack bean. These workers also note that 30 min of pressure cooking in the presence of calcium hydroxide (0.45% by seed weight) is more beneficial than longer periods of pressure cooking in water, or roasting. Barbar *et al.* (1988) found that

Table 6.1 Proximate composition of jack bean seed

Nutrients	*Rodrigues and Torne (1990)*	*Bressani* et al. *(1987)*	*Revilleza* et al. *(1990)*
Moisture (%)	not shown	13.5	8.40–8.56
Total soluble sugar (%)	4.20	not shown	not shown
Starch (%)	34.94	not shown	not shown
Protein (%)	32.23	26.9	27.82–29.41
Fat (%)	2.9	1.8	2.46–2.66
Fibre (%)	9.4	8.5	5.30–5.50
Ash (%)	3.2	3.2	3.95–4.59
Total carbohydrates	not shown	46.1	49.48

Table 6.2 Mineral composition of jack bean (dry matter basis)

Mineral (mg/100 g)	*Source 1*	*Source 2*
Phosphorus	338	360
Calcium	247	170
Magnesium	144	170
Zinc	43	2.87
Copper	10	1.22
Nickel	4	not shown
Magnesium	not shown	0.87
Iron	not shown	trace

Source 1: Rodrigues and Torne, 1990
Source 2: Bressani *et al.*, 1987

application of dry heat is not effective in destroying trypsin inhibitors or reducing the content of polyphenols in jack bean. However, soaking the beans in water for 24 h followed by cooking for 20 min is quite effective. Germination of the seeds for 40 h reduces the content of trypsin inhibitors by 31% and polyphenols by 35%. Abbey and Ibem (1987) in an investigation of the functional properties of raw and pressure cooked jack bean, found that the flours have good water- and oil-absorption properties but poor gel-forming capability. They also note that processing affects such functional properties of jack bean flour as foaming and emulsion capacity.

6.3 NUTRIENT COMPOSITION AND NUTRITIVE VALUE OF JACK BEAN

Table 6.1 shows the proximate composition of jack bean. The bean is a

Table 6.3 Amino acid content (g/g N) of raw and processed jack bean

Amino acid	*Jack bean (raw)*	*Jack bean (cooked)*	*Jack bean (roasted)*	*FAO reference*
Lysine	0.46	0.38	0.38	0.34
Histidine	0.23	0.18	0.22	
Arginine	0.34	0.35	0.32	
Aspartic acid	0.64	0.26	0.64	
Threonine	0.26	0.28	0.23	0.25
Serine	0.30	0.29	0.30	
Glutamic acid	0.65	0.66	–	
Proline	0.25	0.23	0.25	
Glycine	0.26	0.22	0.24	
Alanine	0.27	0.23	0.25	
Valine	0.34	0.30	0.35	0.31
Methionine	0.06	–	–	0.22
Isoleucine	0.39	0.43	0.44	0.25
Leucine	0.43	0.47	0.48	0.44
Tyrosine	0.21	0.20	0.20	0.38
Phenylalanine	0.26	0.27	0.26	0.38
Trytophan	0.05	not shown	not shown	0.06

Source Bressani *et al.* (1987).

good source of protein. Indeed, the protein content of jack bean has been estimated at between 25.6% (Bressani and Sosa, 1990) and 32.2% (Rodrigues and Torne, 1990), depending on the cultivar. The seed contains 46–49% or more, of total carbohydrates, also depending on the cultivar analysed. Of this, starch is about 35%, crude fibre 5–9% and soluble sugars about 4%. As in most edible dry beans, the seed is low in oil content. Table 6.2 shows the concentration of some important mineral elements in jack bean. There is a high content of phosphorus and calcium, although a significant fraction of the phosphorus is most probably present as the largely unavailable phytate phosphorus. There is a moderate content of magnesium, a low content of zinc, and a very low content of copper. Most other minerals are present in only trace amounts.

Table 6.3 shows the content of essential amino acids for raw, pressure-cooked and roasted jack bean. The amino acid profile is similar to that of many dietary oilseeds and legumes, and as is common in these legumes and oilseeds, there is a low content of the sulphur amino acids methionine and cystine. It is obvious from Table 6.3 that pressure cooking or roasting does not significantly affect the amino acid content of jack bean. Jack bean used as sole source of protein meets and surpasses the FAO reference pattern of amino acids in respect of lysine, threonine, valine, isoleucine and leucine, but is deficient in respect of methionine, tyrosine phenylalanine and possibly tryptophan.

Starch content (3.5%) and total carbohydrate content (46–49.5%) are

Table 6.4 Non-reducing sugars in jack bean, lima bean and rice bean

	Sucrose (%)[a]	*Raffinose* (%)[a]	*Stachyose* (%)[a]	*Verbascose* (%)[a]	*Total sugars*
Jack bean A.1	1.49	0.34	1.44	0.01	1.79
Jack bean 8.8	2.47	0.86	1.74	0.11	2.71
Lima bean	1.68	0.93	1.87	0.09	2.89
Lima bean	2.02	1.11	1.99	0.12	3.22
Rice bean	1.86	1.14	0.37	0.74	2.25
Rice bean	3.50	1.73	0.37	0.45	2.55

[a] % dry matter of seed.
Revilleza *et al.* (1990).

high as is common and expected in edible beans. Jack bean has a lower content of all non-reducing sugars than the rice bean.

Content of total oligosaccharides in jack bean is lower than in lima bean and rice bean (Table 6.4). Jack bean also has a lower content of the individual non-reducing sugars than lima bean, and, with the exception of stachyose, has a lower content of these non-reducing sugars than rice bean.

6.4 ANTINUTRITIONAL SUBSTANCES IN JACK BEAN

Jack bean contains canavanine, 2-amino-4-(guanido-oxy) butyric acid, a potentially toxic analogue of arginine. Canavanine has been extracted from the jack bean and is used mainly as a powerful insecticide. Canavanine is a non-protein amino acid but because it is an arginine analogue, canavanine-sensitive organisms, mainly insects, metabolize it by incorporating it into protein macromolecules. Synthesis and utilization of such aberrant and impaired canavanine-containing proteins adversely affects various developmental processes in insects, manifesting itself as toxicity to these organisms. Content of canavanine has been reported to be as high as 26.5 g/kg (Natelson, 1985). It is likely that this high content of canavanine may interfere with the metabolism of arginine in the body (Rosenthal, 1977, 1991).

Three haemagglutinating proteins, canavalin and concanavalin A and concanavalin B, have been identified in the jack bean, which are responsible for the high haemagglutinating activity of jack bean. Other antinutritional components which have been identified/isolated from the jack bean include a concavalin-like lectin (Carlini and Guimares, 1981; Raychaudhuri and Singh, 1986), canatoxin (Ribeiro-Da Silva *et al.*, 1986) and a non-haemagglutinating factor (Dennison *et al.*, 1971). Most of these, like other haemagglutinating proteins, are destroyed by moist heat and are of no consequence when jack bean is properly cooked or autoclaved.

Table 6.5 Content and inactivation of antinutritional substances

	Trypsin inhibitor activity (TIU/g sample)		*Polyphenols (%)*	
Treatment	*Meal*	*Seed*	*Meal*	*Seed*
Dry heat				
Control	12.4	12.4	1.31	1.31
50 °C	12.0	12.4	1.11	1.18
100 °C	9.4	11.5	0.71	0.94
Cooking time				
Control	–	12.4	–	1.31
10 min	–	11.6	–	1.11
30 min	–	8.6	–	0.65
60 min	–	6.4	–	0.53
Autoclaving time				
Control	12.4	–	1.31	–
10 min	3.3	–	1.03	–
20 min	undetectable	–	0.37	–
30 min	undetectable	–	0.12	–
Germination				
Control	–	11.6	–	1.03
12 h	–	10.9	–	0.94
24 h	–	10.2	–	0.85
48 h	–	8.6	–	0.65

Source: Barbar *et al.* (1988).

In addition to haemagglutinins, jack bean contains high levels of trypsin inhibitors, polyphenols and oligosaccharides. Barbar *et al.* (1988) conducted an extensive study of the various processing practices (dry heat, cooking, autoclaving and germination) on trypsin inhibitor activity and polyphenol content of jack bean. They found that the most effective procedure for destroying trypsin inhibitors and reducing polyphenol content is to autoclave at not less than 120°C for about 10 min (Table 6.5). Barbar *et al.* (1988) report that jack bean trypsin inhibitor is resistant to dry heat treatment, similar to the trypsin inhibitor of winged bean (Kadam and Smithard, 1987) and moth bean (Kadam *et al.*, 1986). Moist heat (120°C for 20 min) on the other hand causes rapid destruction and elimination of both trypsin inhibitors and polyphenols in the seed. An even more dramatic loss of trypsin inhibitors and polyphenols occurs when jack bean is soaked in water overnight prior to cooking. Since polyphenols are primarily located in the seed coat, soaking prior to cooking may cause increased leaching of these compounds from the seed.

Bressani and coworkers at the Institute of Nutrition of Central America and Panama (INCAP) demonstrated that the nutritive value of *Canavalia* seeds may be enhanced by alkaline cooking and extrusion cooking more than by regular cooking and roasting. Although the average weight gain and protein efficiency ratio of rats fed jack bean seeds was much lower

Table 6.6 Effect of processing methods on protein quality of jack bean

Processing	*Average gain (g)*	*Feed intake (g)*	*Protein efficiency*
Roasting	24±5	205±22	1.12±0.20
Alkaline cooking	46±9	270±43	1.61±0.11
Cooking	32±4	228±14	1.36±0.12
Extrusion cooking	45±6	264±28	1.66±0.16
Casein control	133±20	448±42	2.81±0.23

Source: Bressani *et al.* (1987).

than in rats fed casein, these experiments have established that jack bean can be an important source of nutrients when properly cooked or processed (Vaidehi and Shivaleeva, 1984; Bressani *et al.*, 1987; Bressani and Sosa, 1990).

6.5 HUMAN AND ANIMAL NUTRITION EXPERIMENTS

The jack bean is not currently used to any significant extent in human nutrition, but it should be. In spite of the reported high toxicity of the raw bean, it is really a bean which can be easily processed into an edible high quality protein for human and animal consumption. When subjected to moist heat for 120°C for about 30 mins, there is complete destruction and elimination of antinutritional constituents in the bean. That level of moist heating is consistent with regular cooking of most edible beans. Indeed, feeding experiments with rats in which low nutritional quality has been reported with raw jack bean or other beans are in practice meaningless, because no bean is consumed raw.

In experiments in which Bressani *et al.* (1987) fed raw, pressure-cooked or roasted jack bean, protein efficiency ratios (PER) were 0.1, 1.21 and 1.18, respectively (Table 6.6). Protein digestibility coefficients were 47.9, 76.4 and 78.7% respectively. A control diet of casein had a PER of 2.89 and a protein digestibility of 90.2%. Bressani *et al.* (1987) report a protein value of 41.5% for jack bean, a value similar to that for commonly eaten legumes. The poor protein quality of jack bean and other edible grain legumes is due to the low content of sulphur amino acids in these legumes. To verify that the growth-limiting amino acid was indeed methionine, Bressani *et al.* (1987) supplemented a jack bean-based diet (10% protein) with 0.3% DL-methionine. The PER rapidly increased from 0.70 to 1.94, a PER equal to 69.5% of the protein value of casein. Bressani *et al.* (1987) suggest that for optimal utilization of jack bean, it is advisable to supplement with both methionine and tryptophan which could also be limiting. If indeed processed jack bean supplemented with methionine and tryptophan has a nutritional value approximately 70% of that of casein, then we as nutritional and agricultural scientists have a duty to provide this information to a hungry world.

REFERENCES

Abbey, B.W. and Ibem, G.O. (1987) Functional properties of raw and heat processed brown bean (*Canavalia rosea* D.C.) flour. *J. Food Sci.* **52**, 406–8.

Barbar, V.S., Chavan, J.K. and Kadam, S.S. (1988) Effects of heat treatments and germination on trypsin inhibitor activity and polyphenols in jack bean (*Canavalia ensiformis* (L.) D.C.). *Plant Foods Hum. Nutr.* **38**, 319–24.

Bressani, R. and Sosa, J.L. (1990) Effect of processing on the nutritive value of *Canavalia* jack beans (*Canavalia ensiformis*. L.). *Plant Foods Hum. Nutr.* **40**, 207–14.

Bressani, R., Brenes, R.G., Garcia, A. and Elias, L.G. (1987) Chemical composition, amino acid content and protein quality of *Canavalia* spp. seeds. *J. Sci. Food Agric.*, **40**, 17–23.

Carlini, C.R. and Guimares, J.A., (1981) Isolation and characterization of a toxic protein from *Canavalia ensiformis* (jack bean) seeds distinct from concanavalin A. *Toxicon (Oxford)*, **19**, 667–75.

Dennison, C., Stead, R.H. and Quicke, G.V., (1971) A non-hemagglutinating toxic factor from the jack bean (*Canavalia ensiformis*). *Landbouwet S. Afric Agric. Sci-S. Afric Agroplantae*, **3**(2), 27–9.

Gopalam, C., Sastri, B.V.R. and Balasubramanian, S.C. (1982) *Nutritive Value of Indian Foods*. National Institute of Nutrition, Hyderabad, India.

Kadam, S.S. and Smithard, R.R. (1987) Effect of heat treatment on trypsin inhibitor and he magglutinating activities in winged bean. *Plant Foods Hum. Nutr.*, **37**, 151–7.

Kadam, S.S., Ghorpade, V.M., Adsule, R.N. and Salunkhe, D.K. (1986) Trypsin inhibitors in moth bean: thermal stability and changes during germination and cooking. *Plant Foods Human Nutr.*, **36**, 43–6.

Kessler, C.D.J. (1990) An agronomic evaluation of jack bean (*Canavalia ensiformis*) in Yucatan, Mexico. III. Germplasm. *Expr. Agric.*, **26**, 31–40.

Natelson, S. (1985) Canavanine to arginine ratio in alfalfa (*Medicago sativa*), clover (*Trifolium*) and jack bean (*Canavalia ensiformis*). *J. Agric Food Chem.*, **33**, 413–19.

National Academy of Sciences (1979) *Tropical Legumes: Resources For the Future.* National Academy of Sciences, Washington DC.

Raychaudhuri, M. and Singh, M. (1986) A Concavalin-A-like lectin in the developing seed of *Canavalia ensiformis. Phytochemistry*, **25**, 793–7.

Revilleza, M.J.R., Mendoza, E.M.T. and Raymundo, L.C. (1990) Oligosaccharides in several Philippine indigenous food legumes: determination, localization and removal. *Plant Foods Hum. Nutr.*, **40**, 83–93.

Ribeiro-DaSilva, G., Carlini, C.R., Pires-Barbosa, R. and Guimaraes, J.A. (1986) Blood glucose alterations induced in rats by canatoxin, a protein isolated from jack bean (*Canavalia ensiformis*) seeds. *Toxicon*, **24**, 775–82.

Rodrigues, B.F. and Torne, S.G. (1991) A chemical study of seeds in three *Canavalia* species. *Trop. Sci.*, **31**, 101–3.

Rosenthal, G.A. (1977) The biological effects and mode of action of L-canavanine, a structural analogue of L-alanine. *Q. Rev. Biol.*, **52**, 155–78.

Rosenthal, G.A. (1991) The biochemical basis for the deleterious effect of L-canavanine. *Phytochemistry*, **30**, 1055–8.

Vaidehi, M.P. and Shivaleeva, H.B. (1984) The protein efficiency ratio of extrusion-cooked blends of jack bean and semolina. *Nutr. Rep. Int.*, **30**, 1173–82.

CHAPTER SEVEN

Chickpea (*Cicer arietinum* L.)

A. Sotelo and R.N. Adsule

7.1 INTRODUCTION

The chickpea is a member of the family Fabaceae (or Leguminosae). The genus *Cicer* to which it belongs, contains some 27 species. It is a small herbaceous annual with a semispreading growth habit and profuse branching. The flowers are typically papilionate and the colour of the corolla may vary from purple, reddish blue or pink to white. Pods are oblong and contain one or two seeds. Seed colour ranges from white through light-brown, dark-brown, brownish-yellow, pinkish-brown or green to black. The seed may be smooth or wrinkled and has a characteristic prominent beak.

The chickpea is a crop initially from the Middle East and can easily be produced in other areas of the world which experience a Mediterranean type of climate. It is important that the crop is able to establish itself in a moist environment and mature in dry conditions. It is an important cool season crop in the Indian subcontinent, and besides the Mediterranean basin it is cultivated in suitable areas of Latin America and to some extent also in Africa. The Indian subcontinent is by far the greatest producer and is responsible for the production of approximately three quarters of the world crop. Mexico is the major producer in the New World. The chickpea is known under a variety of names: Bengal gram, garbanzo, chana, chhole and boot. Two distinct types of chickpea are recognized. The smaller darker desi type is favoured in India and the kabuli type, which is larger and pale in colour, is preferred in the Mediterranean and Middle East (Tables 7.1–7.3).

Legumes and Oilseeds in Nutrition
Edited by E. Nwokolo and J. Smartt
Published in 1996 by Chapman & Hall. ISBN 0 412 45930 2

Table 7.1 The major world producers of chickpea in 1989

Country	*Chickpeas (MT)*
India	5 062 000
Turkey	683 000
Pakistan	456 000
Myanmar	240 000
China	175 000
Mexico	170 000
World	7 430 000

Table 7.2 Production of chickpea in Mexico in 1985

Product	*Production (MT)*	*State where produced*
Chickpea (white)	116 698	Jalisco, Sinaloa, Sonora
Chickpea (forage)	22 774	Jalisco, Michoacan

Table 7.3 Exports of Mexican chickpea in 1988

Country	*Amount (MT)*	*Country*	*Amount (MT)*
Brazil	75	India	4183
Costa Rica	58	Italy	3027
Cuba	4001	Puerto Rico	100
Spain	9 361	Sweden	17
United States	11 281	Uruguay	18
Philippines	34	Venezuela	1136
Total exported	33 296		
Total production	50 875		
Percentage exported	65%		

7.2 UTILIZATION

The chickpea is not only the source of a major pulse for human consumption, it also plays a very significant role in animal nutrition. It is useful in livestock rations for swine and other monogastric animals. Chickpeas used in animal feeding are of the dark seeded, small grained

Table 7.4 Chemical composition of chickpea seeds

Constituent	*Content (/100 g dry seeds)*
Protein (g)	22.0
Fat (g)	4.5
Carbohydrates (g)	70.8
Ash (g)	2.7
Calcium (mg)	114
Phosphorus (mg)	387
Iron (mg)	6.2
Magnesium (mg)	168
Sodium (mg)	30
Copper (mg)	2.3
Zinc (mg)	3.6
Thiamine (mg)	0.3
Riboflavin (mg)	0.2
Niacin (mg)	2.0

Compiled from Gopalan *et al.* (1982), Maga and Sizer (1978) and Salunkhe *et al.* (1985).

Table 7.5 Essential amino acid composition of chickpea seed proteins (g/16 g N)

	Content (g/16 g N)	
Amino acid	*Chickpea*	*FAO reference protein*
Lysine	7.3	5.5
Isoleucine	5.3	4.0
Leucine	9.6	7.0
Valine	5.1	5.0
Methionine + cystine	2.6	3.5
Phenylalanine + tyrosine	9.0	6.0
Threonine	3.7	4.0
Tryptophan	0.8	1.0

Compiled from Gopalan *et al.* (1982) and FAO/WHO (1973).

desi type. There is no significant difference in chemical composition and nutritive value of this type as compared with kabuli chickpeas. The seed can be used in compounding concentrates for feeding, ruminants, non-ruminants and poultry.

7.2.1 Utilization as human food (see also Chapter 2)

The chickpea is an excellent pulse and it is used as such in most if not all areas of production. Preparation presents few difficulties as it contains

Table 7.6 Protein content, protein efficiency ratio (PER) and digestibility of Mexican varieties of chickpea

Sample	*Protein (%)*	*PER*	*Digestibility (%)*
Surutato	16.9	2.31 ± 0.30	84.4 ± 1.7
Sonora	20.6	1.79 ± 0.19	78.7 ± 3.8
Porquero	20.1	2.00 ± 0.35	71.2 ± 2.9
Macarena	20.7	1.65 ± 0.35	73.2 ± 4.0
Breve Duro	19.2	2.22 ± 0.22	77.8 ± 4.0
Surutato 77	18.9	1.99 ± 0.36	76.5 ± 4.0
Sonora 80	18.7	1.91 ± 0.22	77.5 ± 1.6
Quebrado	20.2	1.81 ± 0.14	77.5 ± 1.9
Commercial	19.9	1.74 ± 0.17	83.1 ± 1.6

From Sotelo *et al.* (1987a).

only low concentrations of antinutritional factors (Cravioto *et al.*, 1954). It is a feature of Asian (especially Indian), Mediterranean and Mexican cuisines and as a result of tourism such dishes are gaining popularity outside areas of production. The chickpea also has a high reputation as a health food and is an important component of vegetarian diets. It is very palatable and can be consumed on its own or mixed with other pulses and vegetables. It also is popular in salads and may also be canned in brine or with meats.

Chickpea flour can be used to enrich cereal flours without impairing flavour, baking quality and the like in preparation of bread, cookies and tortillas (Hernandez and Sotelo, 1984, 1987). Ulloa *et al.* (1988) have prepared a chickpea-based concentrate for infant feeding formulas. The nutritional value of this was enhanced by supplementation with methionine and its digestibility was significantly higher than raw chickpea.

7.3 CHEMICAL COMPOSITION AND NUTRITIONAL VALUE

The chemical composition of chickpea seeds is given in Table 7.4. Composition can vary significantly as a result of genetic and environmental factors (Chavan *et al.*, 1989). Chickpea seeds contain relatively less protein than those of many other legume species. Globulin is the major component of the protein fraction followed by glutelin, albumin and prolamin. Non-protein nitrogen is present in significant amounts in the seed. Methionine is the first limiting amino acid in chickpea protein (Table 7.5). The biological value (BV) of chickpea seed protein is reported to be in the range 52–85%, protein efficiency ration (PER) varies between 1.2 and 2.6, the digestibility coefficient (DC) is 78–93% and the net protein utilization (NPU) is 87–92% (Pak and Barja, 1977; Chandrasekharappa,

Table 7.7 Protein content, protein efficiency ratio (PER) and digestibility of chickpea and chickpea + cereal mixtures

Sample	*Protein (%)*	*PER*	*Digestibility (%)*
Chickpea	22.0	1.45 ± 0.15	71.0 ± 0.4
Corn	9.0	1.10 ± 0.15	74.5 ± 3.2
Corn–chickpea (80:20)[a]	10.9	2.09 ± 0.11	69.6 ± 4.2
Chickpea	17.2	2.11 ± 0.35	80.0 ± 4.2
Wheat	10.6	0.63 ± 0.14	90.1 ± 0.6
Wheat–chickpea (60:40)	13.1	2.31 ± 0.30	83.9 ± 1.6
Chickpea	19.9	1.74 ± 0.17	77.9 ± 1.6
Rice	9.0	1.89 ± 0.34	83.9 ± 1.6
Rice–chickpea (77:23)	11.6	2.19 ± 0.22	80.1 ± 2.2

[a]Ratio of chickpea to grain is by weight.

Table 7.8 Amino acid content in wheat, rice, corn and chickpea (g/100 g protein)

	Content (g 100g protein⁻)			
Amino acid	*Wheat*	*Rice*	*Corn*	*Chickpea*
Methionine + cystine	3.63	3.32	3.04	2.61
Tryptophan	1.29	1.23	0.69	1.00
Lysine	1.93	3.25	2.75	7.47
Isoleucine	3.39	3.07	3.14	4.76
Phenylalanine + tyrosine	3.69	8.19	6.34	9.43
Valine	4.36	4.88	4.88	4.84
Threonine	2.52	3.06	3.17	4.23
Leucine	6.79	8.46	14.80	7.18
Chemical score[a]	35 (Lys)	60 (Lys)	51 (Lys)	60 (Sulphur)

[a]According to FAO (1973). Amino acids in parentheses are limiting.

1979). Protein content, protein efficiency ratios and digestibilities of various chickpea varieties are presented in Table 7.6, and the nutritive values of chickpeas, cereals and mixtures are given in Table 7.7.

Supplementation of cornflour with chickpea leads to an improvement in the nutritive value of the tortilla product. In the proportion of 80:20 corn:chickpea, the normal physical characteristics of tortillas are not affected. Indeed the PER is improved from 1.10 in an all-corn tortilla diet to 2.09 when a corn–chickpea mixture is used (Hernandez and Sotelo, 1987). A similar improvement is obtained in a mixture of wheat flour and chickpea. The nutritive value of wheat flour with a PER of 0.63 is remarkably increased in a 60:40 mixture of wheat and chickpea to a PER of 2.31 (Hernandez and Sotelo, 1984). However in a rice–chickpea mixture, the protein quality of rice is not significantly enhanced since the intrinsic protein quality of rice is already high. PER values of rice alone and a 77:23 rice–chickpea mixture are 1.89 and 2.19, respectively. Nutritive value of chickpea protein is improved by addition of methionine. For chickpea alone the PER is 2.06, the addition of 0.3% methionine raises this to 3.04% (Sotelo *et al.*, 1987a, c). Table 7.8 compares the essential amino acid profiles of chickpea, wheat, rice and corn. A comparison of their chemical scores indicates that chickpea protein is superior to those of corn and wheat and equivalent to that of rice.

Starch is the major carbohydrate reserve in chickpea seeds. Sugars present are non-reducing for the most part. Crude fibre is largely concentrated in the seed coat. Desi cultivars of chickpea are higher in crude fibre content than kabuli types. Linoleic, oleic and palmitic acids are the major fatty acids in chickpea lipids. Seeds are a good source of calcium, phosphorus, thiamin, riboflavin, niacin and ascorbic acid.

7.4 ANTINUTRITIONAL FACTORS

The quality of chickpea seed protein is one of the highest in the legume family (FAO, 1970). Sotelo *et al.* (1987c) studied nine Mexican cultivars and found their chemical composition to be very similar with very low contents of trypsin inhibitors and undetectable levels of haemagglutinins. It is important to note that PER values and digestibility coefficients were the same in both raw and cooked seeds. In two samples studied (Surntato and Brevo Duro) the PER values obtained were similar to that of casein. This feature makes chickpea particularly suitable as feed for monogastric mammals and also as the basis for infant feeding formulas. The work of Sotelo *et al.* (1987a, b) showed that chickpea based infant formula was not nutritionally inferior to a soybean based one and had the advantage that infant diarrhoea was much better controlled on the chickpea- than on the soybean-based formula. In addition the chickpea formula could be recommended for use by bottle-fed babies suffering from lactose intolerance.

Other antinutritional factors identified from chickpea include oligosaccharides, polyphenols, phytic acid, α-amylase inhibitors and saponins. The level of flatulence factors (oligosaccharides) is, unlike that of protease inhibitors, higher than in other legumes (Rao and Belavady, 1978). Polyphenols are more abundant in cultivars with dark seedcoats than in those which are pale. Polyphenols have fungistatic properties and can protect seed from fungal invasion. In general, processing procedures, such as cooking, fermentation and germination, reduce levels of antinutritional factors. Dehulling can very effectively reduce levels of polyphenols.

Other desirable effects of feeding high protein foods containing legume flours, such as those of the chickpea, with cereal flours, apart from the obvious one of improving overall nutritional status, is improved retention of nitrogen, calcium and phosphorus in children (Guthkar *et al.*, 1965; Doraiswamy *et al.*, 1969).

The future of the chickpea is promising, it is the subject of a joint breeding programme between ICARDA in Syria and ICRISAT in India. There would seem to be potential for development of consumption well outside areas of production with the burgeoning of the health food markets. It is reasonable to hope that the activities of plant breeders will not only enhance the yielding capacity of the crop but incorporate genetic protection through resistance to realize and safeguard increased yields.

REFERENCES

Chandrasekharappa, G. (1979) Nutritional quality of the protein and blends of wheat and rice with Bengal gram, red gram or black gram. *Nutr. Rep. Int.*, **18**, 401–3.

Chavan, J.K., Kadam, S.S. and Salunkhe, D.K. (1989) Chickpea, in *Handbook of World Food Legumes*, Vol. I, (eds. D.K. Salunkhe and S.S. Kadam), CRC Press, Boca Raton, FL.

Cravioto, J.O., Massieu, H.G., Cravioto, R.O. and Figueroa, F. de M. (1954) Estudios sobre proteinas y amino acidos en dietas Mexicanas. *Ciencia*, **14**, 93–7.

Doraiswamy, T.R., Daniel, V.A. and Swaminathan, M. (1969) Effect of supplementary protein food based on a blend of cotton seed, groundnut and Bengal gram flours, fortified with vitamins and minerals on the growth, nutritional status and nitrogen balance in children subsisting on poor kaffir corn diet. *J. Nutr. Diet.*, **6**, 336–40.

FAO (1970) *Amino Acid Content of Foods and Biological Data on Proteins*. FAO Nutritional Studies, no. 24, Rome.

FAO/WHO (1973) Energy and Protein Requirements. Report of a Joint FAO/WHO Committee. Nutr. Rep. Ser. 52 FAO, Rome.

Gopalan, C., Sastri, B.V.R. and Balasubramanian, S.C. (1982) *Nutritive Value of Indian Foods*. National Institute of Nutrition, Hyderabad, India.

Guthikar, N., Myna, P., Doraiswamy, T.R. *et al.*, (1965) Effects of supplementary protein food based on a blend of groundnut, Bengal gram and sesame on the retention of nitrogen, calcium and phosphorus in undernourished children subsisting on an inadequate diet. *J. Nutr. Dietet.*, **2**, 75–9.

Hernandez, M. and Sotelo, A. (1984) Nutritional evaluation of wheat flour cookies supplemented with chickpea, cheese whey and amino acids. *Nutr. Rep. Int.*, **29**, 845–57.

Hernandez, M. and Sotelo, A. (1987) Protein quality of masa and tortillas supplemented with chickpea. *Nutr. Rep. Int.*, **36**, 213–21.

Maga, J.A. and Sizer, C.E. (1978) Mineral levels in various legume products. *Lebensm. Wiss. Technol.*, **11**, 257–60.

Pak, N. and Barja, I. (1974) Composition, content of toxic substances, protein quality and protein value of green peas, chick peas and lentils grown in Chile. *Cienc. Invest. Agraria.*, **1**, 105–8.

Rao, P.U. and Belavady, B. (1978) Oligosaccharides in pulses: varietal differences and effect of cooking and germination. *J. Agric. Food Chem.*, **28**, 316–19.

Salunkhe, D.K., Kadam, S.S. and Chavan, J.K. (1985) *Postharvest Biotechnology of Food Legumes*, CRC Press, Boca Raton, FL.

Sotelo, A., Arenas, M.L. and Hernandez, M. (1987a) Utilization del garbanzo (*Cicer arietinum* L.) en formulas no lacteas. I. Composicion quimica y nutriva del garbanzo y su comparacion con formulas infantiles comerciales. *Arch. Latinoam. Nutr.*, **37**, 551–9.

Sotelo, A., Hernandez, M., Larracilla, J. *et al.* (1987b) Utilizacion del garbanzo (*Cicer arietinum*) en formulas no lacteas, II. Balance de nitrogeno en ninos con intolerancia a la lactosa, alimentados con una formula a base de garbanzo y un producto comercial de soya. *Arch Latinoam. Nutr.*, **37**, 468–79.

Sotelo, A., Flores, F. and Hernandez, M. (1987c) Chemical composition and nutritional value of Mexican varieties of chickpea (*Cicer arietinum* L.) *Qual. Plant Foods Hum. Nutr.*, **37**, 299–306.

Ulloa, J.A., Valencia, M.E. and Garcia, Z.H. (1988) Protein concentrate from chickpea: nutritive value of a protein concentrate from chickpea (*Cicer arietinum*) obtained by ultrafiltration and its potential use in an infant formula. *J. Food Sci.*, **53**, 1396–8.

CHAPTER EIGHT

Soybean (*Glycine max* (L.) Merr.)

E. Nwokolo

8.1 INTRODUCTION

The soybean is probably the world's most valuable crop, used as feed by billions of livestock, as a source of dietary protein and oil by millions of people, and in the industrial manufacture of thousands of products. It is believed to have been domesticated in China from its probable wild ancestor *Glycine usuriensis*, a wild plant common in eastern Asia and has been used as food and in folk medicine in China and Japan for thousands of years and has been assigned to the China–Japanese Centre of Diversity. Introduced into the United States in the early 1800s it gradually became very popular in the southern and midwestern states. The United States has remained by far the world's largest producer of soybean for more than two decades. Soybean is such an extremely rich source of protein and fat, and such a good source of energy, vitamins and minerals that it has become the standard to which other vegetable food ingredients are compared.

It is, however, a crop of temperate and subtropical climates and has been difficult to introduce and acclimatize to most tropical countries. As a result, production and utilization of soybean is at best marginal in those countries where it can do the greatest good in combating endemic malnutrition.

8.2 BOTANY/ECOLOGY OF THE CROP

Glycine max belongs to the subgenus *Soja*, a member of the genus *Glycine*,

Legumes and Oilseeds in Nutrition
Edited by E. Nwokolo and J. Smartt
Published in 1996 by Chapman & Hall. ISBN 0 412 45930 2

in the family Leguminosae. It is a herbaceous annual with determinate as well as indeterminate growth habits. It is a small, erect, hirsute plant usually 30–70 cm in height, with trifoliolate and alternate leaves and ovate to lanceolate leaflets. The mature flower may be purple or white. The pods are oblong, slightly elongated, about 3–8 cm long and 8–15 mm broad. Pods are hairy, green when immature but turning to a yellowish-brown colour on maturity. There are two or three seeds per pod; the seeds are ovoid to subspherical in shape, a few seeds being irregular. Seeds are smooth. Seed colour varies, common colours being tan, brown, yellow and to a lesser extent black and green. The cotyledons are usually yellow or green. Soybean is a crop of mild climates, growing well and yielding profusely in subtropical and temperate regions, and although it has been introduced not quite successfully to tropical regions of the world, it has, however, been very successful in Brazil where it performs exceptionally well. The soybean performs optimally in the temperature range of 25–32°C. It has a moderate moisture requirement of 400–800 mm rainfall and can therefore be cultivated in a variety of ecological zones. It is a short-day plant yet thrives in day-neutral environments as occurs in the immediate environment of the equator. Although it grows best on fertile, well-drained soils, it tolerates various soil types, preferring mildly acidic soils. Its major requirement is that if soil fertility is low, certain specific nitrogen-fixing rhizobia need to be present in the soil to ensure adequate growth and productivity.

8.3 PRODUCTION

The soybean is a highly productive crop, with an average production cycle of 90–110 days from planting to harvesting. Plants are harvested when the leaves have dropped, the stems have dried and the pods have turned brown and become dry. Harvesting in the United States and other industrialized countries is by combine harvesters, whereas in the developing countries it is usually by hand, followed by threshing. Seed yield varies from continent to continent and is lowest in Africa and highest in North America. Table 8.1 shows the major producers of soybean in the world, while Table 8.2 shows world exports, imports and consumption of soybean products.

8.4 CURRENT AND POTENTIAL USAGE AS FOOD

How soybean is used as food varies from one continent to the other. In North America, its is used primarily as an edible oil source and as a protein concentrate for animal feed manufacture. Only about 5% of total production in the United States is processed into human food, the rest of

Table 8.1 World production of soybean in 1990

Region	*Area harvested (1000 ha)*	*Yield (kg/ha)*	*Production (1000 MT)*
Country			
United States	22 865	2287	52 303
Brazil	11 483	1732	19 888
China	7624	1509	11 508
Argentina	4985	2141	10 672
India	2250	1000	2250
Italy	515	3282	1690
Paraguay	928	1616	1500
Indonesia	1268	1125	1427
Canada	835	1018	850
The former USSR	835	1018	850
Continents			
Africa	446	1038	463
N and C America	33 643	2291	54 175
South America	17 757	1843	32 728
Asia	12 593	1356	17 070
Europe	10 009	2381	2403

Source: FAO Production Yearbook (1991).

Table 8.2 World production, exports, imports and consumption of soybean, soybean protein meal and soybean oil in 1990–1991 (million MT)

	Soybean	*44% protein meal*	*Soybean oil*
Production	103.94	69.67	15.92
Exports	25.05	26.74	3.58
Imports	26.12	27.47	3.64
Crush/consumption	87.98	70.86	15.94

Source: USDA Foreign Agricultural Service Circular (1992).

the crop being crushed for oil and animal feed. The small amount of soybean production which is processed into food is turned into full-fat protein meat, low-fat protein flour, defatted flakes and grits, defatted flour, protein concentrate and protein isolate. These soybean products are processed into a variety of soups and chilli products as well as meat substitutes and meat extenders for incorporation into baby foods, bolognas, luncheon meats and other manufactured food products. Protein isolates may be given a meat-like texture and added to meat products (like hamburgers) to increase the cooked yield of such products. Some of the full-fat soybean is processed into soymilk, a synthetic-milk product which has little or no carbohydrate content and is therefore ideal for diabetics. In the production of soybean oil, the mature dry seed is cleaned, steamed to obtain the appropriate moisture content, ground and passed through a hydraulic or screw press and finally extracted using an azeotropic mixture of solvents.

The oil is refined through a succession of steps which include degumming, alkali refining, bleaching, hydrogenation and deodorization. Degumming yields lecithin, important phospholipids which are very important as wetting and stabilizing agents in the food, cosmetic and pharmaceutical industries and in the manufacture of paints, plastics, soaps and detergents. Alkali soap stocks from the alkali refining stage are useful as food additives. The refined oil is a semi-drying oil useful as a salad oil and for the manufacture of margarine and shortening. It may also be used in the industrial manufacture of soaps, paints, printing ink, linoleum and oil cloths, and as a carrier in the manufacture of insecticides and pharmaceuticals. In the Orient, especially China and Japan, soybean is processed into soy sauce, soy milk and soy curd (tofu) and in Indonesia it is processed into tempeh. Soy sauce is used in the preparation of virtually every meal in China, Japan and many countries in the far East. Soybeans are also sprouted and consumed as bean sprouts, an important and ubiquitous component of daily food in south-east Asia, and increasingly in other parts of the world. There is also the consumption of immature green soybean as a vegetable, although this is somewhat restricted.

In Africa, there is very little human consumption of soybean or soybean products. This is primarily because soybean is alien to the culture and the food systems of the people. Additionally, it takes a very long time to cook and when cooked has an objectionable beany flavour and taste caused by the activity of the enzyme lipoxidase. The enzyme may be inactivated by soaking dry beans in water for 4–6 h followed by blanching for 10–15 min at 100°C. Therefore, in order to encourage the consumption of soybean as a grain legume, it is important that blanched, dry seeds, which have been treated to eliminate any objectionable off-flavour following cooking, are made available commercially and at low cost. There are virtually no large-scale facilities in most African countries for processing soybean into protein concentrates and isolates. Consequently, the extremely high cost of imported protein concentrates and protein isolates in Africa makes any recommendation for increased use of these products not feasible. Additionally, hamburgers, bolognas, luncheon meat and other manufactured meats do not feature in the traditional diets of most Africans, making it very difficult to introduce the use of processed soybean products.

8.5 NUTRIENT COMPOSITION AND NUTRITIVE VALUE

Soybean is an extremely rich source of nutrients, raw soybean containing almost twice the protein content of cowpea and dry beans and about ten

Table 8.3 Proximate composition of soybean and soy products (g/100 g)

Constinuents	*Raw soybean*	*Defatted flour*	*Protein concentrate*	*Protein isolate*
Water (g)	8.54	7.25	5.80	4.98
Food energy (kcal)	416	329	332	338
Protein (g)	36.49	47.01	58.13	80.69
Lipids (g)	19.94	1.22	0.46	3.39
Carbohydrate (g)	30.16	38.37	31.21	0.36
Crude fibre (g)	4.96	4.27	3.77	0.26
Ash (g)	4.87	6.15	4.70	3.58

Source: USDA (1986).

Table 8.4 Vitamin content of soybean and soy products (mg/g)

Vitamin	*Raw soybean*	*Defatted flour*	*Protein concentrate*	*Protein isolates*
Ascorbic acid (mg)	6.0	0.0	0.0	0.0
Thiamin (mg)	0.87	0.70	0.31	0.17
Riboflavin (mg)	0.87	0.25	0.14	0.10
Niacin (mg)	1.62	2.61	0.71	1.43
Pantothenic acid (mg)	0.79	2.00	0.05	0.06
Pyridoxine (mg)	0.37	0.57	0.13	–
Folic acid (mg)	375.1	305.4	340	176.1
Vitamin B_{12} (mg)	0	0	0	0
Vitamin A (IU)	24	40	0	0

Source: USDA (1986).

Table 8.5 Mineral content of soybean and soy products (mg/g)

Minerals	*Raw soybean*	*Defatted flour*	*Protein concentrate*	*Protein isolates*
Calcium (mg)	277	241	363	178
Iron (mg)	15.70	9.24	10.78	14.58
Magnesium (mg)	280	290	315	39
Phosphorus (mg)	704	674	839	776
Potassium (mg)	1797	2384	2202	81
Sodium (mg)	2	20	3	1005[a]
Zinc (mg)	4.89	2.46	4.4	4.03
Copper (mg)	1.65	4.06	0.97	1.60
Manganese (mg)	2.51	3.01	4.19	1.49

Source: USDA 1986.
[a] Sodium was used in preparation of isolates.

Table 8.6 Fatty acid content of soybean and soybean products (g/100 g)

Fatty acids	*Raw soybean*	*Defatted flour*	*Protein concentrate*	*Protein isolates*
Saturated (g)	2.88	0.13	0.05	0.42
14:0 (g)	0.05	0.00	0.0	0.0
16:0 (g)	2.11	0.10	0.03	0.31
18:00 (g)	0.71	0.03	0.01	0.10
Monounsaturated (g)	4.40	0.21	0.08	0.64
16:1 (g)	0.05	0.0	0.0	0.0
18:1 (g)	4.35	0.21	0.08	0.64
Polyunsaturated (g)	11.25	0.53	0.20	1.65
18:2 (g)	9.92	0.47	0.18	1.45
18:3 (g)	1.33	0.06	0.02	0.20
Phytosterols (mg)	161	–	–	–

Source: USDA (1986).

times their oil content. In spite of its low carbohydrate content (30%), its food energy is high due to an abundance of oil in the seed.

The proximate composition of soybean and soy protein products is shown in Table 8.3. The degree to which the oil content of soybean has been extracted from a meal or flour is reflected in a corresponding increase in the protein content of that protein meal or flour. Thus low-fat flours contain more oil but less protein than defatted flour. The protein isolate which is prepared by acid precipitation of soybean protein is very high in protein content as a result of its low fat and carbohydrate content and its almost total absence of fibre. Like other legumes, soybean is a good source of thiamine, riboflavin and niacin (Table 8.4). It is also a fairly good source of ascorbic acid and pantothenic acid. As in most legumes, there is an abundance of potassium but not sodium, and soybean is a very good source of phosphorus, although a significant proportion of it is present as phytic acid phosphorus which has only partial biological availability. Mineral content of soybean and other processed soybean products is shown in Table 8.5. Soybean is a good source of calcium and magnesium but a poor source of iron. It also has minimal amounts of zinc, manganese and copper. Although it has long been assumed that phytates in soybean and other beans depress zinc and iron availability, more recent research shows that phytates in soybean may have no inhibitory effect on iron availability.

Content of fatty acids in soybean oil is shown in Table 8.6. Soybean oil is a semidrying oil which has found popularity as a salad oil, a cooking oil, a baking oil and in the preparation of many culinary delights. The oil is composed of approximately 16% saturated fatty acids, 24% monounsaturated fatty acids and 60% polyunsaturated fatty acids. The major fatty acid is linoleic acid which by itself comprises 53% of the total fatty

Table 8.7 Amino acid content of soybean and soy products (g/100 g)

Amino acids	*Raw soybean*	*Defatted flour*	*Protein concentrate*	*Protein isolates*
Tryptophan (g)	0.53	0.68	0.83	1.11
Threonine (g)	1.58	2.04	2.47	3.13
Isoleucine (g)	1.77	2.28	2.94	4.25
Leucine (g)	2.97	3.82	4.91	6.78
Lysine (g)	2.42	3.12	3.92	5.32
Methionine (g)	0.49	0.63	0.81	1.13
Cystine (g)	0.58	0.75	0.88	1.04
Phenylalanine (g)	1.90	2.45	3.27	4.59
Tyrosine (g)	1.38	1.77	2.30	3.22
Valine (g)	1.82	2.36	3.06	4.10
Arginine (g)	2.83	3.64	4.64	6.67
Histidine (g)	0.98	1.26	1.57	2.30
Alanine (g)	1.71	2.21	2.68	3.59
Aspartic acid (g)	4.58	5.91	7.24	10.20
Glutamic acid (g)	7.06	9.10	12.01	17.45
Glycine (g)	2.13	2.75	3.29	4.96
Proline (g)	2.13	2.75	3.29	4.96
Serine (g)	2.11	2.72	3.36	4.59

Source: USDA (1986).

acids in soybean oil. Another important fatty acid component of soybean oil is linolénic acid, representing 7% of total fatty acids. Because of the high content of these polyunsaturated fatty acids in soybean oil, it is considered to be unstable in use, given to foaming when used as a frying oil, and not recommended for that purpose unless further treated to stabilize it and reduce foaming. Research findings indicate that the presence of fairly high amounts of polyunsaturated fatty acids in edible oils, may enhance the nutritional value of those oils as a source of eicosanoids, some of which have been associated with a reduction in the incidence of ischaemic heart disease in man and experimental animals. Phospholipids are usually present at 2–3.5% of the oil and are removed during the refining of soybean oil. Soy lecithin contains approximately 21% phosphatidylcholine, 22% phosphatidylethanolamine, 19% phosphatidylinositol, 10% phosphatidic acid, 1% phosphatidylserine and 12% glycolipids (Schneider, 1986).

Table 8.7 shows the amino acid profile of the protein of soybean and the various soy products. Soybean has double the protein and amino acids present in most dry beans, lentils, peas and cowpeas used as human food. With the exception of the sulphur amino acids in which soybean is deficient, there are sufficient other essential amino acids. The very high lysine concentration is noteworthy because it places soybean protein in a distinct class of vegetable proteins. Since lysine is limiting in most cereal proteins, it is possible that soy protein may be able to improve total protein status in populations dependent on cereal grains for food.

8.6 ANTINUTRITIONAL FACTORS IN SOYBEAN

Raw soybean, in common with other legumes, contains various antinutritional factors including antiproteolytic agents which, to some extent, interfere with the utilization of the protein of soybean. Although some of these antinutritionals are heat labile, there are others which are neither inactivated nor destroyed by heat. Liener (1981) lists the heat-labile antinutritional factors in soybean as: protease inhibitors, haemagglutinins, goitrogens, antivitamins and phytates. The heat-stable antinutritional agents in soybeans are listed as: saponins, oestrogens, flatulence factors, allergens, and lysinoalanine.

8.6.1 Protease inhibitors

The term protease inhibitor is preferable to the more common term trypsin inhibitor because this family of protein inhibitors interferes with a wide range of proteases including trypsin and chymotrypsin. Protease inhibitors are usually destroyed by application of moist or dry heat (Liener, 1979,1981; Liener and Kuhn, 1979). In the past it was assumed that these antinutritional factors were primarily responsible for the poor growth of rats fed raw soybeans. However, Kakade *et al.* (1973) have clearly demonstrated that destruction of protease inhibitors is responsible for only about 40% of the increased growth which results from heat treatment of raw soybeans. The other 60% of the increased growth is attributable to improved digestion following heat treatment of legume proteins, most of which are in the raw state, refractory to digestive action of proteolytic enzymes. Liener (1979) further suggests that low levels of protease inhibitors are harmless to humans, that human trypsin is only weakly inhibited by protease inhibitors and that the pancreatic hypertrophy associated with protease inhibitors has little effect in humans. Thus, normal cooking reduces the content of protease inhibitors in soybean to 10–20% of the level in raw beans, and it has been suggested that attempts to reduce the level to below 10–20% usually results in heat damage to proteins and a reduction in their nutritional quality (Erdman, 1988).

8.6.2 Haemagglutinins

It has been known for nearly 40 years that raw soybean contains haemagglutinins which bind on to the carbohydrate moiety in certain tissues. It has been shown that with red blood cells, the haemagglutinins bind on to the glycoproteins on the cell surface causing agglutination of the red cells. Some lectins, such as ricin in the castor plant are highly toxic, whereas others are relatively innocuous. Soybean contains several lectins which represent 1–3% of the total protein in defatted soybean flour. Soybean haemaglutinins, like trypsin inhibitors are highly heat labile and are destroyed by any standard moist heat application. This is usually accompanied by an improvement in the nutritive value of the heated protein. What is even more interesting is the observation that soybean lectins play only a minor part in the poor nutritive value of raw soybeans (Liener, 1979).

8.6.3 Flatulence factors

Liener (1981) found that a major impediment to the use of soybean in the human diet is its tendency to cause flatulence. It is known that certain low-molecular-weight oligosaccharides (raffinose and stachyose) are associated with flatus formation. These oligosaccharides are soluble in 80% ethanol and are part of the soy whey solids. Therefore, any processing of soybean which involves extraction with ethanol, or fermentation usually results in a product with low flatus potential.

8.7 PROCESSING OF SOYBEANS

The nutritional quality of soybean and soy products depends to a large extent on conditions during storage and processing. Lester (1988) describes the processing of soybean as involving an initial process of cleaning the beans to remove foreign matter, and a subsequent tempering stage which functions to equilibrate the moisture in the seed and facilitate the removal of the seed hull. The seeds are then cracked into about eight pieces over corrugated rollers, passed over shakers and aspirated to remove the seed hulls. The dehulled soybean pieces (cracks or meat) are conditioned by moist heating to 165°C to soften them for flaking, a process in which the cracked soybeans are passed between rollers which turn them into very thin flakes and ruptures the oil cells. The flakes are solvent-extracted using hexane and subsequently may be flash-desolventized or toasted. The former is a very rapid application of very high heat to remove the hexane, without damaging the protein. Flours processed by this method are specifically for human nutrition. Toasting involves the use of steam to drive off the hexane while cooking the soybean flakes at 210–212°C. Soybean products processed in this manner are termed meals and are specific for animal and poultry nutrition.

Soy concentrates are produced by washing soy flours with water or ethanol to remove soluble sugars. This results in an increase in protein content from about 45% to about 55%, and almost complete elimination of oligosaccharides and their flatulence effects from soy concentrates. Soy isolates are prepared by dissolution of soy flour in alkali followed by acid precipitation of the protein.

8.8 INDUSTRIAL USES OF SOYBEAN PRODUCTS

The established industrial uses of soybean protein and oil are currently numerous and keep multiplying as researchers find newer ways to utilize this nutritional wonder. Soy protein concentrates are used primarily for the manufacture of sausages and other processed meat whereas protein isolates may be texturized and used as meat extenders and substitutes in specialized meat applications. Protein isolates have other industrial non-food uses in a wide range of adhesive applications. Various oil processing by-products which have industrial uses include lecithin, soapstocks, and deodorizer distillates which yield sterols and tocopherols. Refined soybean oil may be converted into a wide range of important industrial oil products. These include blown oils, epoxy resins, esters, fractionated oils and maleic acid oils, alkyd resins, urethane and co-polymer oils, and printing inks. Soybean oil is also used industrially in the manufacture of lubricating oils, lubricating greases and as plasticizers.

8.9 FOOD USES OF SOYBEAN PRODUCTS

8.9.1 Food uses of soybean oil

Food uses of soybean oil include its use as salad and cooking oils. Partial and selective hydrogenation of soybean oil to reduce its linolenic acid content to about 3%, considerably enhances its oxidative stability and flavour. Occasionally, soybean oil may be blended with an oil which is poor in linolenic acid (for example cotton seed oil) to produce an excellent cooking oil. In some cases, an antifoam agent, usually a silicone compound, may be added to soybean oil to improve the frying quality of the oil.

Soybean oil is the major oil of choice in manufacture of mayonnaise and salad dressings. It is also used extensively in margarine production, and constitutes about 64% of the fat in the manufacture of shortening (Mounts, 1988). Lecithin as a by-product of vegetable oil refining has many applications in the food industry. Because lecithin possesses both a hydrophilic and a hydrophobic side chain, it serves to facilitate the bringing together of immiscible materials, through for example, the formation of emulsions. The primary functions of lecithin in food are in wetting, colloidal dispersion, lubrication and release, control of crystallization and starch complexing (Mounts, 1988).

8.9.2 Food uses of soybean protein

The high popularity of soybean products in food systems in the industrialized world, is attributable to its excellent functional properties in food applications, its high nutritional quality, as well as the abundance, availability and low cost of processed soy products (Sipos, 1988). Thus, on a local, nationwide or global scale, soy products hold out the hope that school lunches, feeding of military personnel, hospital feeding and other institutionally prepared feeding programmes can be as nutritious as possible while being reasonable in cost.

Sipos (1988) notes that currently, soy concentrates may be used at a level of 5% of the mixture in the manufacture of emulsified meats like frankfurters. This level of use is expected to increase because the final product has excellent eye appeal, good texture, no off flavours and results in substantial savings to the manufacturer. According to Sipos, hydrated soy proteins may replace up to 20% of the ground beef in pattie-type products without flavour adjustments being necessary. Soy proteins have also been utilized at this level in such ground meat products as meat balls, pizza toppings, taco fillings and sloppy joe mixes. Sipos notes that textured soy proteins are used in canned meats to ensure a firm product while preventing fat separation. Soy products may also be used in bakery products, usually in breads and rolls (1–3% of flour), speciality breads (5–10% of flour), cakes and cake mixes (2–15% of flour), doughnuts (3–5% of flour) and various pastas (up to 15% of flour) without in any way detracting from the sensory and physical characteristics of the final product (Sipos, 1988).

8.9.3 Soy protein for infants and children

Soy proteins are useful in weaning formulas for infants and as special supplemental formulas for undernourished preschool and school-age children. Excellent reviews of the protein quality of soybean products processed for human nutrition have been published (Young and Scrimshaw, 1979; Torun, 1981; Torun *et al.*, 1981). For full-term infants, it has been noted that whereas mother's milk is best for infants, artificial infant formulas based on cow's milk and soybean protein permit optimum growth and development for term-sized well infants (LeLeiko, 1981). The use of artificial formulas primarily containing soybean protein as a sole food for infants is not recommended because of problems such as metabolic acidosis, nutritional inadequacy and gastrointestinal intolerance in some infants (Finberg, 1980). A large-scale experiment in China on the development of a fortified soybean-based formula to replace those based on cow's milk and human milk has produced some interesting results. A formula which in addition contained egg yolk, bone meal and millet was evaluated on 100 human infants (Chou, 1983). Results indicated excellent utilization of this formula, excellent growth response and no digestive problems. In general, it can be concluded that soy protein is an excellent source of dietary amino acids and that when thoroughly processed, can meet the short- and long-term protein and essential amino acid requirements of the human infant, children and adults. Young and Scrimshaw (1979) note that although methionine supplementation of soy protein may be necessary in the nutrition of the preterm infant, such supplementation is unnecessary for full-term infants, children and adults. Thus at protein intake levels similar to current dietary protein allowances, well-processed soy protein products can successfully replace meat and fish proteins in diets of adults and older children. This is also similar to the conclusions of Torun (1981) who, from a review of the available literature, concluded that the nutritional quality of processed soy protein is almost equivalent to that of milk for infants, preschool and school-age children. With preterm infants, there is usually a problem of low weight gain and therefore a need to supplement the basic cow's milk or human milk with processed soy proteins to ensure optimal intake of nutrients (Torun, 1981).

8.9.4 Soy products in adult nutrition

In a review of the role of soybean in food systems, Bressani (1981) points out that in human beings, animal proteins have a higher digestibility coefficient than soy protein in general, although the more the degree of processing of soy products the higher the digestibility coefficient. Values given for true digestibility of most foods are as follows: whole egg 93–100%, milk 90–98%, beef 91–99%, casein 94–97%, soyflour 75–92%, defatted extruded soy flour 90–94%, isolated soy protein 93–97%. Thus, soy isolates and concentrates have a higher digestibility than soy flour. Bressani (1981) shows that in humans as in experimental animals, at low levels of protein intake, supplementation with methionine enhances the nutritional quality of soy protein. Thus, as protein intake increased from 3 to 6 g of nitrogen per day, nitrogen balance was improved. Supplementation with methionine improved nitrogen balance at all levels of nitrogen intake. Bressani (1981) notes that corn, rice and whole wheat have protein efficiency ratios (PER) of 1.00, 1.87 and 1.32, respectively. However, supplementation of these grains with equal amounts of soy flour (8%), increased their PER values to 2.25, 2.88 and 1.91, respectively.

8.9.5 Soy meals in animal nutrition

As mentioned earlier, 95% of the soybean crop in the United States is crushed for the production of oil and soybean meal. Soybean meal and corn are the backbone of the United States' livestock and poultry industry, enabling it to feed a population of almost 250 million people in 1990, and still have a huge surplus for export. For further information on the use of soybean meals in animal and poultry nutrition, the reader is referred to the many books and periodicals and thousands of scientific and review papers which have been published on this topic.

REFERENCES

Bressani, R. (1981) The role of soybeans in food systems. Proceedings of the World Conference on Soya Processing and Utilization. Acapulco, Mexico November 9–14 1980. *J. Am. Oil Chem. Soc.*, **58**, 392–400.

Chou, C. (1983) Studies on the use of soybean food in infant feeding in China and the development of Formula 5410. *Food Nutr. Bull.*, **5**, 45–52.

Erdman, J.W. Jr (1988) Nutritive value of soybean protein in *Proceedings of a World Conference on Emerging Technologies in the Fats and Oils Industry*, (ed. A.R. Baldwin), American Oil Chemists Society, Champaign, IL.

FAO (1991) Production Yearbook. Food and Agriculture Organization of the United Nations, Rome.

Finberg, L. (1980) One milk for all–not ever likely and certainly not yet. *J. Pediatr.*, **96**, 240–1.

Kakade, M.L., Hoffa, D.E. and Liener, I.E. (1973) Contribution of trypsin inhibitors to the deleterious effects of soybean fed to rats. *J. Nutr.*, **103**, 1772–8.

LeLeiko, N.S. (1981) Effects of various diets on the growth and development of premature and full-term infants, in *Textbook of Gastroenterology and Nutrition in Infancy*, (ed. E. Lebanthal), Raven Press, New York.

Lester, W.C. (1988) An overview of soybean processing, in *Proceedings of a World Conference on Emerging Technologies in the Fats and Oils Industry*, (ed. A.R. Baldwin), American Oil Chemists Society, Champaign, IL.

Liener, I.E. (1979) Significance for humans of biologically active factors in soybeans and other food legumes. *J. Am. Oil Chem. Soc.*, **56**, 121–9.

Liener, I.E. (1981) Factors affecting the nutritional quality of soya products. Proceedings of the World Conference on Soya Processing and Utilization. Acapulco, Mexico. November 9–14 1980. *J. Am. Oil Chem. Soc.*, **58**, 406–15.

Liener, I.E. and Kuhn, G.L. (1979) Dry roasting of beans, in *Soybean Utilization Alternatives*. Proceedings of a Symposium by The Center for Alternative Crops and Products, University of Minnesota, St Paul, (ed. L. McCann).

Mounts, T.L. (1988) Edible soybean oil products, in *Soybean Utilization Alternatives*. Proceedings of a symposium by The Center for Alternative Crops and Products, University of Minnesota, St Paul. (ed. L. McCann).

Schneider, M. (1986) Phospholipids, in *Proceedings of a World Conference on Emerging Technologies in the Fats and Oils Industry*, (ed. A.R. Baldwin), American Oil Chemists Society, Champaign, IL.

Sipos, E.F. (1988) Edible uses of soybean protein, in *Proceedings of a World Conference on Emerging Technologies in the Fats and Oils Industry*, (ed. A.R. Baldwin), American Oil Chemists Society, Champaign, IL.

Torun, B. (1981) Soybeans and soy products in the feeding of children. *J. Am. Oil Chem. Soc.*, **58**, 460–4.

Torun, B., Viteri, F.E. and Young, V.R. (1981) Nutritional role of soy protein for humans. *J. Am. Oil Chem. Soc.*, **58**, 400–6.

USDA (1986) *Composition of Foods: Legumes and Legume Products*. USDA Human Nutrition Information Service. Agriculture Handbook No. 8–16, United States Department of Agriculture, Washington, DC.

USDA (1992) Foreign Agricultural Service Circular. United States Department of Agriculture, Washington, DC.

Young, V.R. and Scrimshaw, N.S. (1979) Soybean protein in human nutrition: an overview. *J. Am. Oil Chem. Soc.*, **56**, 112–20.

CHAPTER NINE

Hyacinth bean (*Lablab purpureus* (L.) Sweet)

M. Akpapunam

9.1 INTRODUCTION

The hyacinth bean is one of the lesser known tropical and subtropical food legumes. Though little known, it has, however, been cultivated for many years in Asia, the Middle East, Africa, the Americas and Europe. It is a good source of protein and can be used in the form of unripe pods, green mature and dry seed. The dry seed, leaves, haulm (green or as hay or silage) and straw are excellent livestock feeds (Purseglove, 1968; Duke, 1983).

The world production of this species is not known accurately since it is not a significant commodity in world trade. It is widely used as a subsistence crop and has considerable unrealized potential for feeding hungry peoples of the world either directly or indirectly through livestock. This chapter considers its botany, ecology, production, current and potential usage as human food and animal feed as well as nutrient composition and nutritive value. The presence or absence of toxic materials and antinutritional factors is also considered as are their chemical constitution and methods of detoxification.

9.2 BOTANY, PRODUCTION AND USE AS HUMAN FOOD

The hyacinth bean exists in two very distinctive forms, one with deeply pigmented flowers, stems and foliage is a popular ornamental vine

Legumes and Oilseeds in Nutrition
Edited by E. Nwokolo and J. Smartt
Published in 1996 by Chapman & Hall. ISBN 0 412 45930 2

whereas the form with white flowers and unpigmented stems and foliage is the common pulse form. The name hyacinth bean is derived from the purple flowered form, it is also known as bonavist bean, lablab, seim, lubia, Indian bean and Egyptian bean (Purseglove, 1968). Synonyms include, *Dolichos lablab* L., *Lablab vulgaris* Savi and *Lablab niger* (Medik.). It is a member of the tribe Phaseoleae in the legume family which includes the majority of the common pulse species. The hyacinth bean is a native of South and East Asia from which area it has been introduced to many others in the tropics. It is usually grown as a rain-fed crop. The crop is drought tolerant and can be grown in areas with low annual rainfall of 60–90 cm. It will grow on marginal soils if these are well drained and can be grown from sea level to altitudes up to approximately 2300 m. The range of latitudes in which it can be grown is from 30°N to 30°S with day temperatures of 20–35°C. Photoperiodic sensitivity occurs but both short-day and day-neutral cultivars are reported. Seed yields of 450 kg/ha have been reported from mixed cropping systems and yields as high as 1460 kg/ha have been reported from pure stands (Duke, 1983). Fodder yields of 5–10 tonnes/ha have been recorded.

Like other pulse species the hyacinth bean is a good source of proteins, vitamins, minerals and calories. The young pods and tender green mature beans are popular vegetables in India and other tropical countries. Mature, dry seeds are consumed as dhal or split peas in India. The beans may also be sprouted, soaked in water, de-hulled, boiled and ground to a paste which can be fried with spices and condiments.

9.3 CURRENT AND POTENTIAL USE AS AN ANIMAL FEED

The stem, leaf, root, pod and seed of the hyacinth bean, in common with most legume species, contain high amounts of protein. This applies even to stem, leaf and root harvested at an advanced state of maturity. They also contain high levels of calcium and phosphorus and are good sources of vitamins A and D. They can, therefore, play an important role in complementing other animal feeds. Hyacinth bean grown for livestock forage can be fed to goats, cattle and swine; mixed with oats it can be fed to horses. Herbage production by the individual hyacinth bean is approximately twice that of the cowpea plant. Although the stems are tough and more fibrous the leaves are more succulent than those of the cowpea. There is considerable potential for use in livestock feeding which is not at present being realized. Research is needed to determine how best to use this species and to develop more suitable and adapted genotypes for use in seed and forage production. Improving the volume of production of satisfactory quality could allow expansion of production of pigs and poultry, if levels of concentrates fed to cattle can be reduced (that is assuming some are fed!).

Table 9.1 Proximate composition of hyacinth beans[a]

	Content (%)
Moisture	9.6
Protein	24.9
Ether extract	0.8
Ash	3.2
Crude fibre	1.4
Total carbohydrate	60.1

[a]Compiled from Purseglove (1968) and Duke (1983).

Table 9.2 Amino acid composition of hyacinth beans

Amino acid	*Content (g/16 gN)*
Lysine	6.8
Methionine	0.9
Cystine	–
Arginine	6.6
Glycine	4.6
Histidine	3.2
Isoleucine	4.4
Leucine	8.5
Phenylalanine	4.9
Tyrosine	3.6
Threonine	4.2
Valine	5.2
Alanine	4.5
Aspartic acid	12.0
Glutamic acid	15.7
Hydroxyproline	0.6
Proline	4.3
Serine	5.4
Protein (N × 6.25)	23.4

Source: Duke (1983).

9.4 NUTRIENT COMPOSITION AND NUTRITIVE VALUE

Hyacinth bean, like most pulses, is high in protein (Table 9.1), this ranges from 21.5 to 24.9%. Amino acid composition is given in Table 9.2. The amino acid profile is adequate for most amino acids with the exception of the sulphur amino acids methionine and cystine which may be nutritionally limiting. This is common in most edible legume seeds. Hyacinth beans are also a good source of soluble carbohydrate which comprises 60.1%. Some varieties contain up to 6.8% crude fibre (Duke,

Table 9.3 Mineral and vitamin contents of hyacinth beans

	Content (mg/100 g)
Calcium	98
Phosphorus	345
Iron	3.9
Sodium	–
Potassium	–
Thiamine	0.7
Riboflavin	0.1
Niacine	2.1
Ascorbic acid	–

Source: Duke (1983).

1983) whereas others contain only 1.4% (Table 9.1). Contents of phosphorus and potassium are high whereas those of calcium, iron and sodium are low (Table 9.3). Vitamin contents are also given in Table 9.3.

9.5 PRESENCE, CHEMICAL NATURE AND DETOXIFICATION OF ANTINUTRITIONAL FACTORS

In common with many pulse crops the hyacinth bean contains antinutritional substances which can adversely affect biological value. These may include protease inhibitors, haemagglutinins, cyanogens, phytic acid, oligosaccharides and saponins. A brief summary is given below of those factors which have been identified in the hyacinth bean.

9.5.1 Protease inhibitors

These are proteins which inhibit the action of proteolytic enzymes by binding on to them. The most important are the inhibitors of trypsin and chymotrypsin. Their action in preventing protein hydrolysis results in impaired growth, poor food utilization and pancreatic hypertrophy. Although concentration levels have not been established, they are thought to occur in hyacinth beans at considerably lower levels than are found in the soyabean (Elkowicz and Sosulski, 1982).

9.5.2 Haemagglutinins

Although high levels of haemagglutinins have been recorded for other members of the tribe Phaseoleae such as *Phaseolus* species, their presence in seed of the hyacinth bean has not been confirmed. Haemagglutinins are destroyed by moist heat and so normal cooking will eliminate these factors when and if they occur.

9.5.3 Cyanogens

These have been reported from the hyacinth bean as cyanogenic glucosides. Three such compounds have been identified in plants; the one commonly found in legumes is linamarin. The toxic effect is produced by hydrolysis of the glucoside to release hydrogen cyanide which is exceedingly toxic. Concentration levels are apparently very much lower than those found in *Phaseolus lunatus* beans which have been responsible for some fatalities, the lethal dose for humans is of the order of 0.5–3.5 mg/kg body weight. Detoxification can be achieved by fermentation, presoaking and in the course of adequate cooking when the relatively volatile HCN is removed. This effect is enhanced, if the water is also changed in the course of cooking.

9.5.4 Phytic acid

Phytic acid binds calcium, magnesium and iron making them unavailable. It is less in evidence in germinated seed and in fermented products. It is not, however, heat labile and is unaffected by cooking.

9.5.5 Oligosaccharides

The α-galactosides raffinose, stachyose and verbascose are found in many pulses especially those of the Phaseoleae. They produce flatulence in humans and animals consuming them (Akpapunam and Markakis, 1979). The human digestive system does not produce α-galactosidase which is essential to hydrolyse oligosaccharides; these enter the large intestine where anaerobic fermentation occurs to produce gas and abdominal discomfort. They may also impede nutrient absorption. They are not labile and are not destroyed by normal cooking. They are water soluble and can be eliminated by adequate presoaking. Germination and fermentation are known to reduce their concentration. Their concentration has been reduced by selective breeding in *Phaseolus vulgaris* and this might be feasible in other species.

9.5.6 Saponins

Saponins are glycosides which are implicated in growth depression, mineral binding and *in vitro* haemolysis of erythrocytes. They are not known to be a problem in hyacinth beans. Where they do occur they can be removed by aqueous extraction with hot water.

9.6 NUTRITIONAL STUDIES

From its amino acid profile the hyacinth bean appears to be a quite typical pulse. Its amino acid profile is deficient in sulphur amino acids (cystine and methionine) and the nutritional quality of the seed protein could well be improved by supplementation with amino acids to improve the nutritional quality (Kon *et al.*, 1974; Onayemi and Potter, 1976). Adequate cooking would probably be sufficient to remove any adverse heat-labile factors, coupled with presoaking if saponins or oligosaccharides are a problem. Studies of the incidence of these compounds have not been exhaustive and among the wide diversity of material known there could be some genotypes in which low concentrations of antimetabolites might occur in the seed.

The hyacinth bean is a pulse which merits further study. It is underexploited, for no very good reason it appears, and some efforts in encouraging production and utilization are indicated. It has appreciable drought tolerance and could be of value in areas suffering from low levels of rainfall.

REFERENCES

Akpapunam, M.A. and Markakis, P. (1979) Oligosaccharides of 13 American cultivars of cowpea (*Vigna sinesis*). *J. Food Sci.*, **44**, 1317–18.

Duke, J.A. (1983) *Handbook of Legumes of World Economic Importance*. Plenum Press, New York, pp. 102–6, 275–8, 293–6.

Elkowicz, K. and Sosulski, F.W. (1982) Anti-nutritional factors in eleven legumes and their air-classified protein and starch fractions. *J. Food Sci.*, **47**, 1301–4.

Kon, S., Wagner, J.R. and Booth, A.N. (1974) Legume powders: Preparation and some nutritional and physiochemical properties. *J. Food Sci.*, **39**, 897–9.

Onayemi, O. and Potter, N.N. (1976) Cowpea powders dried with methionine. Preparation, storage stability, organoleptic properties, nutritional quality. *J. Food Sci.*, **41**, 48–53.

Purseglove, J.W. (1968) *Tropical Crops: Dicotyledons*, Vols. 1 and 2. Longman, London, pp. 273–6, 290–4, 318–21.

CHAPTER TEN

Lentil (*Lens culinaris* Medik.)

R.N. Adsule

10.1 INTRODUCTION

Lentil has made a significant contribution to the human diet since ancient times. India and Turkey together account for about half the total production of world lentils.

Lentil (*Lens culinaris* Medik., synonym *L. esculenta* Moench.), also known as red dhal, masur, massar, masuri, tillseed and split pea, is a member of the family Leguminosae (Fabaceae). There are two subspecies of lentil: *Lens culinaris* ssp. *macrosperma* with large pods and seeds and *L. culinaris* ssp. *microsperma* with small pods and seeds. Lentils are thought to have originated from the Mediterranean region (Cubero, 1984).

Lentil is a much-branched erect or suberect, slightly pubescent annual bushy herb. The flowers may be white, pink, red, purple, pale blue or violet in colour. The pods are oblong, flattened, smooth and small and contain one to three seeds. The seeds are small, round or flat and biconvex. The colour of the seed may range from pale pink or buff to dark reddish brown, sometimes mottled with grey or dark green. The cotyledons may be red, orange, yellow or green in colour.

Lentil is grown as a winter season crop. It can be grown on a wide range of soils and can tolerate moderate alkalinity.

10.2 UTILIZATION

Lentils are mainly used for human consumption in the form of cooked whole seeds or split cotyledons (dhal). The methods of preparing lentils

Legumes and Oilseeds in Nutrition
Edited by E. Nwokolo and J. Smartt
Published in 1996 by Chapman & Hall. ISBN 0 412 45930 2

Table 10.1 Chemical composition of lentil seeds

Constituent	*Content (per 100 g dry seeds)*
Protein (g)	28.6
Fat (g)	0.8
Carbohydrates (g)	68.2
Ash (g)	2.4
Calcium (mg)	80
Phosphorus (mg)	336
Iron (mg)	5.5
Magnesium (mg)	94
Sodium (mg)	40
Copper (mg)	0.7
Zinc (mg)	2.1
Thiamine (mg)	0.45
Riboflavin (mg)	0.2
Niacin (mg)	3.0

Compiled from Gopalan *et al.* (1982) and Salunkhe *et al.* (1985).

for consumption vary widely among the different ethnic groups. The Indian Council of Medical Research (ICMR) has developed recipes for weaning foods such as 'khicheri' and 'keera pongal' using lentil dhal (ICMR, 1977).

Lentil straw can be utilized as a quite nutritious cattle feed (Ranjhan, 1980).

10.3 NUTRITIONAL COMPOSITION

The chemical composition of lentil seeds is given in Table 10.1. The composition varies significantly with genetic and environmental factors (Adsule *et al.*, 1989).

Globulin is the major protein fraction in lentils and is further characterized as 7S (vicilin) and 11S (legumin) fractions on the basis of sedimentation coefficients. The essential amino acid profile of lentil seed protein is presented in Table 10.2, and is of inferior quality to the FAO reference protein primarily because of its low content of sulphur amino acids. The digestibility coefficient (DC), biological value (BV) and protein efficiency ratio (PER) of 14 lentil samples were in the ranges 78–93%, 32–58% and 0.1–1.2, respectively (Patwardhan, 1962).

Most of the carbohydrate in lentil seeds is present in the form of starch, which is composed mainly of amylopectin. Sucrose is the major sugar in lentil seeds. Dietary fibre content is relatively low.

Lentil lipids contain a high proportion of oleic, linoleic and palmitic acids. Lentils are a good source of iron, zinc and magnesium and a fair source of thiamine, riboflavin and niacin.

Table 10.2 Essential amino acid composition of lentil seed proteins

Amino acid	*Lentil (g/16 g N)*	*FAO reference protein*
Lysine	7.0	5.5
Isoleucine	4.3	4.0
Leucine	7.5	7.0
Valine	4.9	5.0
Methionine + cystine	1.9	3.5
Phenylalanine + tyrosine	7.5	6.0
Threonine	3.5	4.0
Tryptophan	1.0	1.0

Compiled from Gopalan *et al.* (1982) and FAO/WHO (1973).

10.4 ANTINUTRITIONAL FACTORS

Lentil seeds contain relatively low levels of antinutritional factors. These factors include trypsin inhibitors, haemagglutinins, polyphenols, phytic acid, flatulence factors, α-amylase inhibitor, cyanogens and saponins (Adsule *et al.*, 1989). The trypsin inhibitor activity in lentils is significantly reduced after germination and soaking and is almost completely destroyed after cooking.

Haemagglutinating activity of lentils is completely destroyed by autoclaving for 20 min (Tannous and Ullah, 1969). Dehulling significantly reduces the polyphenols of lentils as they are mainly present in the seed coat. Germination and cooking significantly reduce the phytate phosphorus and oligosaccharides in lentil seeds.

10.5 HUMAN AND ANIMAL STUDIES

The human and animal studies with lentil are mostly related to the evaluation of their nutritional quality. Studies with rats have been conducted to evaluate the protein digestibility, biological value, protein efficiency ratio, net protein utilization and amino acid supplementation of lentil proteins.

REFERENCES

Adsule, R.N., Kadam, S.S. and Leung, H.K. (1989) Lentil, in *Handbook of World Food Legumes*, Vol. III, (eds D.K. Salunkhe and S.S. Kadam), CRC Press, Boca Raton, FL.

Cubero, J.I. (1984) Taxonomy, distribution and evolution of lentil and its wild relatives, in *Genetic Resources and Their Exploitation – Chickpeas, Faba Beans and Lentils*, (eds J.R. Witcombe and W. Erskine), Martinus Nijhoff Publishers, The Hague, The Netherlands.

FAO/WHO (1973) *Energy and Protein Requirement, FAO Nutrition Meeting Rep. Ser. No. 52/WHO Technical Rep. Ser. No. 522*, Food and Agriculture Organization, Rome, Italy.

Gopalan, C., Sastri, B.V.R. and Balasubramanian, S.C. (1982) *Nutritive Value of Indian Foods*. National Institute of Nutrition, Hyderabad, India.

ICMR. (1977) *Studies on Weaning and Supplementary Foods. ICMR Tech. Rep. Ser. No. 27*. Indian Council of Medical Research, New Delhi, India.

Patwardhan, V.N. (1962) Pulses and beans in human nutrition. *Am. J. Clin. Nutr.*, **11**, 12–16.

Ranjhan, S.K. (1980) *Animal Nutrition and Feeding Practices in India*, 2nd edn, Vikas Publishing House, New Delhi, India.

Salunkhe, D.K., Kadam, S.S. and Chavan, J.K. (1985) *Postharvest Biotechnology of Food Legumes*. CRC Press, Boca Raton, FL.

Tannous, R.I. and Ullah, M. (1969) Effects of autoclaving on nutritional factors in legume seeds. *Trop. Agric.*, **46**, 123–5.

CHAPTER ELEVEN

Lupin (*Lupinus* spp.)

N.A. Todorov, D.C. Pavlov and K.D. Kostov

11.1 GENERAL BOTANICAL, ECOLOGICAL AND PRODUCTION INFORMATION

Lupin (*Lupinus* sp. L.) is an ancient crop that is well known in many countries. It is a member of the Papilionaceae (Fabaceae) family. In total there are 300–400 different species of lupin (Popov *et al.*, 1966), although only about 150 of them are widely propagated (Allen and Allen, 1981). Only a few are actually cultivated. Those cultivated species include the white lupin (*Lupinus albus* L.), the yellow lupin (*Lupinus luteus* L.) the blue lupin (*Lupinus angustifolius* L.), the perennial lupin (*Lupinus polyphyllus* Lind.) and the sweet lupin (*Lupinus mutabilis*, Sweet).

Lupin possesses great adaptability and can be cultivated in widely different ecological conditions. These range from tropical environments like Brazil and the Sudan, to Mediterranean climates like Egypt, to low temperature ecologies like Alaska and the Canadian Yukon. The blue lupin and the perennial lupin varieties are usually well adapted to low temperature conditions whereas the white and yellow varieties have higher temperature requirements. White and yellow lupins are adapted to dry conditions, whereas blue and perennial lupins have higher water requirements. All lupin species have a high requirement for light and solar energy. Blue and yellow lupins are long-day plants, whereas the white lupin is day-length neutral. Most lupin species do not have specific requirements for soil type and fertility, most of them doing quite well on sandy soils, well-drained sands over clay, or sandy loam soils with acid to neutral pH.

Legumes and Oilseeds in Nutrition
Edited by E. Nwokolo and J. Smartt
Published in 1996 by Chapman & Hall. ISBN 0 412 45930 2

They do not have any specific requirements as to their location in a crop rotation system. However, they usually come after wheat or other grains in low land rotations, and after oats, rye or potatoes in high land rotations.

11.1.1 Production

Lupins can be used as a highly productive source of green material or grain for animal fodder and human nutrition. Some strains of lupin are capable of very high yields of grain and green forage; for example, a lupin cultivar, the Kievskij mutant, has been reported to yield 100 800 kg/ha green matter and 6800 kg/ha grain. It also yields more than 2000 kg/ha of digestible protein (Gvozdev *et al.*, 1983). Of the annual lupin varieties, the white lupin is most productive, yielding about 34 560 kg/ha green matter and 7350 kg/ha dry matter. The yellow lupin is the next most productive species, yielding 23 760 kg/ha green matter and 5440 kg/ha dry matter. The blue lupin has the lowest level of productivity, about 20 060 kg/ha green matter and 4720 kg/ha dry matter.

11.2 CURRENT AND POTENTIAL USAGE AS HUMAN FOOD

Lupins were being used as food more than 4000 years ago, probably first in Greece and subsequently in Egypt and Italy from where its use spread to France, Spain, Portugal and South Russia. Initially, all lupin grain used as food had to be processed to remove the bitter seed component, a process of debittering (Alekseev, 1968). However, plant breeders have now succeeded in producing non-bitter varieties of lupin, the so-called sweet lupin varieties, which are not extensively used as human food.

Lupin seed flour, like other legume flours, can be profitably used as a supplement to improve the dietary protein quality of local foods. Lupin flour, protein isolates and concentrates can probably be used in the future as supplements to weaning food mixtures, arabic bread, biscuits, and other foods. Processed alkaloid-free lupin isolate contains up to 62% protein and is suitable for incorporation into many prepared foods (Feldheim, 1975). Lupin isolates can successfully be included in bread to replace 7–10% of wheat flour. The quality is good and the protein content increases from 11.1% to 13.9% with 10% lupin meal. Bread lipids increase from 3.7% to 4.6% and carbohydrates decrease from 83.2% to 79.3%. Additionally, the protein efficiency ratio is increased from 28% to 76%. Lupin-based dishes were well accepted by children aged 8–15 years over a 3-year period of evaluation in Bulgaria. Alkaloid-free full-fat lupin meal containing about 38% protein, 12.7% fat and 36% nitrogen-free extract is used in bread making, replacing 3–12% of the wheat. Bread made from

100% wheat contains about 10.3% protein. Use of lupin flour significantly increases the protein content, and improves the amino acid balance. Experiments in Bulgaria suggest that sweet lupin flour may be used largely as a wheat flour supplement in bread-making, thus saving wheat and enhancing the protein quality of bread.

Alkaloid-free full-fat sweet lupin flour has also been successfully included in the preparation of biscuits, replacing up to 20% of the wheat meal. Results show good acceptability and high protein efficiency ratio (Ballester *et al.*, 1986).

Fortification of wheat meal with 10% full-fat sweet lupin meal is recommended in preparing biscuits or other foods to improve the nutritional quality of school feeding programmes (Wittig de Penna *et al.*, 1987). Sweet lupin flour or alkaloid-free lupin flour can be used as a supplement in the preparation of macaroni, spaghetti and noodles. Flour from dehulled lupin seed contains about 47% crude protein, 9% lipids, 34% total carbohydrates, 5% ash and 5% fibre. The flour may be included in macaroni to replace up to 6% of the flour, and in spaghetti replacing up to 10–12% of the flour. Macaroni supplemented with soybean flour reaches its maximum weight at 2% of supplementation. With lupin flour, the maximum weight is not reached until 6% of supplementation, thus achieving a higher protein content in macaroni. Additionally, macaroni supplemented with lupin flour has very good cooking quality, and a desirable amber colour, eliminating the addition of eggs or artificial colours in macaroni manufacture.

Supplementation of spaghetti with lupin flour improves colour, appearance and cooking qualities, while enhancing protein efficiency ratio and biological value. Lupin flour can successfully replace 5–10% of the wheat flour in noodles and can be used in the production of imitation milk. Alkaloid-free lupin meal can also be included in the preparation of products such as soups, purees and creams for human consumption (Oliva and Ballester, 1984). The flour has been incorporated into some commonly eaten dishes like chocolate cake, cream puree, apple pie, various sauces and soups and other specialties. When the flour is included at 4–11 g per portion, the products are of good appearance, flavour and high acceptability (Gross *et al.*, 1976b). Seeds of many lupin varieties contain up to 18% fat and can therefore, be used as a source of edible oils. After removal of alkaloids, the oil is an excellent dietary oil, the main fatty acid being oleic acid and the main tocopherol being α-tocopherol (Hatzold *et al.*, 1983).

11.3 CURRENT AND POTENTIAL USAGE AS ANIMAL FEED

The great advantage of lupin as animal feed lies in the high crude protein content of the seed and vegetative parts, its high productivity and its great

Table 11.1 Chemical composition of some legume grains

Legume	*Water (%)*	*Protein (%)*	*Fat (%)*	*NFE (%)*	*Ash (%)*	*Fibre (%)*	*Energy (Kcal)*
Lupin	9.0	40.0	13.0	26.0	3.0	9.0	420
Fenugreek	8.6	29.6	5.2	50.0	3.3	7.2	365
Broad bean	10.6	25.0	1.8	53.7	3.0	5.9	354
Lentil	12.2	23.7	1.3	57.4	2.2	3.2	351
Chickpea	11.5	19.2	6.2	56.7	2.2	3.2	376

From Tannous (1978). NFE = Nitrogen-free extract.

adaptability to various climatic environments. The seed is fed as a grain, whole or ground, to all types and ages of livestock, being fed at a daily rate of 1.5–2.0 kg to bulls, and 2.0–2.5 kg to milking cows. Horses may be fed at a daily rate of 2.0–2.5 kg. For swine, adult pigs should be fed at a daily rate of 0.5–0.7 kg, growing pigs at 0.3–0.4 kg, and young pigs at 0.5 kg. For chickens, lupin flour may be incorporated into diets at 11–20% of the diet, it should not be used as the sole protein source, but rather mixed with soybean meal. The seed and vegetative parts of the lupin contain alkaloids which are toxic to domestic animals. These toxic components have in the past been removed by soaking in water, steam treatment and high temperature drying. The entire vegetative part of the sweet lupin can be used as green fodder for ruminants, horses and pigs. The plants are cut at flowering or when the first pods are formed. The green feed is good forage for cows and can be included in the daily diet at a rate of 20–25 kg/head. Green lupin forage may also be included in adult swine diets at up to 10–11 kg daily, and at 4 kg for young pigs. Green fodder may also be fed to rabbits, goats and chicken.

11.4 NUTRIENT COMPOSITION AND NUTRITIVE VALUE

Lupin grain is a good source of dietary protein and fat for animals and humans.

11.4.1 Proteins

Lupin grain has considerably higher protein content compared to broad beans, lentils, chickpea, fenugreek and other pulses (Table 11.1). Only soybean approaches lupins in this respect. There is considerable variation among lupin species in protein content of lupin meal, from 24 to 61%; yellow lupin contains 39–47% and blue lupin 31–38%. Lupin protein has a relatively good amino acid profile with high content of arginine (4.1–11.2%), leucine (7.5–9.4%), lysine (4.3–5.2%) and phenylalanine (3.0–6.8%). The protein is poor in sulphur amino acids, (Szelenyine-Galantai *et al.*,

1984). Among pulses, lupin ranked third in protein quality after soybean and chickpea. According to the study by Tannous (1978) the chemical score is 61 for chickpea, 55 for lupin, 44 for lentils and 34 for broad beans. There are differences in amino acid composition among the protein fractions of lupin. The albumin fraction has a better amino acid composition than the globulin fraction. All three fractions (albumins, globulins and prolamines) are rich in glutamic acid, aspartic acid, arginine and leucine.

11.4.2 Carbohydrates

Lupin grains are rich in structural (cell wall) polysaccharides and have a very low content of starch, 0.4%. Total furfural generators are about 20%, most of them being soluble in either water or dilute acid. Ethanol-soluble sugars are more than 10% (Cerning-Beroard and Filiatre, 1976). The cell wall material of lupin cotyledon is characterized by a low content of cellulose and high content of pectic substances. The main sugar component of cell wall material is galactose, followed by arabinose, fucose, xylose, and glucose (Carre *et al.*, 1985).

11.4.3 Fat

The content of fat extracted by ether in most lupin species is 4–12%, and varies a lot between the different species. Seeds of *L. albus* contain 5–14%, *L. luteus* 5–7% and *L. angustifolius* 4–8.5%. The richest in fat are the seeds of *L. mutabilis* Sweet with 13–23% (Hudson, 1979). The fat content of lupin is relatively high, only a few pulses like soybean exceed lupin in this respect. The high fat content confers a high energy value on lupin meal as food and feed. The fatty acid profile of *L. mutabilis* is as follows: oleic acid 37.7–58%, linoleic acid 22.9–40.1%, palmitic acid 9.5–11.6%, stearic acid 6.6–7.0%, linolenic acid 1.9–2.0%, arachidonic acid 0.6–1.0% and behenic acid 0.5–0.7% (Eckard and Feldheim, 1974). The oil also contains α- and β-tocopherols, sitosterol, campesterol, stigasterol and squalene. As a dietary oil, lupin compares favourably with soybean and rapeseed oils.

11.4.4 Minerals

Lupin grain is a good source of macro and micro mineral elements as shown in Table 11.2. It is an excellent source of phosphorus, although much of it is present as phytates, salts of phytic acid, which are utilized poorly by domestic species. Lupin contains an exceptionally high quantity of the trace element manganese, in white lupin manganese content is as high as 3 g/kg, much of it concentrated in the cotyledons. Lupin seeds have various contents of carotenoids. The content of β-carotene and zeaxanthin is low in white lupin (23–44.7 mg/100 g) but higher in yellow

Table 11.2 Mineral content of lupin grain (at 87% dry matter)

Macro elements	*g/kg*	*Micro elements*	*mg/kg*
Calcium	1.8	Iron	55
Phosphorus, total	4.0	Copper	6
Phosphorus, avail.	0.8	Zinc	3.5
Sodium	0.1	Manganese	800–3000
Potassium	9.0	Cobalt	0.04
Chlorine	0.2	Selenium	0.07
Magnesium	1.5	Molybdenum	3
Sulphur	2.4	Nickel	5

and blue lupins, about 142 mg/100 g carotene, and 1110 mg/100 g zeaxanthin. They are therefore good sources of provitamin A in human and animal nutrition.

11.5 PROCESSING OF LUPIN SEEDS

Lupin seeds which are not of the sweet type need to be processed to remove the various toxic components commonly present in these seeds. After processing, a good quality flour containing 60–62% protein results. Seeds of the lupin are milled and solvent extracted to remove the fat. The oil cake resulting from solvent extraction of *L. mutabilis* contains 93.0% dry matter, 65.3% crude protein, 0.6% ether extract, 4.0% crude fibre, 6.57% ash and 23.5% nitrogen-free extracts (Schoeneberger *et al.*, 1987). During extraction of fat, the alkaloids are concentrated in the oilcake and can be removed by washing the oilcake with an ethanol–water mixture. An oilcake produced in this way contains about 72% protein and 0.6% residual alkaloids, a level which is quite safe for use. Most commercially produced protein concentrates from lupin seeds contain 1.4% moisture, 72.8% protein, 17.9% crude fat, 0.7% ash, 2.3% total sugars and 0.26% reducing sugars (Blaicher *et al.*, 1981; Sathe *et al.*, 1982). The chemical composition of various lupin products is presented in Table 11.3.

After alkaline extraction and precipitation at pH 5.1, 4.2 or 4.9, a lupin protein isolate is obtained which has an extremely high protein content (95–96%) and no fibre. Lupin protein isolates have better solubility than soybean isolates and have a similar emulsification capacity. They may, therefore, be considered as potential substitutes for soybean proteins in food manufacture (King *et al.*, 1985).

Table 11.3 Chemical composition of lupin protein products

Nutrients	*Undefatted meal*	*Defatted meal*	*Protein isolates*		
			pH 4.9	*pH 5.1*	*pH 4.2*
Moisture (%)	6.2	9.7	1.7	2.1	1.9
Protein (%)	39.3	43.5	95.7	95.7	95.6
Fat (%)	12.0	0.9	–	–	–
Ash (%)	3.8	4.1	2.9	2.2	2.5
Crude fibre (%)	2.7	3.4	–	–	–

Source King *et al.* (1985).

11.6 CHEMICAL NATURE AND DETOXIFICATION OF ANTINUTRITIONAL SUBSTANCES

It is recognized that lupin plants and seeds contain different antinutritional substances, which are capable of being destroyed using various processing procedures. These include alkaloids, glucosides, trypsin inhibitors, etc.

11.6.1 Alkaloids

The alkaloid content of lupin plants ranges from 0.0001–5.0%. Lupin feeds which contain alkaloids at a level of 0.1–15 mg/100 g are considered as being without alkaloids. Low alkaloid lupin feeds contain 50–200 mg/100 g, and high alkaloid feeds contain 1600–3200 mg/100 g. Lupin species which belong to the first group are usually referred to as sweet. Flowers and leaves of most lupin species have a high ability to synthesize and accumulate alkaloids and all aerial parts and seeds of poisonous species are toxic to some extent. With animals, death occurs when large amounts of high alkaloid lupin are consumed over a short period. Signs of poisoning include nervousness, depression, loss of muscular coordination and a reluctance to move, twitching leg and other muscles, difficulty in breathing, frothing at the mouth, convulsions, coma and death. For human consumption, lupin grain, flour and other products can only be used as food when they contain less than 0.002% alkaloids (Gross *et al.*, 1976a). When the alkaloid content is higher, there is a danger of accidental poisoning. Agid *et al.* (1988) reported a progressive motor neurone disease in a woman following normal pregnancy and consumption of high-alkaloid seeds of *L. albus*. Within 2 months after ceasing to consume the lupin meal, the patient had begun to recover. After 20 months, the clinical picture remained stable, allowing an almost normal life despite the incomplete recovery.

11.6.2 Glucosides

The different lupin species contain various amounts of glucosides. Cyanogenic glucoside content in *L. luteus* is 0.42 mg HCN/100 g. This is much lower than in *Phaseolus vulgaris* which contains 1.52–1.83 mg HCN/100 g (Contreras *et al.*, 1974). Hydrocyanic acid content in lupin species varies between 0.53 and 2.89 mg/100 g seed compared to 9.75 mg/100 g in *P. lunatus* (Schoeneberger *et al.*, 1981).

11.6.3 Trypsin inhibitors

Different lupin species contain different amounts of trypsin inhibitors. *L. albus* and *L. angustifolius* do not show any trypsin and amylase inhibitor activity (Rahma and Rao, 1984), whereas trypsin inhibitor activity (0.4 mg/g) is found in *L. mutabilis* Sweet (Hove and King, 1979). *L. termis* has been shown to have a trypsin inhibitor activity of 1.16 TIU/ml compared to 30.1 TIU/ml in soybean (Schoeneberger *et al.*, 1981). Lupin grain contains much less trypsin inhibitor than pea and horse bean (Korol *et al.*, 1986).

11.6.4 Detoxification

Seeds and aerial parts of plants may be soaked in running water for 48 h to remove toxic components. Similarly drying the aerial parts in high heat, or pelleting at high temperatures (about 800°C) significantly reduces alkaloid content, thus rendering such feeds harmless. Pressed cakes may also be extracted with 70% methanol in water to make the cakes harmless. Detoxification by any of these processes also reduces trypsin and amylase inhibitor activities, and haemagglutinin activity.

11.7 ANIMAL STUDIES WITH LUPIN SEED OR MEAL

Many experiments have been conducted with rats to evaluate the digestibility, nutritive value and the antinutritional effects of lupin meal. Experiments with rats have demonstrated that it is important that lupin-based diets are adequately supplemented with methionine and cystine. Thus methionine supplementation significantly increases efficiency of protein utilization in protein isolates and concentrates from 39 and 19% to 84 and 89%, respectively (Elmadfa *et al.*, 1982). PER of lupin protein isolate is increased from 0.47 to 2.90 compared to a casein value of 3.12 in control rats. A similar result on the positive influence of supplementation with methionine was noted by Sgarbieri and Galeazzi (1978) who observed that addition of 0.15% DL-methionine to lupin meal increases PER from 1.03 to 2.02, compared to 2.1 with casein.

Pigs tolerate lupin meals containing low levels of alkaloids, diets containing 0.3–0.34% alkaloids causing toxicity. Godfrey *et al.* (1985) consider that free-feeding growing pigs may tolerate dietary lupin alkaloid levels of 0.2%. Unprocessed yellow and white lupin meals significantly depress weight gain while increasing cost of gain. When a small quantity is included for a brief period, lupin meal can replace soybean meal in swine diets. However, when 50% of soy protein is replaced by lupin meal, growth is significantly reduced and feed/gain ratio increased (Szelenyine-Galantai *et al.*, 1988).

Poultry are very sensitive to lupin alkaloids in the grain. When lupin meal containing 0.1% alkaloid is fed to chickens, there is reduced live weight gain in growing chickens and depressed productivity in laying hens.

Lupin meal can be used as a supplementary feed to young animals and pregnant ewes during the dry seasons of many countries. Lupin grain and meal can be used to maintain high productivity and normal growth during periods when fresh fodder or grazing is not available. Supplementation of diets of pregnant ewes with 400 g daily for 2 weeks prior to lambing increases lamb birth weight and lamb growth rates. Even better benefits are obtained when the ewes are in poor condition during gestation. Lupin grain and meal may also be fed successfully to beef steers and dairy cows to enhance productivity.

REFERENCES

Agid, Y., Pertuiset, B. and Dubois, B. (1988) Motor-neurone disease as manifestation of lupin toxicity. Letter to *Lancet*, 1347.

Alekseev, E.K. (1968) *Annual Forage Lupin*, Colos, Moscow (in Russian).

Allen, N.O. and Allen, E.K. (1981) *The Leguminosae. A Source Book of Characteristics, Uses and Nodulation*. Macmilan, Wisconsin.

Ballester, D., Carreno, P., Urrutia, X. and Yanez, E. (1986) Chemical composition and nutritional quality of sugar cookies containing full-fat sweet lupin flour (*L. albus* cv. Multolupa). *J. Food Sci.*, **51**, 645–8.

Blaicher, F.M., Nolte, R. and Mukherjee, K.D. (1981) Lupin protein concentrates by extraction with aqueous alcohols. *J. Am. Oil Chem. Soc.*, **58**, 761–5.

Carre, B., Brillouet, J.M. and Thibault, J.F. (1985) Characterization of polysaccharides from white lupin (*Lupinus albus* L.) cotyledons. *J. Agric. Food Chem.*, **33**, 285–92.

Cerning-Beroard, J. and Filiatre, A. (1976) A comparison of the carbohydrate composition of legume seeds: horsebean, peas and lupins. *Cereal Chem.*, **53**, 968–78.

Contreras, S., Araya, H., Pak, N. and Tagle, M.A. (1974) Toxic factors in legumes growing in Chile. 1. Cyanogenetic glucosides. *Arch. Latinoam. Nutr.*, **23**, 251–9.

Eckard, R.W. and Feldheim, W. (1974) Lupins as new oilseed for South America.

Z. Lebensm. Unters Forsch., **155**, 92–3.

Elmadfa, I., Schoeneberger, H. and Gross, R. (1982) Evaluation of the quality of protein concentrates and protein isolates from the oil cake of *Lupinus mutabilis*. *Fette Seifen Anstrichmittel*, **84**, 61–70.

Feldheim, W. (1975) Lupin seeds as food. *Umschan Wiss. Technic*, **75**, 280 (in German).

Godfrey, N.W., Mercy, A.R., Emms, Y. and Payne, H.G. (1985) Tolerance of growing pigs to lupin alkaloids. *Aust. J. Exp. Agric. Anim. Husb.*, **25**, 791–5.

Gross, R., Morales, E., Gross, U. and Baer, E. von. (1976a) The lupin: contribution to the supply of food in the Andes. 3. A nutritional physiological study with flour from the sweet lupin (*Lupinus albus*). *Z. Ernahrungswiss.*, **15**, 391–5.

Gross, U., Reyes, J., Gross, R. and Baer, E. von. (1976b) The lupin: contribution to the supply of food in the Andes. 4. Acceptability of *Lupinus albus* flour. *Z. Ernahrungswiss.*, **15**, 396–402.

Gvozdev, V.A., Taaev, E.A., Golovtchenko, V.I. and Plotkin, A.L. (1983) *Producing Grasses and Lupin Seeds on the Industrial Base.* Colos Moscow (in Russian).

Hatzold, T., Elmadfa, I. and Gross, R. (1983) Edible oil and protein concentrate from *Lupinus mutabilis*. *Qual. Plant. Plant Foods Hum. Nutr.*, **32**, 934–8.

Hove, E.I. and King, S. (1979) Trypsin inhibitor content of lupin seed and other grains. *NZ J. Agric. Res.*, **22**, 41–2.

Hudson, B.J.F. (1979) The nutritional quality of lupin seed. *Qual. Plant.*, **29**, 15–21.

King, J., Aguirre, C. and Pablo, S. (1985) Functional properties of lupin protein isolates (*Lupinus albus* cv. Multolupa). *J. Food Sci.*, **50**, 82–7.

Korol, W., Barczynzka-Niedzialek, A. and Matyka, S. (1986) Antitrypsin activity of locally available legume seeds. *Bull. Infor, Przemyslu Paszowego*, **25**, 19–24.

Oliva, P. and Ballester, D. (1984) Chemical composition and biological quality of precooked legumes commonly eaten in Chile: lentils, beans, chickpeas alone or mixed with sweet lupin (*Lupinus albus* cv. Multolupa) flour. *Rev. Chil. Nutr.*, **12**, 51–6.

Popov, A., Pavlov, K. and Popov, P. (1966) *Crop Production Zemizdat*, Sofia.

Rahma, E.H. and Rao, M.S.N. (1984) Effect of debittering treatment on the composition and protein components of lupin seed (*Lupinus termis*) flour. *J. Agric. Food Chem.*, **32**, 1026–30.

Sathe, S.K., Deshpande, S.S. and Salunkhe, D.K. (1982) Functional properties of lupin seed (*Lupinus mutabilis*) proteins and protein concentrates. *J. Food Sci.*, **47**, 491–502.

Schoeneberger, H., Moron, S. and Gross, R. (1987) Safety evaluation of water-debittered Andean lupin (*Lupinus mutabilis*)–12-week rat feeding study. *Qual. Plant. Plant Foods Hum. Nutr.*, **37**, 169–82.

Schoeneberger, H., Sam, O., Gross, R. *et al.* (1981) The protein quality of water debittered lupins (*Lupinus mutabilis*) in combination with other protein sources. *Nutr. Rep. Int.*, **25**, 763–71.

Sgarbieri, V.C. and Galeazzi, M.A.N. (1978) Some physico-chemical and nutritional properties of a sweet lupin (*Lupinus albus* var. Multolupa) protein. *J. Agric. Food. Chem.*, **26**, 1438–42.

Szelenyine-Galantai, M., Jecsai, G., Juhasz, B. and Bodis, L. (1984) Determination of the feeding value of yellow and white flowered varieties of sweet lupin and of soya varieties. *Allattenyesztes es Takarmanoyovas*, **33**, 281–7.

Szelenyine-Galantai, M., Jecsai, G. and Juhasz, B. (1988) Replacement of extracted soya protein by various sweet lupin species in pig feed. *Acta Agron. Hung.*, **37**, 101–9.

Tannous, R., Shakra, S.A. and Hallab, A.H. (1978) Nutritional quality and importance of food legumes in the Middle Eastern diet. Food Leg Improv. Develop. Proceedings of a Workshop held at the University of Alepo Syria 2–

7 May 1978. ICARDA, IDRC, Canada, pp. 29–32.
Wittig-de Penna, E., Carreno, P., Urrutia, X. *et al.* (1987) Sensory evaluation and acceptability of cookies enriched with sweet lupin (*Lupinus albus* cv. Multolupa). *J. Food Sci.*, **52**, 1434–5.

CHAPTER TWELVE

African locust bean (*Parkia filicoidea* Welw.)

A.C. Uwaegbute

12.1 INTRODUCTION

The African locust bean is a tree legume found all over West Africa, southeast Asia, and tropical South America. It is a medium-sized tree found in open savanna woodlands where it may stand as the only tree species in a location, and characterized by its fruits, which are elongated pods, 5–11 inches long, and found in clusters. The immature fruit is green, assuming a brown colour as it matures. The mature pod is made up of a husk which encloses a powdery yellow material in which are embedded dark brown seeds. Seeds comprise about 25.4% of the weight of the dry pod. The whole fruit contains on a dry matter basis 12.7% protein, 6.8% fat, 6.2% ash and 18% fibre. The mature seed contains about 30% digestible carbohydrates (starch and sugars), 30% protein, 20% fat, 9% crude fibre and 5% minerals. Not only is the fat content higher than in most grain legumes, it is made up predominantly of unsaturated fatty acids (54% of total fatty acids). The unsaturated fatty acids are linoleic acid (42.5% of total), oleic acid (8.8% of total) and palmitoleic acid (2.7% of total). The saturated fatty acids which contribute 46% of total fatty acids, comprise palmitic acid (31% of total), stearic acid (7.7% of total), arachidic acid (4.2% of total) and behenic acid (3.1% of total).

The powdery yellow pulp contains more carbohydrates than the seed, the carbohydrates being primarily reducing sugars (19%), non-reducing sugars (9%) and other complex carbohydrates (36%).

Legumes and Oilseeds in Nutrition
Edited by E. Nwokolo and J. Smartt
Published in 1996 by Chapman & Hall. ISBN 0 412 45930 2

12.2 USE OF LOCUST BEAN

Almost every part of the tree serves a useful purpose in the household. The leaves are rich in nitrogen and ash, serving as an important source of green manure for homestead agriculture. The whole pod provides valuable fodder for animals particularly in the dry season when animal feed shortages are common. The yellow pulp may be used as an ingredient in the preparation of various stews, soups and sauces for the consumption of rice, cereals and grain legumes. The pulp may also be pressed into cakes and preserved for later use, or used in the preparation of various indigenous drinks. The dark brown seeds are the most valuable part of the crop. The seeds are not eaten in their natural state probably because of their high content of indigestible components and antinutritional factors (Odunfa, 1983). Seeds are fermented into a most important flavouring condiment in West Africa which is used in much the same way as bouillion cubes of the western world. In northern Togo, some of the local people grind the seeds to a fine meal and steep it in water to make a refreshing gruel.

12.3 FOOD VALUE OF LOCUST BEAN

Fermentation significantly improves the nutritional value of locust bean and makes it one of the most important sources of dietary protein in those areas of the world where it is consumed as a food ingredient. The fermented product is known by various local names in different countries, and constitutes a very important food flavouring agent. Fermented locust bean cakes are eaten in various countries as a meat substitute, especially during periods of famine and starvation which are endemic to many countries in Africa and Asia. It may be combined with pepper and salt and turned into sauce for consumption of cereals, cassava, yam, and other roots and tubers. For example, in Northern Nigeria it has been estimated that the locust bean ranks third in order of importance as a source of dietary protein intake, after guinea corn (*Sorghum* spp.) and millet. In general, it contributes 1.5% of the caloric intake and 5% of the protein intake of the Hausa people of Northern Nigeria (Simon, 1976). Even though it is recognized that things have changed, over 25 years ago it was estimated in Ilesha, Western Nigeria, that fermented locust bean represented the second most important source of dietary protein after cowpeas, cowpeas having been acknowledged as providing as much as 80% of total protein intake (May, 1965).

12.4 PROCESSING/FERMENTATION OF LOCUST BEAN

Locust bean seed has a very hard, water-impermeable testa and therefore prolonged cooking is necessary during processing to soften the seed coat prior to dehulling. According to Odunfa (1985), presoaking for 12 h and cooking for 30 min in a pressure cooker is sufficient to soften the seed, but removal of the seed coat is more difficult. Pressure cooking for 90 min without presoaking makes the seed coat easy to remove while sufficiently softening the seed. Seeds may be dehulled manually or by machine. Machine dehulling of the seeds before cooking is practised as an easier alternative since the hydrophobic seed coat which prolongs cooking is eliminated. After the initial cooking regime, the dehulled seeds are cooked for an additional 30 min and gently pounded to further soften the seeds. Spoonfuls of the pounded seeds are wrapped in leaves and stored in various sized containers for fermentation to occur. Piling of the wrapped seeds in storage baskets causes the temperature of storage to rise to a level adequate for fermentation to be initiated.

Fermentation occurs in 48–72 h at an ambient temperature of 25–35°C. Fermentation depends on chance inoculation from the environment by organisms identified by Odunfa (1981) as belonging to *Bacillus subtilis*, coagulase negative *Staphylococcus*, *Bacillus licheniformis* and *Staphylococcus saprophyticus*. Other microorganisms involved have been identified as *Pseudomonas aeruginosa*, *Enterobacter aerogenes* and *Coryneformis* bacteria (Kuboye, 1985). Chance inoculation with these various microorganisms makes the quality and flavour of the final product inconsistent. The bacteria produce viscous strands of mucilaginous material which link the beans together and produce a strong ammoniacal odour. During fermentation, there is extensive hydrolysis of proteins to amino acids and of oligosaccharides to simpler sugars. Hydrolysis of phytates and oxalates during the fermentation process also releases minerals rendering them more available, especially calcium and phosphorus. Protein and fat content of the fermented product is much higher than in the locust bean seed, the protein of the fermented cake being higher by 40% and the fat by 35%. Riboflavin and thiamine may increase threefold and sixfold, respectively, following fermentation. However, fermentation causes a one-third decrease in niacin content. Fermented seeds are partially dried in the sun during which they become darker in colour. When partially dry, the beans are moulded into balls or made into flat cakes of about 5 cm in diameter and further dried to improve their shelf life and make them stable for at least one year (Campbell-Platt, 1980).

Table 12.1 Nutrient content of locust bean

	Content (g/100 g dry matter)	
Nutrient	*Unfermented*	*Fermented*
Ash	5.14	6.75
Protein	30.57	38.50
Fat	15.20	31.20
Fibre	7.83	6.23
Carbohydrates	40.09	23.55

Source: Eka (1980).

Table 12.2 Mineral and vitamin content of locust bean

	Content (mg/100 g dry matter)	
Nutrient	*Unfermented*	*Fermented*
Minerals		
Potassium	2.50	5.50
Sodium	2.40	2.50
Zinc	15.00	18.00
Magnesium	80.50	83.50
Calcium	3.30	3.60
Copper	1.50	2.00
Iron	22.50	2.80
Phosphorus	2.80	3.20
Vitamins		
Thiamine	0.65	1.35
Riboflavin	0.45	1.30
Ascorbic acid	7.50	5.20

Source: Eka (1980).

12.5 NUTRITIONAL VALUE OF LOCUST BEAN

Tables 12.1–12.3 demonstrate that fermentation improves the nutritional value of locust bean. In fact, the increase in protein and fat is relative, attributable to a corresponding loss in the digestible carbohydrate fraction as a result of fermentation. Availability of minerals is affected by the presence of such chelators as phytic acid and oxalic acid. Fermentation substantially reduces the concentration of these chelators in locust bean. Eka (1980) found that after fermentation, cyanide is reduced from 26×10^{-4} to 12×10^{-4} and oxalic acid from 0.21 ± 0.1 to 0.12 ± 0.12 whereas total phytic acid phosphorus is reduced from 15% in unfermented beans to 7.5% in the fermented bean.

Table 12.3 Amino acid content of locust bean

	Content (mg/100 g dry matter)	
Amino acids	*Fermented*	*Unfermented*
Lysine	4.24	6.72
Histidine	1.60	1.68
Arginine	2.84	2.88
Tryptophan	0.12	0.24
Aspartic acid	8.40	12.00
Threonine	3.80	4.48
Serine	3.22	5.62
Glutamic acid	16.00	19.60
Proline	4.24	6.11
Glycine	6.84	9.20
Alanine	5.60	7.84
Cystine	1.20	2.10
Methionine	0.12	0.24
Isoleucine	0.88	0.40
Leucine	1.44	0.80
Tyrosine	1.12	0.80
Phenylalanine	0.32	0.16

Source: Eka (1980).

Locust bean protein shows deficiencies of leucine, isoleucine, phenylalanine and tryptophan. Content of tyrosine may also be low. Amino acid profile of locust bean protein shows a characteristic deficiency of the sulphur-containing amino acids. The content of some other amino acids is high and increased substantially in the fermented bean. However, the low content of many essential amino acids indicates that the overall protein quality is low, an observation supported by experiments with rats. In a study in which fermented locust bean was fed to weaning rats, dry matter digestibility was higher (97.6%) than in casein (85.6%) (Umoh and Oke, 1974). However, protein efficiency ratio, biological value and net protein utilization were 0.97, 62.93 and 61.73, respectively, for locust bean, as compared with values of 2.8, 85.1 and 73.3 for casein diets.

REFERENCES

Campbell-Platt, G. (1980) African locust bean (*Parkia* spp) and its West African fermented food product, *Dawada. Ecol. Food Nutr.*, **9** 123–32.

Eka, O.U. (1980) Effect of fermentation on the nutrient status of locust beans (*Parkia filicoidea*). *Food Chem.*, **5**, 303–8.

Kuboye, M.A. (1985) Traditional fermented foods and beverages of Nigeria, in *Provisional Report no. 20.* Proceedings of the IFS/UNU Workshop on Development of Indigenous Fermented Foods and Food Technology in Africa, Douala, Cameroun, October 1985. International Foundation for Science (IFS), Stockholm, Sweden.

May, J.M. (1965) Ecology of malnutrition in middle Africa. Studies in medical geography. Vol 1, (eds G. Charalambous and G. Inglett), Academic Press, London, pp. 61–92.

Odunfa, S.A. (1981) Microorganisms associated with fermented African locust bean during 'iru' preparation. *J. Plant Foods*, **25**, 245–50.

Odunfa, S.A. (1983) Carbohydrate changes in fermenting locust bean during 'iru' preparation. *Qual. Plant. Plant Foods Hum. Nutr.*, **32**, 3–10.

Odunfa, S.A. (1985) African fermented foods: from art to science, in *Provisional Report no. 20*. Proceedings of the IFS/UNU Workshop on Development of Indigenous Fermented Foods and Food Technology in Africa, Douala, Cameroun, October 1985. International Foundation for Science (IFS), Stockholm, Sweden.

Simon, E.B. (1976) Calorie and protein intake in three villages of Zaria Province, May 1970–July 1981. *Samaru Miscellanous Paper no. 55*. Institute of Agricultural Research, Ahmadu Bello University, Zaria, Nigeria.

Umoh, I.B. and Oke, O.L. (1974) Nutritional value of some lesser known oilseeds in rats. *Nutr. Rep. Int.*, **9**, 453–60.

CHAPTER THIRTEEN

The African oil bean (*Pentaclethra macrophylla* Benth.)

S.C. Achinewhu

13.1 INTRODUCTION

The African oil bean, popular in Nigeria (its local name is Ugba) is a tropical tree in the family Leguminosae (Mimosoideae). It is native to tropical Africa, although a representative of the genus exists in tropical South and Central America. The tree grows to a height of 21 m, is well branched and forms a crownlike canopy. The oil bean tree is found in the southern rain forest zone of West Africa where it grows wild, having never been cultivated to any real extent by individuals around homesteads or commercially in plantations. The tree produces flowers between March and April and also between June and November. The flowers are yellow and pinkish-white in colour, sweet smelling, and attract myriads of insects including the honey bee. The fruit is a long green pod which slowly darkens with maturity, 36–46 cm long and 5–10 cm broad. Each pod contains up to 10 seeds and at maturity, the pod splits open explosively scattering its seeds up to a distance of 20 m from the tree. The seeds are flat in shape, hard but smooth in texture, brown in colour and about 6 cm long.

13.2 PRESENT AND POTENTIAL USAGE IN NUTRITION

Seeds of the African oil bean are edible and are normally boiled for up to 12 h to facilitate the removal of the hard seed coat. After removal of the

Legumes and Oilseeds in Nutrition
Edited by E. Nwokolo and J. Smartt
Published in 1996 by Chapman & Hall. ISBN 0 412 45930 2

Table 13.1 Proximate composition of fermented and unfermented oil bean

Nutrient	*Unfermented bean*	*Fermented bean*
Moisture	2.42	3.90
Crude protein	39.90 + 0.80	40.70 + 0.06
Ether extract	36.33 + 0.50	38.55 + 1.47
Crude fibre	3.56 + 0.80	2.51 + 0.16
Total ash	2.50 + 0.80	2.52 + 0.20
Carbohydrates	15.30	12.1

Results are expressed as a percentage of dry matter, and are means of four determinations. Carbohydrates determined by difference.

seed coat, the embryo is sliced into smaller pieces with a sharp knife, and boiled for a further couple of hours. The sliced seeds are washed thoroughly in water with four or five changes of water, to reduce/remove the bitter taste of the seed. The washed slices are wrapped in blanched banana leaves, packed together in baskets and left to ferment for 3 days at room temperature (29–32°C). Fermentation is purely by chance inoculation with desirable fermentation agents which are ubiquitous. Products left to ferment for longer periods or allowed to mature after fermentation have better flavours than those consumed immediately after 3 days. The fermented product is very popular especially in the eastern states of Nigeria where it is eaten alone or with other ingredients like stockfish, garden egg and sliced tapioca as a mixed vegetable salad. This is the extremely popular 'African salad', a favourite snack at masquerade and other festivals.

13.3 CHEMICAL COMPOSITION AND NUTRITIVE VALUE

Proximate composition of oil bean seed (Table 13.1) is similar to those of most oilseeds. Protein and ether extract are very high (40% and 36–38%, respectively), carbohydrate content is low (12–15%), fibre and ash are very low (2.5–3.5%, 2.2–2.3%, respectively). There is great similarity in nutrient content (except for carbohydrates) between fermented and unfermented oil bean indicating that fermentation is not associated with any significant loss of nutrients except the carbohydrate fraction. Achinewhu and Riley (1986) have shown that fermentation significantly increases the content of thiamine, riboflavin and niacin in the African oil bean. Thiamine increased from 1.1 to 2.2 mg/100 g dry sample weight, riboflavin from 0.11 to 0.3 mg/100 g dry weight, and niacin increased from 2.0 to 3.2 mg/100 g dry weight.

Table 13.2 Carbohydrate composition of fermented and unfermented oil bean

Sugars	*Concentration (mg/g dry wt)*	*% of total carbohydrates*
Fermented oil bean		
Fructose	8.30 ± 0.8	24.5
Galactose	5.24 ± 0.1	15.5
Glucose	1.74 ± 0.0	5.1
Glucose + galactose	7.98 ± 0.9	23.6
Total unknown monosaccharides	6.86	20.3
Total monosaccharides	30.0 ± 1.9	88.5
Sucrose	2.62 ± 0.2	7.7
Maltose	0.64 ± 0.0	1.9
Raffinose	0.15	0.4
Stachyose	0.16 ± 0.0	0.5
Unknown disaccharides	0.32 ± 0.0	0.9
Total di/oligosaccharides	3.89 ± 0.2	11.5
Total carbohydrate of extract	33.90 ± 2.1	
Unfermented oil bean		
Fructose	0.22 ± 0.0	0.3
Galactose	0.34 ± 0.0	0.5
Glucose	0.34 ± 0.0	0.5
Glucose + galactose	0.54 ± 0.0	0.9
Unknown monosaccharide	0.66 ± 0.0	1.0
Total monosaccharides	1.76 ± 0.0	2.8
Sucrose	33.40 ± 1.8	52.8
Maltose	0.22 ± 0.0	0.3
Raffinose	2.10 ± 0.1	3.3
Stachyose	20.90 ± 1.3	33.1
Verbascose	4.41 ± 0.2	7.0
Unknown	0.05 ± 0.0	0.1
Total di/oligosaccharides	61.10 ± 3.4	96.7
Total carbohydrates of extract	63.2 ± 3.4	

13.3.1 Carbohydrates

The carbohydrate composition of the oil bean is presented in Table 13.2. The carbohydrate fraction of unfermented bean contains mostly sucrose (52%), followed by stachyose (33.1%), verbascose (6.9%), and raffinose (3.3%). The content of these sugars decreases after fermentation which causes a corresponding increase in total monosaccharides, from 2.7% to 88.5% of total carbohydrates. Fructose (24.5%) and galactose (15.5%) are the predominant sugars in fermented oil bean. The increase in fructose and galactose is probably due to the breakdown of raffinose, stachyose and verbascose by the appropriate fermenting organisms. Raffinose, stachyose, and verbascose are tri-, tetra- and pentasaccharides, respectively. The source of fructose could be sucrose. There is no doubt that the fermenting organism also derives metabolic energy from the breakdown of oligosaccharides. Glucose concentration is lower than fructose and galactose probably because the fermenting organisms preferentially utilize glucose more than other sugars. It is known that glucose enters the fermentation pathway directly (Vedamuthu, 1982), whereas galactose goes through a series of reactions first. Fermentation considerably reduces the concentration of raffinose and stachyose and eliminates verbascose. These oligosaccharides are flatulence-forming in many persons who consume legumes; this is because alpha-galactosidase, the enzyme necessary to cleave the galactose unit, is not present in the human digestive system. Fermenting microorganisms possess this enzyme (Fleming, 1980). These oligosaccharides are known to cause raised levels of hydrogen excretion in humans (Calloway *et al.*, 1971) and rats (Wagner *et al.*, 1976, Fleming, 1980). The human intestine has microflora which can metabolize them to carbon dioxide, hydrogen and methane.

Other legumes contain some of these flatus-forming oligosaccharides. Fleming (1981) showed that the navy bean, red kidney bean, garbanzo bean and mung bean contain 0.37–0.67% raffinose, 1.67–4.0% stachyose, and 0.5–1.7% verbascose. Similarly, Odunfa (1983) demonstrates the presence of sucrose, raffinose, and stachyose at levels of 31.0, 12.0 and 28.0 mg/g dry weight, respectively, in unfermented African locust bean. He found a reduction in these sugars after fermentation. Similarly, Chompreeda and Fields (1984) noted that soybean contains raffinose and stachyose at 1.48 and 5.03 g/100 g of dry matter, respectively, and that fermentation considerably reduces or eliminates these oligosaccharides. Vishalakshi *et al.* (1980) have also found sucrose, raffinose, stachyose and verbascose in three varieties of *Phaseolus vulgaris* and raw kidney bean.

13.3.2 Fatty acid composition

The fatty acid composition of fermented and unfermented oil bean is presented in Table 13.3. Unsaturated fatty acids comprise 77–78% of total fatty acids whereas saturated fatty acids account for the remaining 22–23%. In unfermented beans, the major unsaturated fatty acids are linolenic acid (40.9–42.8%) and oleic acid (30.3–31.2%). The saturated fatty acids of oil bean contain fairly high quantities of lignoceric and behenic acids. Achinewhu (1982) noted that these fatty acids contribute 18.1 and 17.2%, respectively, of total fatty acids, an observation which cast doubt on the industrial potentials of lipid extracts of the African oil bean. This was borne out by earlier unsuccessful attempts to use oil from the African oil bean in soap manufacture. Insoluble soaps were produced, probably due to the presence of these fatty acids. The unsaturated fatty acids, oleic and linoleic acids, together formed about 73% of total fatty acids. The presence of large amounts of unsaturated fatty acids in this oil (77% of total fatty

Table 13.3 Fatty acid composition of the oil bean seed (% total)

Fatty acids	*Unfermented bean*	*Fermented bean*
Saturated		
8:0 Caprilic	–	0.02
10:0 Capric	–	0.05
12:0 Lauric	–	0.02
14:0 Myristic	–	0.03
16:0 Palmitic	5.03 + 0.03	4.52 + 0.03
17:0 Margaric	–	–
18:0 Stearic	4.35 + 0.2	3.39 + 0.3
22:0 Behenic	3.78 + 0.3	3.93 + 0.4
24:0 Lignoceric	10.0 + 0.6	10.0 + 0.8
Total saturated	23.2 + 1.4	22.0 + 1.8
Unsaturated		
18:1 Oleic	30.30 + 1.3	31.20 + 1.8
18:2 Linoleic	42.80 + 1.1	40.90 + 1.5
18:3 Linolenic	2.75 + 0.1	4.51 + 0.5
20:1 Gadoleic	0.66 + 0.0	1.10 + 0.0
Total unsaturated	76.6 + 2.5	77.8 + 3.8
Total fatty acids	99.7 + 3.9	99.7 + 5.6
Percentage saturated	23.2	22.0
Percentage unsaturated	76.8	78.0
Ether extract	46.3 + 1.6	46.8 + 2.1

acids), enhance its potential use following refining, as a dietary oil. The oil could therefore be used as a linolenic-rich salad oil, or could be hydrogenated to an edible fat or margarine (Achinewhu, 1986).

Fermentation does not seem to have much of an effect on the fatty acid composition of oil bean lipids, except for a slight decrease in total unsaturated fatty acids. The fermenting organisms probably do not utilize as much of the lipids as carbon or energy sources as they do carbohydrates. Beuchat and Worthington (1974) found that microorganisms do not appear to utilize lipids during the fermentation of peanuts.

13.3.3 Protein and amino acids

Table 13.4 shows the protein and amino acid composition of the African oil bean seed. When compared with isolated soybean protein, the oil bean protein is fairly high in all the essential amino acids except methionine. The high content of essential amino acids makes the oil bean a potential source of protein. Following fermentation of oil bean, there is an increase in soluble nitrogen, indicating enhanced action of proteolytic enzymes. Predigestion by enzymes may improve the protein quality of oil bean products.

Table 13.4 Amino acid composition of African oil bean seed

Amino acid	*mg/g*
Aspartic acid	618.9
Threonine	236.5
Serine	323.9
Glutamic acid	660.0
Proline	252.2
Glycine	343.1
Alanine	377.7
Valine	419.5
Isoleucine	281.6
Leucine	432.0
Phenylalanine	425.2
Tyrosine	305.6
Histidine	122.1
Lysine	395.9
Arginine	280.2
Methionine	44.3
Cystine	86.7
Tryptophan	59.0
Protein (N × 6.25)	22.5
Total nitrogen (mg)	925.6

Fermentation, however, does not have much of an effect on total essential amino acid composition (Achinewhu, 1983a). Other workers (Zamora and Fields, 1979; Au and Fields, 1981; Chompreeda and Fields, 1984) have shown an improvement in microbiologically available lysine, methionine and tryptophan following fermentation of soybean and sorghum. The oil bean, with its high protein content and good essential amino acid profile, has a potential for increased utilization as an important vegetable protein source in diets of southern Nigerians and others who already consume it as food.

13.3.4 Energy and protein quality evaluation

The energy and protein quality of African oil bean seed have been determined by Isichei and Achinewhu (1988) and other researchers. Achinewhu (1987) determined the protein quality of fermented oil bean seed based on *in vitro* digestibility experiments with four enzymes, amino acid scoring procedures, and a comparison with amino acid requirements of human infants. Results showed that the digestibility was 80%. Some essential amino acids especially methionine and tryptophan were most limiting when compared with the FAO/WHO scoring patterns for infants (FAO/WHO, 1973; FAO/WHO/UNU, 1985). Despite the limitations, the oil bean could meet the protein, amino acid and energy needs of babies up

Table 13.5 Energy values of fermented and unfermented oil bean (dry matter basis)

Oil bean	*Gross energy (kcal/g)*	*P^e(%)*	*NDpCal (%)*
Unfermented	6.67	29.3	10.5
Fermented	6.48	29.4	15.0

P^e is the concentration of protein in terms of energy. NDpCal represents the utilizable protein calories.

to one year of age, when fed at an appropriate level of 200 g/day.

Table 13.5 shows the gross energy values, the protein-energy ratio(P^e%) and the net dietary protein caloric percentage (NDpCal%) of oil bean-based diets (Isichei and Achinewhu, 1988). These values were calculated using the procedure of Miller and Payne (1961). There is little difference between the values of P^e%, although these values are very much higher than recommended 'safe' values of 11–12 for protein utilization of about 60% (Beaton and Swiss, 1974). There is, however, an appreciable difference in NDpCal% values between fermented and unfermented oil bean seed. Fermented oil bean has a much higher value than unfermented oil bean, although these NDpCal% values are much higher than a recommended value of 8 which is the NDpCal% of human breast milk. This is considered optimum for infant growth. Okeiyi and Futnell (1983) had earlier obtained NDpCal% values of 10 and above in their study of formulated cereal and legume mixtures for child rearing. The gross energy values determined by Isichei and Achinewhu are similar to the value of 6.26 kcal/g obtained for pumpkin seeds (Kamel *et al.*, 1982). A gross energy value of 5.0 kcal/g reported for cooked, defatted fluted pumpkin (Longe *et al.*, 1983) also compared favourably with gross energy of African oil bean. The high energy content of oil bean is noteworthy as protein utilization and energy intake are closely interrelated. If energy intake is seriously inadequate, it is unlikely that protein would be effectively utilized by the body.

13.4 NUTRITIONAL EXPERIMENTS WITH ANIMALS

In rats fed on diets containing fermented or unfermented oil bean, there was a consistent, significant depression in weights at the end of the study. Feed intake was similarly depressed throughout the duration of the study, although the depression was very severe in the first two weeks. Isichei and Achinewhu (1988) suggested that the low feed intake of rats fed on diets containing oil bean seed was due to extreme unpalatability of the oil bean or the presence of an as-yet-unidentified toxic compound. These workers in the first 14 days observed a 20% mortality in rats fed on the unfermented oil bean, and a 10% mortality in those fed on the fermented oil bean. It was likely that the duration of fermentation in these experiments was not enough to fully inactivate or destroy the toxic components in oil bean.

The protein efficiency ratio (PER), digestibility and feed efficiency ratio of diets containing fermented and unfermented oil bean highlight differences between oil bean-based diets and casein diets. Consumption of unfermented oil bean was associated with greater weight loss than consumption of fermented oil bean. The inadequate feed consumption and therefore low protein intake, combined with continuous weight loss resulted in negative PER values for unfermented and fermented oil bean. Apparent digestibility coefficients of fermented and unfermented oil bean are low. These studies demonstrate that although oil bean is rich in protein, it suffers severe nutritional drawbacks. Oil bean could not promote or maintain growth in rats. The poor performance of rats fed on oil bean diets may not be due to inadequacies in amino acid profile, or overall poor digestibility of the bean. It is most probably due to the presence of toxic components in the oil bean. The presence of these toxic components in the oil bean may affect its nutritive value by impairing protein utilization.

13.5 PRESENCE AND CHEMICAL NATURE OF ANTINUTRITIONAL SUBSTANCES

Some toxic substances have been identified in the oil bean, toxic constituents which are not destroyed by boiling or fermentation. Achinewhu (1983) showed the presence of saponins, whereas Duke (1981) reported the existence of a poisonous alkaloid, pancine in the African oil bean. Other workers (Fetuga *et al.*, 1974; Osaniyi and Eka, 1978; Achinewhu, 1983b, 1984) have observed a depression in weight, together with negative protein utilization in rats fed various tropical legume seeds. They attribute these adverse effects to the presence of heat-stable toxicants in these seeds.

Processed oil bean products are widely consumed as snacks or used as food condiments in many areas of Nigeria, although any adverse effects the oil bean may have on consumers have not been extensively studied or fully understood. Further investigations are needed on the appropriate processing methods necessary to inactivate these toxic components and to improve the nutritive quality of the oil bean.

REFERENCES

Achinewhu, S.C. (1982) Composition and food potential of the African oil bean seed (*Pentaclethra macrophylla*) and velvet bean (*Mucuna uriens*) *J. Food Sci.*, **47**, 1736–7.

Achinewhu, S.C. (1983a) Chemical and nutrient composition of fermented products from plant foods. *Nigeria Food J.*, **1**, 115–17.

Achinewhu, S.C. (1983b) Protein quality of the African oil bean seed (*Pentaclethra macrophylla*). *J. Food Sci.*, **48**, 1374–5.

Achinewhu, S.C. (1984) Amino acid composition and nutritive value of proteins in horse bean (*Mucuna uriens*). *Qual. Plant Plant. Foods Hum. Nutr.*, **34**, 181–4.

Achinewhu, S.C. (1986) The effect of fermentation on carbohydrate and fatty acid composition of the African oil bean (*Pentaclethra macrophylla*). *Food Chem.*, **19**, 105–16.

Achinewhu, S.C. and Riley, J. (1986) Effect of fermentation on thiamin, riboflavin and niacin content of melon seed (*Citrullus vulgaris*) and the African oil bean seed (*Pentaclethra macrophylla*). *Food Chem.*, **20**, 243–252. *J. Food Sci.*, **47**, 1736–7.

Achinewhu, S.C. (1987) Protein quality evaluation of weaning food mixtures from indigenous fermented foods. *Nigeria J. Nutr. Sci.* **8**, 23–31.

Au, P.M. and Fields, M.L. (1981) Nutritive quality of fermented sorghum. *J. Food Sci.*, **46**, 652–4.

Beaton, G.H. and Swiss, L.D. (1974) Evaluation of nutritional quality of food supplies: prediction of desirable or safe protein: caloric ratios. *Am. J. Clin. Nutr.*, **27**, 485–504.

Beuchat, L.R. and Worthington, R.E. (1974) Changes in the lipid content of fermented peanuts. *J. Agric. Food Chem.*, **22**, 509–12.

Calloway, D.H., Hickey, C.A. and Murphey, E.L. (1971) Reduction of intestinal fermented peanuts. *J. Agric. Food Chem.*, **22**, 509–12.

Chompreeda, P.T. and Fields, M.L. (1984) Effects of heat and natural fermentation on amino acids, flatus producing compounds, lipid oxidation and trypsin inhibitor in blends of sorghum and corn meal. *J. Food Sci.*, **49**, 563–5.

Duke, J.A. (1981) *Handbook of Legumes of World Economic Importance*. Plenum Press, New York.

FAO/WHO (1973) Energy and Protein Requirements. *FAO Nutrition Meeting Report Series No. 52*. Rome, WHO Technical Report Series No. 522 Geneva.

FAO/WHO/UNU (1985) *Energy and Protein Requirements*. WHO, Geneva.

Fetuga, B.L., Babatunde, G.M. and Oyenuga, V.A. (1974) Protein quality of some Nigerian feedstuffs: biological evaluation of protein quality. *J. Sci. Food Agric.*, **24**, 1515–23.

Fleming, S.E. (1980) Measurement of hydrogen production in the rat as an indicator of flatulence activity. *J. Food Sci.*, **45**, 1012–18.

Fleming, S.E. (1981) A study of the relationship between flatus potential and carbohydrate distribution in legume seeds. *J. Food Sci.* **46**, 794–8.

Isichei, M.O. and Achinewhu, S.C. (1988) The nutritive value of African oil bean seeds (*Pentaclethra macrophylla*). *Food Chem.*, **30**, 83–92.

Kamel, B.S., Deman, J.M. and Blackman, B. (1982) Nutritional, fatty acid and oil characteristics of different agricultural seeds. *J. Food Technol.*, **17**, 263–9.

Longe, O.G., Farimu, G.O. and Fetuga, B.L. (1983) Nutritional value of the fluted pumpkin (*Telfaria occidentalis*). *J. Agric. Food Chem.*, **31**, 89–92.

Miller, D.A. and Payne, P.R. (1961) Problems in prediction of protein values of diets: the influence of protein concentration. *Br. J. Nutr.*, **15**, 11–19.

Odunfa, S.A. (1983) Carbohydrate changes in fermenting locust bean (*Parkia filicoidea*) during iru preparation. *Plant Food Hum. Nutr.*, **32**, 3–10.

Okeiyi, C. and Futnell, M.F. (1983) Evaluation of protein quality of formulations of sorghum flour and legume seeds. *Nutr. Rep. Int.*, **28**, 451–61.

Osaniyi, C.B. and Eka, O.U. (1978) Studies on the chemical composition and nutritive value of horse eye bean (*Mucuna uriens*). *W.A. J. Biol. Appl. Chem.*, **21**, 53–61.

Vedamuthu, E.R. (1982) Microbial modification of milk constituents–lactose metabolism, in *Economic Microbiology–Fermented Foods*, Vol. 7 (ed A.H. Rose), Academic Press, London, pp. 2201–3.

Vishalakshi, S.D.K., Sathe, S.K. and Rockland, L.B. (1980) Quick cooking beans (*Phaseolus vulgaris L*) II. Phytates, oligosaccharides and antienzymes. *Qual. Plant. Plant Foods Hum. Nutr.*, **30**, 45–52.

Wagner, J.R., Becker, R., Gumbman, M.R. and Olson, A.C. (1976) Hydrogen production in the rat following ingestion of raffinose, stachyose, and oligosaccharide-free bean residue. *J. Nutr.*, **106**, 466–70.

Zamora, A.F. and Fields, M.L. (1979) Nutritive quality of fermented cowpea. (*Vigna sinensis*) and chickpea (*Cicer arietinum*). *J. Food Sci.*, **44**, 234–6.

CHAPTER FOURTEEN

Escumite bean (*Phaseolus acutifolius* A. Gray)

A. Sotelo

14.1 GENERAL BOTANICAL, ECOLOGICAL AND PRODUCTION INFORMATION

Escumite bean (*Phaseolus acutifolius*) is the least cultivated of Mexican dry beans and grows in those places where environmental or pathological stresses do not permit the cultivation of the common bean. The main advantage is that it grows in arid zones and is resistant to microbiological invasion and to many beetles which attack other beans. This bean is called tepari in Sonora and Sinaloa, tecuzi in Chihuahua and escumite or xmayum in southern Mexico. It is mainly found in the southwestern area of the United States, the north and southeast regions of Mexico and Guatemala and other countries in Central America. The attention of agricultural scientists has been focused on this bean as it can be easily adapted to grow in arid zones and has a potential for use in the transfer of desirable traits from tepari to the common bean (Pratt and Nabhan, 1988).

14.2 CURRENT AND POTENTIAL USAGE AS HUMAN FOOD

The chemical composition of escumite bean is very similar to other *Phaseolus* species (*P. vulgaris, P. lunatus* and *P. coccineus*). Sulphur amino acids are limiting and a very low tryptophan content has been found (Sotelo *et al.*, 1985). In small towns of Chiapas state in Mexico, the

Legumes and Oilseeds in Nutrition
Edited by E. Nwokolo and J. Smartt
Published in 1996 by Chapman & Hall. ISBN 0 412 45930 2

escumite bean is relished, people who eat it finding the flavour excellent. Its use does not, however, seem to be widespread. Escumite bean is toxic in the raw state and needs to be thoroughly cooked before it can be used as human food. Cooking inactivates its antinutritional factors which all seem to be heat-labile. The toxicity of escumite bean in the raw state precludes or limits its usage as animal feed, unless the bean is steamed or heat-treated prior to feeding to animals.

14.3 PRESENCE AND DETOXIFICATION OF ANTINUTRITIONAL SUBSTANCES

The most important antinutritional factor in *P. acutifolius* is the high concentration of lectins which agglutinate cow red blood cells. According to Jaffé and Brücher (1972) the higher the concentration of these lectins, the greater the toxicity of the product to rats. Ingestion of raw escumite bean induces ultrastructural changes in epithelial intestine cells of rat. An *in vitro* test also showed that cell numbers and intestinal epithelial cells are significantly affected (Sotelo and Hernandez, 1980). Studies by Pusztai *et al.* (1979) on the damage induced by raw legumes at the structural level of the intestinal epithelium have demonstrated similar results.

14.4 HUMAN OR ANIMAL STUDIES WITH LEGUMES

Sotelo *et al.* (1983) found that when rats were fed on diets containing raw escumite bean, the animals died two days later. However, since heat treatment destroys the antinutritional factors, no noxious effect is observed when escumite bean is cooked and eaten by humans. Further research is needed to investigate not only the mechanism of action of the lectins on the intestinal epithelium but also their nutritional significance in the diets containing this bean.

14.5 OTHER LOCAL MEXICAN BEANS

Before the arrival of the Spaniards, the area now known as Mexico was inhabited by several ethnic groups with independent political and social organizations with customs and habits different from each other. However, an important commercial nexus existed from the centre of the country to Honduras. Since corn and bean were the primary food staples of all these people, it became imperative that the early inhabitants improved the yield of local food crops. It was also very necessary to obtain new varieties with desirable characteristics, adapted to the different regions, with different climates and soils. The state of Chiapas in Mexico has considerable variations in climate and in soil types, from tropical jungle in the west to cold highlands in the centre of the state. It thus represents a wide climatic spectrum and is suitable for a study of the diversity of grain legumes used as food. Besides, it is a region with numerous ethnic groups with different traditional feeding habits. Based on this, several samples of dry beans belonging to the genus *Phaseolus* were obtained in the markets of different towns in Chiapas state. Samples from markets in Oaxaca and the Yucatan peninsula were included for investigation into the chemical composition of beans in the species *P. lunatus*, *P. coccineus* and *P. vulgaris*. Proximate composition of nearly 32 varieties of these Mexican dry beans have been studied (Sotelo and Hernandez, 1980; de la Vega and Sotelo, 1986). These researchers investigated 19 varieties of *P. vulgaris*, 11 varieties of *P. lunatus* and two varieties of *P. coccineus*. Among the 19 varieties of *P. vulgaris* obtained from various areas of Mexico, protein content varied from a low of 18.6% to a high of 25.7% and fat content varied from a low of 0.7% to a high of 2.6%. Fibre content was very high in the de Enredo variety (13.8%) and in the range 5.2–10.3% in the others. Most of the varieties of *P. vulgaris* had similar ash (3.5–4.5%) and carbohydrate (59.8–71.8%) contents. Among the 11 varieties of *P. lunatus* examined, protein ranged from 19.8 to 31.8%, fat from 1.4 to 3.0%, and, with the exception of the guinze variety which was very high in fibre (12%), fibre ranged from 5.9 to 8.5%. There was no significant variation among the 11 varieties in content of ash (3.4–5.5%) and carbohydrates (50–67%).

In general, protein content in all the beans studied is not significantly different, ranging from 19.0 to 31.8% which is the normal variability range observed in beans. Differences are even observed in essential amino acid profiles from samples of wild Mexican legumes. The greatest variability among the various wild legumes is in the sulphur amino acid content, the range being from 47.4 to 191.1 mg/gN. There is high variability in the trypsin inhibitor content. Haemagglutinins are present in all the bean samples. The content of cyanogenic glucosides is much higher in wild beans than in cultivated ones, the breeding process having contributed to the innocuousness of the cultivated as compared to the wild type bean (de la Vega and Sotelo, 1986).

It is difficult to estimate even roughly the production of the varieties of legumes cultivated in the different areas of Mexico. This is because most of them are used for local consumption and in many cases, just for family consumption. Additionally, there is a lack of coordinated information on legume cultivation in many areas of Mexico, especially in many isolated areas in the arid zones of the Tarahumaras area and the jungles of the Lacandon area, where many other wild legumes are used in the daily diet.

REFERENCES

de la Vega, A. and Sotelo, A. (1986) The nutritional quality and toxin content of wild and cultivated lima beans (*Phaseolus lunatus*). *Qual. Plant. Plant Foods Hum. Nutr.*, **36**, 75–83.

Jaffé, W.G. and Brucher, O. (1972) Toxicidad y especificidad de diferentes fitohemaglutininas de frijoles. *Arch Latinoam. Nutr.*, **22**, 267–81.

Pratt, C.R. and Nabhan, P.G. (1988) Evolution and diversity of *Phaseolus acutifolius*. in *Genetic Resources of* Phaseolus *Bean*, (ed. P. Gepts), Kluwer Academic, New York.

Pusztai, A., Clarke, E.M.W. and King, T.P. (1979) The nutritional toxicity of *Phaseolus vulgaris* lectins. *Proc. Nutr. Soc.*, **38**, 115–21.

Sotelo, A. and Hernandez, M. (1980) Nutritional evaluation of three varieties of beans (*Phaseolus*) using chemical and biological methods. *Nutr. Rep. Int.*, **22**, 607–16.

Sotelo, A., Gonzalez, L.A., Gonzalez, M.T. *et al.* (1983) Ultrastructural changes of epithelial intestinal cells induced by ingestion of raw *Phaseolus acutifolius*. *Nutr. Rep. Int.*, **27**, 329–37.

Sotelo, A., Uvalle, A., Lucas, B. and de la Vega, A. (1985) Composicion quimica y aspectos toxicologicos de siete varietades de frijoles cultivados en el Estado de Chiapas. *Tecol. Aliment.* (*Mex.*), **20**, 18–25.

CHAPTER FIFTEEN

Lima bean (*Phaseolus lunatus* L.)

E. Nwokolo

15.1 INTRODUCTION

Lima bean (*Phaseolus lunatus* L.) is still regarded worldwide as a minor crop in spite of its ubiquity, diversity and yield potential. It is cultivated as far north as Canada, as far south as Argentina, and is widespread in Asia, Africa, Europe and Australia. It is truly a cosmopolitan crop although different varieties are cultivated in these diverse ecological zones. Despite its ubiquity, lima bean production in many countries is so low that its production statistics are combined with other dry beans. In almost all of these countries, very little lima bean enters the international trade in beans, production being primarily for local consumption.

Baudoin (1988) discusses the origin, evolution and domestication of the lima bean. Published literature indicates that Mackie (1943) first hypothesized Central America as a general and Guatemala as a specific area of origin of the lima bean, although archaeological excavations and their arguments have not supported this hypothesis. Rather, it has been proposed that although the origins are obscure, two groups of lima beans were independently domesticated from separate origins (Allard, 1960; Kaplan, 1965; Smith, 1968). One group represents the large flat-seeded varieties (big lima) which are believed to have originated in Peru, whereas the other group represents the small, flat-seeded varieties (sieva) believed to have originated in the more northern parts of Central America. Indeed, the issue of origin(s) of the lima bean has been very murky, Kaplan (1965) proposing the warm humid Andean region and Erickson (1982) the whole humid Amazon basin as possible sources of origin of the big lima types.

Legumes and Oilseeds in Nutrition
Edited by E. Nwokolo and J. Smartt
Published in 1996 by Chapman & Hall. ISBN 0 412 45930 2

Smith (1968) considers Central America and Mexico as probable sites of origin of the small-seeded sieva types. It is known, however, that Peru and Central America are two primary centres of domestication of the lima bean, and remain centres of genetic diversity for both wild and cultivated types of lima bean. It is also believed that South East Asia and Africa are secondary centres of genetic diversity also for cultivated types of lima beans.

15.2 DISTRIBUTION

Lima beans are distributed widely from sea level to mountainous regions of the Americas, Europe, Asia and Africa. It is believed that early European explorers took the lima bean from the Americas and spread it worldwide. Indeed, Mackie (1943) singles out Spanish explorers as having taken the lima bean to Europe and Asia and were instrumental in taking it to Africa through slave ships. Following introduction to these diverse ecological zones, the lima bean underwent considerable adaptation and hybridization to produce the various local strains that exist in these vastly different areas. For production purposes, only three types of lima beans are recognized. These are the **big lima type** with large-sized flat seeds, the **sieva type** with medium-sized flat seeds and the **potato type** with small-sized globular seeds. Plant breeders in the USA have developed various commercial cultivars which are extremely high yielding under intensive agronomic management, and are designed to satisfy specific consumer requirements.

15.3 BOTANY OF THE CROP

The lima been is a herbaceous species which includes both annual determinate bush types and perennial indeterminate climbing types, perennial due to the presence of large swollen tap roots. The bush types are short, about 0.6 m tall, whereas the climbing types are up to about 2–5 m. Each plant on maturity produces a large number of pods. Each pod is oblong, often curved and terminates in a sharp beak facing the dorsal suture. Pods contain two to four seeds and dehisce on drying. Seeds may be small, medium or large-sized depending on the lima bean type. Seed weight varies from about 0.5 g/seed in the small-seeded types to as much as 3–4 g per seed in the large-seeded types. Seed shape may range from kidney, through rhomboid to round. Seed colour is variable and may be speckled, mottled, or full-coloured. Colours vary and usually include white, grey, yellow, brown, purple or black. Wild varieties usually have small pods, are small-seeded, seed colours being predominantly grey, black or speckled.

15.4 PRODUCTION

The world's largest producer of lima bean is the USA. The second largest producer is the Malagasy Republic, in the Indian ocean off the coast of Africa, and Peru is the third largest producer in the world. In the United States, lima beans are cultivated as sole crops primarily for commercial processing, the small-seeded types being planted and harvested more in northern than in the southern states where the large-seeded varieties are more common. High yields per hectare are achieved with irrigation, fertilization and pest control. In more traditional agricultural systems as seen in Latin America, Africa and Asia, lima beans are not monocrops but are rather intercropped with a whole range of crops including corn and millet. In Nigeria, where climbing types of lima bean are very common, the bean is intercropped with yam, corn, cocoyam and other vegetables in homestead farms. Under traditional farming systems, dry seed yield is low, about 200–600 kg/ha whereas under intensive agronomic management in the United States, dry seed yield may be as high as 1080–1350 kg/ha. Dry seed productivity of improved seeds under minimal chemical input in experimental plots in Columbia has been estimated at an average of 2.6 MT/ha, with a range of 1.7–4.8 MT/ha (Lyman *et al.*, 1985). As the average yield of 2.6 MT/ha may not be achieved in farmers' plots, this high level of productivity using improved genotypes is indicative of the tremendous potential that exists with this crop. In the US, yield of green beans is about 2.7–3 MT/ha. Commercial cultivars developed in the US are both of the bush and the vining types. Bush cultivars developed to produce small seeds include Kingston, Dixie Butter, Bridgetown and Jackson Wonder. Bush cultivars which produce large seeds include the well-known Fordhook 242 and Fordhook 1042. Vining cultivars which produce small seeds include King of the Garden, Florida Speckled Butter, Wilbur and Western, and vining cultivars which produce large seeds include the well-known Ventura. Duke (1981) suggests that King of the Garden and Florida Speckled Butter are vining types which would be suitable for cultivation in the tropics.

15.5 CURRENT AND POTENTIAL USAGE AS FOOD

In many developing countries, grain legumes are currently contributing an even more significant fraction of dietary protein intake than they did a decade ago. In the mid 1970s, even before the present serious decline in the economies of most developing countries, field workers had estimated that grain legumes accounted for about 12–20% of the protein intake of West Africans. In all likelihood, as meat production declines in these countries, more and more people will turn to grain legumes to meet dietary protein needs. Crops like lima bean which have considerable potential for increased productivity, therefore assume an even greater importance in meeting dietary needs of a hungry world. Lima bean is easily suited to meet this demand. In many areas of the world, lima bean is consumed fresh and green as a vegetable or a pulse. In many African countries, the lima bean is a culinary delight, which is prepared in a great variety of ways, and may be eaten alone or in combination with various cereals or starchy roots and tubers. In Nigeria, there are many cultivars, Ologhobo and Fetuga (1983a, b) have reported the existence of over 100 accessions and 800 different breeding lines of cultivated lima beans. Lima bean usage is so prevalent in Nigeria that virtually every village or clan has a distinct vernacular name for it. There are indications that lima beans, cowpeas and other beans are not only being increasingly consumed by adults and children, but are also being used increasingly in the preparation of weaning diets in Nigeria and other African countries. In the United States, green lima beans are canned or frozen and used as vegetables. There is also considerable consumption of the mature seed, especially the large-seeded types, precooked and canned, or precooked and pickled for use in vegetable salads. Lyman *et al.* (1985) found that leaves, fruits and seeds are eaten as vegetables in India and Asia, and that dry seeds are milled to produce a protein-rich flour which is added to bread or noodles in the Philippines, as well as used for a bean paste in Japan. The seeds are also used in the preparation of bean soups and stews in Central and South America, and in southern states of the US.

Fresh pods may be harvested repeatedly from the same plant, the pods split open and the beans eaten as green beans. In southern Nigeria, lima beans are left to mature, the dry beans being preferred to the green seeds. In some regions of Latin America, lima beans are cultivated primarily as a source of dry beans and/or immature seeds, whereas in other regions they are cultivated as a source of leaves and immature pods, eaten much as french beans (Lyman *et al.*, 1985). In the development of the US cultivars, the content of antinutritional factors was of prime concern to plant breeders. Consequently, these cultivars contain the lowest amounts of antinutritional factors known in lima bean and are totally safe. This is not so with wild types in some tropical countries where HCN content may be as high as 318–400 mg HCN/100 g flour (Hernandez-Xolocotzi *et al.*, 1984). In a study of 18 varieties of lima bean obtained from the Nigerian National Cereals Research Institute, Ibadan, and the International Institute of Tropical Agriculture, Ibadan, Nigeria, the total HCN values in raw seeds were found to range from 26.5 to 55.3 mg/100 g of bean (Ologhobo *et al.*, 1984). In such cases, discarding the cooking water once or twice removes the toxicity as well as any bitterness in the cooked bean. In India, lima bean flour which has been processed to eliminate antinutritional factors, is used as a partial replacement for wheat flour in

Table 15.1 Proximate composition of raw and cooked, large and baby lima beans

	Content (per 100g edible portion)			
	Raw, large	*Raw, baby*	*Cooked, large*	*Cooked, baby*
Water (g)	10.17	12.07	69.79	67.15
Protein (g)	21.46	20.62	7.80	8.04
Fat (g)	0.69	0.93	0.38	0.38
Carbohydrate (g)	63.38	62.83	20.89	23.31
Crude fibre (g)	6.34	5.71	3.09	3.59
Ash (g)	4.30	3.55	1.15	1.12
Food energy (kcal)	338	335	115	126

From USDA (1986).

pastry making and noodle production. In Nigeria, the common dry lima bean types are the unimproved, small to medium-sized varieties which have a high HCN content. As a result, lima bean is usually cooked overnight over a slow fire, with many changes of cooking water, for consumption the following day. Lima beans are added to soups and stews in South America, and they may be boiled, fried or cooked with vegetables in Africa.

15.6 NUTRIENT CONTENT AND NUTRITIVE VALUE

15.6.1 Proximate composition

Because the lima bean is a popular bean in the US and elsewhere, its chemical composition has been determined as part of an investigation of commonly consumed foods (USDA, 1986). Nutrient composition in some 18 lima bean cultivars grown in Africa has also been reported by Ologhobo and Fetuga (1983b). Protein content is about 20–25.9%, with a mean of about 22%, although there may be slight variations in value reflecting differences in genotype and ecology. Lyman *et al.* (1985) note that the major proteins are the alpha- and beta-globulins, with a smaller quantity of albumin. Lima bean has a low fat content (0.6–3.5%), most samples have a fat content of about 1.5%. Ash content is moderate (3.4–6.4%) with a mean value of about 5%. Fibre content is also moderate, in the range 4.3–7.5%, with a mean value of about 5.6%. Composite values of the major nutrients in lima bean is shown in Table 15.1.

15.6.2 Carbohydrates

From an analysis of the carbohydrate constituents of 18 varieties of lima bean, Ologhobo and Fetuga (1982b) concluded that starch comprised

Table 15.2 Mineral and vitamin content of lima bean

	Content (per 100g of edible portion)			
	Raw, large	*Raw, baby*	*Cooked, large*	*Cooked, baby*
Minerals				
Calcium (mg)	81.0	81.0	17.0	29.0
Iron (mg)	7.5	6.2	2.4	2.4
Magnesium (mg)	224.0	188.0	43.0	53.0
Phosphorus (mg)	385.0	370.0	111.0	127.0
Potassium (mg)	1724.0	1403.0	508.0	401.0
Sodium (mg)	18.0	13.0	2.0	3.0
Zinc (mg)	2.8	2.6	1.0	1.0
Copper (mg)	0.7	0.7	0.2	0.2
Manganese (mg)	1.7	1.7	0.5	0.6
Vitamins				
Ascorbic acid (mg)	0.0	0.0	0.0	0.0
Thiamin (mg)	0.51	0.60	0.16	0.16
Riboflavin (mg)	0.20	0.22	0.06	0.06
Niacin (mg)	1.54	1.71	0.42	0.66
Pantothenic acid (mg)	1.36	1.27	0.42	0.47
Vitamin B_6 (mg)	0.51	0.33	0.16	0.79
Folacin (mg)	395	400	83	150
Vitamin B_{12} (mg)	0.0	0.0	0.0	0.0
Vitamin A (IU)	0.0	0.0	0.0	0.0

From USDA (1986).

about 44% of the seed (range 32.1–50.6%) and oligosaccharides comprised about 9.5% (range 7.5–12.8%). Glucose, fructose and sucrose were only minor components of the seed and comprised 0.07, 0.62 and 1.54%, respectively. Cellulose content was only 5.8% of the seed (range 5.0–6.0%) and lignin content was only 2.0% of the seed (range 1.0–2.4%). Meredith *et al.* (1988) analysed lima bean seeds for the monosaccharide fructose, the disaccharide sucrose and the oligosaccharides raffinose, stachyose and verbascose at three stages of growth. They showed that as maturity increased, the concentration of fructose and sucrose decreased whereas the concentration of the oligosaccharides increased. No significant differences were found in the starch concentration. These workers suggested that sufficient variability existed within the various genotypes that breeding for low oligosaccharide content was possible.

15.6.3 Minerals

A 1986 USDA publication presents the content of minerals in common cultivars of lima bean in the USA, whereas data from Ologhobo and Fetuga (1983a, b) are based on 18 cultivars obtained from West Africa.

Table 15.3 Fatty acid profile of lima bean fat

Fatty acids	Content (per 100g edible portion)			
	Raw, large	*Raw, baby*	*Cooked, large*	*Cooked, baby*
Saturated				
14:0 (g)	0.002	0.003	0.001	0.001
16:0 (g)	0.118	0.161	0.065	0.065
18:0 (g)	0.032	0.043	0.018	0.017
Total (g)	0.161	0.219	0.089	0.088
Monounsaturated				
16:1 (g)	0.004	0.006	0.002	0.002
18:1 (g)	0.052	0.071	0.029	0.028
22:1 (g)	0.004	0.006	0.002	0.002
Total (g)	0.062	0.084	0.034	0.034
Polyunsaturated				
18:2 (g)	0.215	0.292	0.118	0.169
18:3 (g)	0.095	0.129	0.052	0.052
Total (g)	0.309	0.420	0.171	0.169

From USDA (1986).

Table 15.4 Lipid content and fatty acid profile of lima bean, soybean and cowpeas

Fatty acids	Fatty acid profile (% of total fatty acids)		
	Lima bean	*Soybean*	*Cowpeas*
12:0	0.04	–	0.06
14:0	0.39	0.10	0.06
16:0	21.30	9.71	25.10
16:1	0.63	0.68	–
18:0	6.04	3.10	5.48
18:1	9.03	27.30	7.98
18:2	40.00	43.80	31.70
18:3	17.80	4.86	18.80
20:0	1.58	2.09	2.09
22:0	0.89	–	5.66
Sat. fatty acids (%)	30.20	15.10	38.20
Unsat. fatty acids (%).	60.50	53.70	52.10
Total lipid content	1.58	26.60	1.96

From Ologhobo and Fetuga (1983b).

According to the USDA data (Table 15.2), potassium content is very high (1724 mg/100 g), phosphorus and magnesium content are fairly high (385 and 224 mg/100 g, respectively) and calcium and sodium content are low

Table 15.5 Amino acid content of lima bean

Amino acids	*Content ($g\ 100g^{-1}$ of edible portion)*			
	Raw, large	*Raw, baby*	*Cooked, large*	*Cooked, baby*
Tryptophan	0.25	0.25	0.09	0.10
Threonine	0.93	0.89	0.34	0.35
Isoleucine	1.13	1.09	0.41	0.42
Leucine	1.85	1.78	0.67	0.69
Lysine	1.44	1.38	0.52	0.54
Methionine	0.27	0.26	0.10	0.10
Cystine	0.24	0.23	0.86	0.09
Phenylalanine	1.24	1.19	0.45	0.46
Tyrosine	0.76	0.73	0.28	0.28
Valine	1.29	1.24	0.47	0.48
Arginine	1.32	1.26	0.48	0.49
Histidine	0.66	0.63	0.24	0.25
Alanine	1.10	1.05	0.40	0.41
Aspartic acid	2.77	2.66	1.01	1.04
Glutamic acid	3.04	2.92	1.10	1.14
Glycine	0.91	0.87	0.33	0.34
Proline	0.98	0.94	0.35	0.37
Serine	1.43	1.37	0.52	0.54

From USDA (1986).

(about 81 and 18 mg/100 g, respectively). Iron, zinc, manganese and copper are present only in trace amounts (7.5, 2.8, 1.7 and 0.7 mg/100 g, respectively). Data on West African cultivars parallels the data on American lima beans. Potassium is very high (mean of 1874 mg/100 g), phosphorus and magnesium content is fairly high (385 and 176 mg/100 g, respectively) whereas iron, zinc, manganese and copper are present in trace amounts (6, 4.2, 2 and 1.3 mg/100 g, respectively).

15.6.4 Fats and fatty acids

Fat content of lima bean is very low and consequently the fatty acids will not contribute in any meaningful way to dietary energy. In spite of that, the fatty acid profile as published by the USDA (1986) is presented in Table 15.3. Saturated fatty acids comprise about 30%, whereas monounsaturated and polyunsaturated fatty acids comprise 12% and 58%, respectively, of total fatty acids. The predominant fatty acids are linoleic acid (0.22 g/100 g of bean) and palmitic acid (0.12 g/100 g of bean). Table 15.4 shows a comparison of the fatty profiles of lima bean, soybean and cowpea.

Table 15.6 Amino acids in lima bean compared with the FAO amino acid standard

	Amino acid content (g/16 gN)	
Amino acids	*Lima bean*[a]	*FAO reference*[b]
Lysine	6.78	4.20
Histidine	3.30	2.40
Arginine	6.59	2.00
Aspartic acid	12.20	–
Threonine	4.29	2.80
Serine	5.17	–
Glutamic acid	19.40	–
Proline	4.51	–
Glycine	5.18	–
Alanine	5.62	–
Cystine	1.07	2.00
Valine	5.77	4.20
Methionine	1.42	2.20
Isoleucine	5.60	4.20
Leucine	8.52	4.80
Tyrosine	4.56	–
Phenylalanine	7.51	2.80
Tryptophan	1.39	1.40

[a]From Ologhobo and Fetuga (1983b).
[b]From FAO/WHO (1973).

15.6.5 Amino acids

The amino acid profile of African lima bean has been studied by Evans and Boulter (1974) and by Ologhobo and Fetuga (1983b) and the profile of cultivars common in the US has been compiled by the USDA (1986). The amino acid profile is good, lima bean containing fairly high amounts of lysine although like other beans, it is deficient in the sulphur amino acids which are the most limiting amino acids in the bean (Evans and Boulter, 1974). During the preparation of this chapter, the author visited about ten villages in various segments of southern Nigeria where lima beans feature prominently in diets. In many of these villages, lima beans are cooked together with dry corn. Fresh vegetables are added if in season. It is apparent that, without being aware of it, the villagers are taking advantage of the complementarity which exists between lima beans which have a low content of sulphur amino acids but a high content of lysine, and corn which has a high content of sulphur amino acids but a low content of lysine. Comparative values of amino acids in various lima bean types are shown in Table 15.5, and Table 16.5 compares the lima bean amino acid profile with that of FAO reference protein (FAO/WHO, 1973).

Table 15.7 HCN content of raw and processed lima bean

	HCN content (mg 100g^{-1})		
Lima bean	*Free*	*Bound*	*Total*
Raw	38.3	3.15	41.5
Autoclaved	16.6	2.25	18.8
Cooked	9.0	1.68	10.7
Soaked (2 days)	27.3	2.67	30.0
Soaked (4 days)	16.2	1.69	17.9
Soaked (6 days)	10.5	1.01	11.5
Germinated (2 days)	29.7	2.54	31.6
Germinated (4 days)	17.3	1.49	18.8
Germinated (6 days)	7.9	0.60	8.51

From Ologhobo *et al.* (1984).

15.6.6 Vitamins

The following vitamins are found in raw lima bean (mg/100 g of bean): thiamine 0.51 mg, riboflavin 0.20, niacin 1.54 mg, pantothenic acid 1.36; pyridoxine 0.51 mg; and folacin 359 μg (USDA, 1986). Lima bean contains no measurable quantities of ascorbic acid, vitamin B_{12} or vitamin A. Studies of the availability of folic acid and pyridoxine in lima bean indicate that although folic acid in cooked lima bean is over 70% available (Clifford *et al.*, 1991), pyridoxine availability in blanched or heat-treated lima beans is reduced by almost 50% (Ekanayake and Nelson, 1990).

15.7 CONTENT OF ANTINUTRITIONAL FACTORS

Antinutritional factors present in lima bean include trypsin and chymotrypsin inhibitors (Birk, 1967; Krahn and Stevens 1971, 1972a, b; Ologhobo and Fetuga, 1983a). They also include haemagglutinins (Galbraith and Goldstein, 1970, 1972a, b) and cyanogenic glucosides (Hernandex-Xolocotzi *et al.*, 1984; Ologhobo *et al.*, 1984). Other anti-physiological agents present include polyphenols and phytic acid (Ologhobo and Fetuga, 1982a) and various oligosaccharides which lead to flatulence when lima beans are consumed (Ologhobo and Fetuga, 1982b; Meredith *et al.*, 1988). Most edible beans contain varying amounts of some or all of these antinutritionals. The winged bean, faba bean and mung bean, although edible, contain much more of these antinutritional factors than cowpeas or pigeon pea.

15.7.1 Protease inhibitors

These are heat-labile proteins which have an ability to inhibit the body's proteolytic enzymes, primarily trypsin and chymotrypsin, following consumption of lima beans. Liener (1976) reported that rats fed raw seeds containing high amounts of protease inhibitors, grew very slowly, suffered pancreatic hypertrophy and probably sustained considerable loss of endogenous sulphur amino acids. Various food-processing methods affect the trypsin inhibitor activity of lima bean differently. Although cooking almost completely destroys trypsin inhibitors, soaking and germination have relatively little effect on inhibitor activity (Ologhobo and Fetuga, 1983a). Raw lima bean has a trypsin inhibitor activity of 32.63 TIU/mg protein, whereas cooked lima bean has no measurable activity. Lima bean soaked in water for 2, 4 or 6 days has an inhibitor activity of 31.02, 26.87 and 21.34 TIU/mg protein, respectively. Those germinated for 2, 4 or 6 days have an inhibitor activity of 29.30, 24.02 and 18.18 TIU/mg protein (Ologhobo and Fetuga, 1983a).

15.7.2 Haemagglutinins

Although the presence of lectins capable of agglutinating type A erythrocytes was suspected, confirmation followed isolation and characterization of two types of anti-A haemagglutinins and other lectins in lima bean (Gould and Scheinberg, 1970; Galbraith and Goldstein, 1970, 1972a,b). The haemagglutinins have been further purified leading to identification of their subunit structures (Gould and Scheinberg, 1970). In its pure form, the lectin is a glycoprotein composed of 31 kDa subunits which are joined together by disulphide bridges to form dimers. These lectins are heat-labile and agglutinate erythocytes by virtue of their ability to bind to specific glycoprotein receptor sites on the surfaces of red blood cells. The lectins show a specificity for agglutination of erythrocyte types A and AB of human blood groups.

15.7.3 Cyanogenic glucosides

Lima beans contain various amounts of cyanogenic glucosides, depending on whether it is an improved or unimproved type or a wild type. HCN content of raw and processed lima bean is presented in Table 15.7. As mentioned earlier, unimproved varieties have HCN contents of 21.5–55.0 mg/100 g, whereas wild types have very much higher HCN contents which may range from 318 to 400 mg HCN/100 g. Although USDA permits a maximum HCN concentration of 20 mg/100 g in lima beans grown for human consumption in the United States, most of the common cultivars in the US have HCN contents much lower than this level.

Cyanogenic glucosides are present as linamarin and are hydrolysed by the action of the enzyme linamarinase or beta-glucosidase, with the release of HCN. Hydrocyanic acid is potentially toxic, even lethal, if sufficient amounts are consumed. The wild Mexican varieties reported by Hernandez-Xolocotzi *et al.* (1984) to contain 300–400 mg HCN/100 g of beans may on consumption lead to severe poisoning and death, since lethal doses of ingested cyanide in human adults are well under 200 mg. Various processing procedures are successful to varying degrees in reducing the HCN content of lima beans. Autoclaving at 105°C, at 15 psi for 30 min results in an average loss of 54% of HCN whereas cooking is even more effective (77.8% loss of HCN). Soaking the beans for 2 days is only marginally effective (29% loss), more effective after 4 days (55.6%) and quite effective after 6 days (73.5%). However, one must doubt the usefulness of a detoxification procedure which involves soaking seeds for 6 days. Germination follows the same trend as soaking and is effective only after 6 days. It is apparent that soaking and germination are not practical procedures to detoxify lima beans.

15.7.4 Oligosaccharides

Intestinal gases formed by microbial fermentation of oligosaccharides in the lower intestine are a major problem when lima beans are consumed. The oligosaccharides responsible for flatulence in lima beans and most other beans are raffinose, stachyose and verbascose. The concentration of these three oligosaccharides in US cultivars of lima beans has been determined by Meredith *et al.* (1988). Cultivars studied included Fordhook 242, Henderson Bush, King of the Garden, Fordhook 1072, Belgreen, Bridgeton, and a Fordhook × Henderson cross. Results indicate that concentration of flatulence-causing oligosaccharides increases with maturity of lima beans. Raffinose content is low in most cultivars (4.9–5.9 mg/g), being lowest in Fordhook 1072 (2.4 mg/g) and highest in Belgreen (27.8 mg/g). Stachyose concentration is low in Belgreen and Fordhook (5.8 and 10.1 mg/g, respectively) but very high in all other cultivars. All cultivars have a very low content of verbascose. A report by Ologhobo and Fetuga (1982b) on the carbohydrate constituents of 18 lima bean varieties from West Africa gives a single value for total oligosaccharides, rather than individual values, this ranges from 7.5 to 12.8 g/100 g of seed, with a mean of 9.5 g/100 g. Meredith *et al.* (1988) suggest that since there is great variability in oligosaccharide content of lima bean, a breeding programme designed to eliminate or reduce these oligosaccharides would probably be successful.

Table 15.8 Nutritional quality assessment of lima bean protein

Diet type	*Wt gain*	*PER*	*FER*	*N intake*	*Body N*	*NUP*
Casein	68.5	2.93	3.37	3.63	3.67	62.5
Soybean	52.5	2.34	4.30	2.85	3.59	50.97
Lima bean protein	26.9	1.68	5.94	2.20	2.54	48.41
Lima bean protein + meth + lys + val + phenylala.	68.4	3.04	3.42	3.63	3.60	72.70

From Maneepun *et al.* (1974).
Weight gain was over 21 days, PER = protein efficiency ratio; FER = feed efficiency ratio; N intake = average daily intake per rat; Body N = body nitrogen content (g); NPU = net protein utilization.

15.7.5 Phytic acid and polyphenols

Most legumes are rich sources of phosphorus, much of which is present as phytic acid. Past research has identified phytic acid as capable of interfering with mineral absorption and induction of rickets in animals fed cereal-based diets. Similarly, polyphenols, such as tannins, have been implicated in the reduction of digestibility of proteins, depression in nutritive value of feedstuffs, and the increase in requirements for sulphur amino acids (Bressani and Elias, 1980). From their study of West African lima beans, Ologhobo and Fetuga (1982a) have shown that phytic acid phosphorus comprises 47% of the total phosphorus in lima bean. Phytic acid content varies from 0.77 to 1.14%, with a mean of 0.88%. The concentration of tannic acid varies from 0.71 to 0.93%, with a mean of 0.77%. These workers suggest that tannin content may be linked with consumer acceptability of lima bean varieties, since consumers always inherently prefer the low tannin varieties.

15.8 NUTRITIONAL EXPERIMENTS

Fairly high yields of lima protein concentrate may be obtained from large lima beans by precipitation of solubilized proteins with phosphates (Maneepun *et al.*, 1974). It is necessary to add the phosphate slowly with continuous stirring and to keep the slurry on the boil for at least 10 min, in order to properly coagulate the proteins and inactivate trypsin inhibitors. It is also necessary to freeze dry the protein curd rather than to oven dry it in order to obtain a light-coloured, light weight, soft-textured product.

Protein efficiency ratio (PER) and net protein utilization (NPU) of the freeze-dried lima protein concentrate are 1.67 and 48.4, and for freeze-dried concentrate supplemented with lysine, phenylalanine and valine, PER and NPU are 3.04 and 72.7, respectively (Table 15.8) (Maneepun *et al.*, 1974). The influence of raw and cooked lima has also been investigated in an attempt to determine the effects of antinutritional factors in lima beans (Aletor and Fetuga, 1984a, b). Feeding of raw lima beans to weaning rats raises the liver enzymes glutamic pyruvate transaminase (GPT) and glutamic oxaloacetic transaminase (GOT) while depressing the urea cycle enzyme ornithine carbamyl transferase (OCT). It also depresses body weight gain, feed consumption, nitrogen retention and protein digestibility. Since these antinutritional factors are easily destroyed by heat treatment and since human beings cook their meals before eating them, experiments in which raw beans are fed to experimental animals do not have much practical relevance except to indicate the presence of antinutritional factors.

REFERENCES

Aletor, V.A. and Fetuga, B.L. (1984a) Effect of graded levels of raw lima bean (*Phaseolus lunatus*) on some liver enzyme activity in the rat. *Nutr. Rep. Int.*, **29**, 565–70.

Aletor, V.A. and Fetuga, B.L. (1984b) Influence of raw and cooked lima bean (*Phaseolus lunatus*) meals on performance, nutrient utilization and disaccharidase activities in growing rats. *Nutr. Rep. Int.*, **29**, 571–8.

Allard, R.W. (1960) Patterns of evolution of cultivated species, in *Principles of Plant Breeding*, Wiley, New York.

Baudoin, J.P. (1988) Genetic resources, domestication and evolution of the lima bean *Phaseolus lunatus*, in *Genetic Resources of* Phaseolus *Beans*, (ed. P. Gepts), Kluwer Academic Publishers, Boston.

Birk, Y. (1967) Lima bean trypsin inhibitors. *Methods Enzymol.*, **45**(pt. B), 707–9.

Bressani, R. and Elias, L.G. (1980) The nutritional role of polyphenols in beans, in *Toxic Constituents of Legumes* 2nd edn, (ed. R. Bressani), Academic Press, New York.

Clifford, A.J., Heid, M.K., Peerson, J.M. and Bills, N.D. (1991) Bioavailability of food folates and evaluation of food matrix effects with a rat bioassay. *J. Nutr.*, **121**, 445–53.

Duke, J.A. (1981) *Handbook of Legumes of World Economic Importance*. Plenum Press, New York.

Ekanayake, A. and Nelson, P.E. (1990) Effect of thermal processing on lima bean vitamin B-6 availability. *J. Food Sci.*, **55**, 154–7.

Erickson, H.T. (1982) Lima bean legacy. *Hortscience*, **17**, 702.

Evans, I.M. and Boulter, D. (1974) Amino acid composition of seed meals of yam bean (*Sphenostylis stenocarpa*) and lima bean (*Phaseolus lunatus*). *J. Sci. Food Agric.*, **25**, 919–22.

FAO/WHO (1973) Energy and Protein Requirements. FAO Nutrition Meeting Report Series No. 52, Rome. WHO Technical Report Services No. 522, Geneva.

Galbraith, W. and Goldstein, W. (1970) Phytohemagglutinins: a new class of metalloproteins. Isolation, purification and some properties of the lectin from *Phaseolus lunatus*. *FEBS Lett.*, **9**, 197–201.

Galbraith, W. and Goldstein, W. (1972a) Lima bean (*Phaseolus lunatus*) lectin. *Methods Enzymol.*, **28**(pt B), 318–23.

Galbraith, W. and Goldstein, W. (1972b) Phytohemagglutinin of the lima bean (*Phaseolus lunatus*). Isolation, characterization and interaction with type A blood group substance. *Biochemistry*, **11**, 3976–84.

Gould, N.R. and Scheinberg, S.L. (1970) Isolation and partial characterization of two anti-A hemagglutinins from *P. lunatus*. *Arch. Biochem. Biophys.*, **137**, 1–11.

Hernandez-Xolocotzi, E., Bruner-Cruz, M. and Ortega-Delgado, M.L. (1984) Composition of seeds of *Phaseolus lunatus* L. from Mexico and South America. *Ann. Rep. Bean Improvement Coop.*, **27**, 175–6.

Kaplan, L. (1965) Archeology and domestication in American *Phaseolus* (Beans). *Econ. Bot.*, **19**, 358–68.

Krahn, J. and Stevens, F.C. (1971) Stoichiometry of the interaction of lima bean protease inhibitor with trypsin and/or chymotrypsin. *FEBS Lett.*, **13**, 339–41.

Krahn, J. and Stevens, F.C. (1972a) Antitrypsin site of lima bean protease inhibitor. *Biochemistry*, **11**, 1804–8.

Krahn, J. and Stevens, F.C. (1972b) Lima bean protease inhibitor: comparative study of the trypsin and chymotrypsin inhibitory activities of four chromatographically pure variants. *FEBS Lett.*, **28**, 313–16.

Liener, I.E. (1976) Legume toxins in relation to protein digestibility; a review. *J. Food Sci.*, **41**, 1076–81

Lyman, J.M., Baudoin, J.P. and Hidalgo, R. (1985) Lima bean (*Phaseolus lunatus*, L), in *Grain Legume Crops*, (eds R.J. Summerfield and E.H. Roberts), William Collins, London.

Mackie, W.W. (1943) Origin, dispersal and variability in the lima bean. *Hilgardia*, **15**, 1–29.

Maneepun, S., Luh, B.S. and Rucker, R.B. (1974) Amino acid composition and biological quality of lima bean protein. *J. Food Sci.*, **39**, 171–4.

Meredith, F.I., Thomas, C.A., Snook, M.E. *et al.* (1988) Soluble carbohydrate, oligosaccharides and starch in lima bean seeds. *J. Food Sci.*, **53**, 768–71.

Ologhobo, D.A. and Fetuga, B.L. (1982a) Polyphenols, phytic acid and other phosphorus compounds of lima bean (*Phaseolus lunatus*). *Nutr. Rep. Int.*, **26**, 605–11.

Ologhobo, D.A. and Fetuga, B.L. (1982b) Carbohydrate constituents of some lima bean (*Phaseolus lunatus*) varieties. *Nutr. Rep. Int.*, **26**, 981.

Ologhobo, D.A. and Fetuga, B.L. (1983a) Trypsin inhibitor activity in some lima bean (*Phaseolus lunatus*) varieties as affected by different processing methods. *Nutr. Rep. Int.*, **27**, 41–7.

Ologhobo, D.A. and Fetuga, B.L. (1983b) Compositional differences in some lima bean (*Phaseolus lunatus*) varieties. *Food Chem.*, **10**, 297–307.

Ologhobo, D.A., Fetuga, B.L. and Tewe, O.O. (1984) The cyanogenic glycoside content of raw and processed lima bean varieties. *Food Chem.*, **13**, 117–28.

Smith, C.E. (1968) The new world centres of origin of cultivated plants and the archeological evidence. *Econ. Bot.*, **22**, 253–66.

USDA (1986) Composition of Foods: Legumes and Legume Products. United States Department of Agriculture, Human Nutrition Information Service. *Agriculture Handbook no. 8–16*. USDA, Washington, DC.

CHAPTER SIXTEEN

Common bean (*Phaseolus vulgaris* L.)

E. Nwokolo

16.1 INTRODUCTION

The common bean *Phaseolus vulgaris* is a member of that immense family of plants, the Leguminosae (Fabaceae) which comprises about 600 genera and about 13 000 species. The genus *Phaseolus* includes 150–200 species of plants many of which are cultivated as food or garden ornamentals. The specific name *Phaseolus vulgaris* refers to hundreds of varieties and cultivars of the common bean which has been in cultivation for thousands of years. Thus the name *Phaseolus vulgaris* refers equally to the kidney bean, navy bean, French bean, haricot bean, pinto bean, field beans, China bean, frijol, marrow bean, snap bean, black bean or white bean. What is important is to specify the type and variety of bean referred to following the use of the specific name. In this chapter, an attempt will be made to present as generalized a discussion of the common bean as is possible.

The common bean has at least 20 well known and generally accepted names which in addition to the ones listed above include: yellow bean, pea bean, wax bean, string bean, poroto, caraoto and feijao. These are in addition to countless local names in myriads of languages and dialects. This multiplicity of names is as a result of the common bean being cultivated in very many countries and ecological zones worldwide. It is a very common crop of most temperate, subtropical and moist-to-dry tropical climates. It is predominantly produced in India, Brazil, China, United States and Mexico. The major producers together with their productivity and yields are presented in Table 16.1.

According to Kaplan (1965) the common bean is native to the New

Legumes and Oilseeds in Nutrition
Edited by E. Nwokolo and J. Smartt
Published in 1996 by Chapman & Hall. ISBN 0 412 45930 2

Table 16.1 Major world producers of dry beans in 1990

Region	*Area harvested (ha)*	*Yield (kg/ha)*	*Production (1000 MT)*
Country			
India	9760	410	4000
Brazil	4478	456	2040
China	1417	1351	1915
USA	844	1742	1471
Mexico	2099	616	1292
Indonesia	350	1314	460
Uganda	450	778	350
Myanmar	400	754	301
Thailand	190	1147	301
Tanzania	450	622	280
Continent			
Africa	2682	665	1782
N & C America	3651	909	3319
S. America	5140	507	2603
Asia	13 617	587	7990
World	26 407	617	16 294

Source: FAO (1990).

World, probably Mexico, Central America and the northern areas of South America. It is postulated that it is from this region that early explorers took it to the Old World, spreading it throughout Africa, Europe, the Middle East, Asia and China. Over 14 000 cultivars of the common bean are known to exist, the International Centre for Tropical Agriculture in Colombia being the main repository of the germplasm. That the bean has been cultivated in most countries for hundreds of years has aided the evolution of common names and a natural selection for seed colour which appeals to local people. Duke (1981) notes that there are strong regional preferences in Latin America and Africa for coat colour and brilliance. Thus, there is a predominance of black beans in Venezuela and Guatemala, red beans in Columbia and Honduras, cream or tan beans in Peru and black or tan beans in Brazil. This author has observed that all over West Africa, there is a predominance of white and red beans, and to a smaller extent black beans. According to Duke (1981) common beans cultivated in the United States are primarily of four types: (1) red kidney beans, long kidney-shaped beans measuring 1.5 cm or more and very popular in Latin American countries; (2) pinto beans, medium-sized beans measuring 1.0–1.2 cm in length, usually of pinkish-buff colour with brown spots, very popular and extensively cultivated in the United States; (3) yellow-eyed white beans also called marrow beans, measuring 1.0–1.5 cm in length, also cultivated in the United States but to a lesser extent than others; (4) navy beans (pea beans), small round white beans measuring 8 mm or less, very popular in the United States and cultivated in California.

16.2 THE CROP

The common bean is a highly polymorphic, annual herbaceous species. It exists in two types, a short bushy erect type usually 25–60 cm high, and a long twining pole type, usually 2–3 m or longer. Flowering occurs about 4 weeks after sowing, the flowers being variegated although the basic colours are white, purple or pink. Pods develop rapidly after flowering and are usually slender and flat, turning nearly cylindrical on maturity. Pods are usually 10–20 cm long and 1.0–1.5 cm wide, coming in a variety of colours which include green, yellow or black, sometimes purple. In early maturing cultivars, maturation of pods occurs about two weeks after flowering, and mature green beans may be harvested every 2–3 days on a regular basis. There are usually 6–8 seeds in a mature pod, although the number may be as low as four less. Seed colour is usually red, white, tan, grey, black or mottled. Seed size varies from 10–30 g/100 seeds in the small, white navy bean cultivars, to 50–70 g/100 seeds in the large-seeded red and white kidney beans and Great Northern beans.

16.3 CURRENT AND POTENTIAL USAGE AS FOOD

The common bean is the most widely cultivated leguminous crop of all, cultivated to a large extent in over 50 countries and to some extent in over 70 countries in Africa, North and South America, Asia and Europe (FAO, 1990). Total worldwide production in 1990 exceeded 16 million MT, of which 8 million MT was produced in Asia, 3.3 million MT in North and Central America, 2.6 million MT in South America, 1.8 million MT in Africa, and nearly 0.6 million MT in Europe (Table 16.1). Wherever it is grown, the common bean is a popular vegetable and a popular component of many dishes. In temperate countries, some of the crop is harvested immature and green and may be marketed fresh, frozen or canned. Fresh green beans are marketed whole, whereas those designated for freezing and canning are cut. Frozen and canned beans are exported worldwide and are available in supermarkets in even the poorest countries of the world. In Africa, Asia and Latin America, there is widespread acceptance and consumption of dry beans, the common bean eaten in a variety of ways constituting a significant proportion of daily food intake. Sgarbieri (1989) found that in countries of North, Central and South America, *P. vulgaris* is by far the most commonly consumed grain legume. In Africa and Asia the most commonly consumed legumes are species of the genera *Dolichos*, *Vigna* and *Cajanus* whereas in the Middle East, it is species of the genera *Pisum* and *Vicia*. In these countries dry beans provide a major fraction of the daily protein intake of millions of

Table 16.2 Major African and Latin American producers of dry beans in 1990

Country	*Area harvested (ha)*	*Yield (kg/ha)*	*Production (1000 MT)*
Africa			
Uganda	450	778	350
Tanzania	450	622	280
Rwanda	300	667	200
Burundi	151	992	149
Zaire	173	581	122
Cameroun	170	700	119
Latin America			
Brazil	4478	456	2040
Mexico	2099	616	1292
Argentina	165	970	160
Colombia	165	803	132
Guatamala	142	834	118

Source: FAO (1990).

low and moderate income families (Duke, 1981). Production and yield data on the major African and Latin American producers of dry beans for 1990 are presented in Table 16.2.

Legumes are a valuable source of protein, calcium, iron, thiamine and riboflavin in diets of people in poorer parts of the world (Walker, 1982). However, rather than capitalize on the excellent nutritional qualities of beans and other legumes, moderate to high income citizens continuously aspire to and emulate western-type diets which have high proportions of animal products, with a corresponding high proportion of total fat, saturated fat and sugar. These western-type diets invariably lead to over nutrition and obesity which are accompanied by diabetes and coronary heart disease (Walker, 1982). Thus, maintaining or increasing the consumption of beans and other legumes without attempting to introduce high levels of animal products would spare huge segments of the population the so-called diseases of affluence. A review of published information on the physiological effects of legumes in the human diet, highlighted the immense health and nutritional advantages of consumption of beans (Walker, 1982). Increased production and consumption of the common bean provides a means of enhancing protein intake at a low cost, through the complementarity which exists between legumes and cereals, in diets of people under marginal nutrition.

Table 16.3 Proximate composition of common beans

	Content (per 100 g raw dried beans)			
Constituents	*Red kidney*	*White*	*Pinto*	*Navy*
Water	11.75	11.32	10.95	12.36
Energy (kcal)	337.00	333.00	340.00	335.00
Protein (g)	22.53	23.36	20.88	22.33
Lipids (g)	1.06	0.85	1.13	1.28
Carbohydrates (g)	61.30	60.27	63.41	60.65
Crude fibre (g)	6.19	5.98	6.01	5.52
Ash (g)	3.37	4.20	3.63	3.37

Source: USDA (1986).

16.4 NUTRIENT COMPOSITION AND NUTRITIVE VALUE

There is overwhelming scientific evidence that consumption of beans and other legumes is 'the right thing to do'. The high fibre content of beans and other indigenous foods of Africa has been linked to the observation that many Africans living in Africa have been spared such diseases as rectal and colon cancer, diverticular disease, coronary heart disease, etc. (Burkitt and Trowell, 1975). Beneficial effects of the consumption of beans and other legumes include the following:

1. The high dietary fibre content of beans which is reported to be about 15.2% in mung beans, 16.7% in peas and 25.4% in kidney beans. Crude fibre contents of these legumes were also reported as 4.4%, 6.0% and 4.8% respectively (Kamath and Belavady, 1980).
2. The high content of unavailable carbohydrates (Kamath and Belavady, 1980) which reduces the magnitude of postprandial hyperglycaemia and therefore possibly prevents or reduces obesity.
3. A hypocholesterolaemic effect following legume consumption which has been demonstrated with chickpea, mung bean and other beans (Mathur *et al.*, 1964; Shurpelakar *et al.*, 1979).

16.4.1 Proximate composition

The common bean is a good source of many dietary nutrients. Compositional data on red kidney bean, white bean, pinto bean and navy bean, which are considered representative of *P. vulgaris* beans are presented in Table 16.3. Total carbohydrates, protein, fibre and dietary energy are uniformly high in these beans, ash content is moderate, lipid content is very low. Values in Table 16.3, which are composites from

Table 16.4 Mineral content of common beans

	Content (mg/100 g raw dried beans)			
Mineral	*Red kidney*	*White*	*Pinto*	*Navy*
Calcium	83	240	121	155
Iron	6.69	10.44	5.88	6.44
Magnesium	138	190	159	173
Phosphorus	406	301	418	443
Potassium	1359	1795	1328	1140
Sodium	12	16	10	14
Zinc	2.79	3.67	2.54	2.54
Copper	0.70	0.98	0.77	0.87
Manganese	1.11	1.79	1.13	1.30

Source: USDA (1986).

various beans analysed and reported on by the USDA (1986), are in all respects very similar to values reported by other workers (Moraes and Angelucci, 1971; Meiners *et al.*, 1976). For example, Duke (1981) reports analytical data of 22.9% protein, 1.3% fat, 60.6% carbohydrates and 3.2% ash for a commercial brand of common bean. Duke also reports that dried mature seeds of white and red beans as well as pinto beans contain the following levels of nutrients: food energy 340, 343 and 349 kcal/100 g; protein 22.3, 22.5 and 22.9%, fat 1.6, 1.5 and 1.2%; total carbohydrates 61.3, 61.9 and 63.7%; fibre 4.3, 4.2, 4.3%; and ash 3.9, 3.7 and 3.9%, respectively.

16.4.2 Mineral and vitamin content

Common beans are a fairly good source of some minerals, especially potassium, phosphorus and magnesium, and a poor source of iron, sodium, zinc, manganese and copper. There is also considerable variation among the various varieties of beans in respect of these minerals (Table 16.4). For example red kidney bean has only a third to a half of the calcium content of white, pinto and navy beans. Analytical data reported by Duke (1981) for minerals in various samples of common beans are (mg/100 g dry beans): calcium 180; phosphorus 309; magnesium 183; sulphur 166; potassium 1160; sodium 43.2; iron 6.6; copper 0.6 and chlorine 1.8. Dry kidney bean has also been reported (Duke, 1981) to contain calcium at 86 mg/100 g and phosphorus at 247 mg/100 g. In general, there is a threefold decrease in mineral content of beans on cooking. This loss in mineral content is partially due to leaching and partially as a result of increased water content of cooked food. Leaching minerals are irretrievably lost when cooking water is discarded.

The vitamin content of red kidney, white, pinto and navy beans is shown in Table 16.5. In general, beans are a fair source of ascorbic acid

Table 16.5 Vitamin content of common beans

Vitamin	Content (per 100 g raw dried beans)			
	Red kidney	*White*	*Pinto*	*Navy*
Ascorbic acid (mg)	4.5	0.0	7.3	3.0
Thiamin (mg)	0.61	0.43	0.55	0.64
Riboflavin (mg)	0.21	0.14	0.23	0.23
Niacin (mg)	2.11	0.47	1.44	2.06
Pantothenic acid (mg)	0.78	0.73	0.76	0.68
Pyridoxine (mg)	0.40	0.31	0.44	0.43
Folacin (μg)	394.1	387.8	506.3	369.7
Vitamin A (IU)	8	0	5	4

Source: USDA (1986).

and niacin and a poor source of most of the other B-complex vitamins. Beans are also a very poor source of fat-soluble vitamins. Sgarbieri (1989) states that in terms of the contribution of vitamins to daily dietary intake, one cup of a cooked bean serving weighing about 170 g and containing about 65% moisture, will supply less than 10% of the niacin and riboflavin, 10–12% of pyridoxine, 25% of the thiamin and 30% of the folic acid required per adult per day. It is likely, according to this author, that the problem is one of bioavailability of bean vitamins, a topic that has not been adequately researched.

16.4.3 Protein and amino acids

Protein content of dry beans ranges from 20 to 30%, although for most beans the average value is about 23%. For example, pinto beans contain about 20.8% protein, navy beans 22.3% and kidney beans 23.5%. The protein is composed of globulins (55–80%) and albumins (10–20%) and minor amounts of glutelins and prolamins, which vary with the type and cultivar of beans (Chang and Satterlee, 1982). Since storage proteins constitute the largest quantity of proteins in beans, attempts have been made to investigate the nature of the storage proteins of beans. In kidney bean, for example, a minor storage protein, glycoprotein I, and a major storage protein, glycoprotein II, have been identified (Pusztai, 1966; Pusztai and Watt, 1970). Both storage proteins contain very little of the sulphur amino acids methionine and cystine which partially explains why *Phaseolus* beans are generally low in sulphur amino acids. Average amino acid content of common beans is shown in Table 16.6. Beans are a good source of most essential amino acids, although as mentioned above, like most legumes they are a poor source of the sulphur amino acids, methionine and cystine. There is a significant variation among varieties and cultivars of common beans, in respect of most amino acids. It is

Table 16.6 Amino acid content of common beans

	Content (mg/100 g raw dried beans)			
Amino acid	*Red kidney*	*White*	*Pinto*	*Navy*
Tryptophan	0.27	0.27	0.24	0.26
Threonine	0.95	0.98	0.87	0.94
Isoleucine	1.00	1.03	0.92	0.98
Leucine	1.80	1.86	1.66	1.78
Lysine	1.55	1.60	1.43	1.53
Methionine	0.34	0.35	0.31	0.33
Cystine	0.24	0.25	0.22	0.24
Phenylalanine	1.22	1.26	1.12	1.20
Tyrosine	0.63	0.66	0.58	0.62
Valine	1.18	1.22	1.09	1.16
Arginine	1.39	1.44	1.29	1.38
Histidine	0.63	0.65	0.58	0.62
Alanine	0.94	0.97	0.87	0.93
Aspartic acid	2.72	2.82	2.52	2.70
Glutamic acid	3.43	3.56	3.18	3.40
Glycine	0.88	0.91	0.81	0.87
Proline	0.95	0.99	0.88	0.94
Serine	1.22	1.27	1.13	1.21

Source: USDA (1986).

important therefore to bear in mind that *Phaseolus* beans are not one homogeneous entity but rather varieties of beans which, although having many properties in common, still differ in certain significant aspects. Therefore, for effective utilization of these beans adequate heat treatment is required coupled with the provision of a complementary source of sulphur amino acids through, for example, judicious supplementation with cereals, as in corn–bean diets which are very popular in Latin America and Africa.

16.4.4 Carbohydrates

Carbohydrates comprise 60–65% of the dry weight of the common bean. Starch is the predominant carbohydrate, representing 38–45% of the dry weight (Moraes and Angelucci, 1971) or 63–75% of the carbohydrate fraction. Sucrose accounts for 3–5% and crude fibre 5–7% of bean weight. There are also smaller quantities of oligosaccharides such as stachyose (1.2–3.5%), raffinose (0.2–0.6%) and verbascose (0.2–0.4%) which are responsible for the occurrence of flatulence when these beans are consumed (Tobin and Carpenter, 1978). Most beans contain variable amounts of amylose and amylopectin, the two major components of starch. Digestibility of the starch present in bean is relatively high in many bean varieties although Shurpelakar *et al.* (1979) observed fair-to-moderate digestibility in mung bean and black gram starch. However, Geervani and Theophilus (1981) showed that starch from *Phaseolus* beans (black gram and mung beans) supported as much growth as corn starch in casein diets fed to rats. These workers also showed that test diets containing starch from *Phaseolus* beans had higher lysine and methionine availability than diets containing starch from pigeon pea and chickpea. The influence of mung bean starch on protein utilization was investigated by Rao and Rao (1978) who showed that raw mung bean starch depressed the protein quality of casein.

16.4.5 Oligosaccharides

Most legume seeds contain variable amounts of low-molecular-weight carbohydrates which not only include sucrose but also other sucrosyl oligosaccharides, such as raffinose, stachyose and verbascose (Cerning-Beroard and Filiatre, 1977). In some beans these oligosaccharides may be as high as 10% of seed weight (Cerning-Beroard and Filiatre, 1977) although they are generally much lower (Tobin and Carpenter, 1978). Raffinose, stachyose and verbascose are indigestible following ingestion because the human body lacks the α-galactosidase enzyme necessary for their digestion. Intestinal microorganisms ferment the oligosaccharides producing larger volumes of such gases as methane, hydrogen and carbon dioxide causing intestinal distension and discomfort, nausea and diarrhoea. Although raffinose and stachyose occur in all legume seeds, including dry beans (Kosson and Bakowski, 1986), other oligosaccharides, like verbascose and ajugose, have been identified only in specific legumes like broad beans, soybean, mung bean and peas (Cerning-Beroard and Filiatre, 1977). Only a few varieties of dry bean contain verbascose. Cooking in plenty of water has been a well-known practical approach to reduce effectively the concentration of flatulence-producing sugars in beans, as has soaking dry beans in water prior to cooking. A combination of lower temperature soaking (20° C) and cooking has been reported by Kosson and Bakowski (1986) to be more effective than cooking or soaking alone.

16.5 ANTINUTRITIONAL FACTORS

Common beans can only be effectively utilized in the body, after they have been subjected to considerable heat treatment. This is partially because raw bean proteins are refractive to the action of digestive enzymes, but also because most beans have some native proteins which have toxic or antinutritional properties. Antinutritional factors which have been identified in varieties of the common bean include lectins (haemagglutinins), protease inhibitors (trypsin and chymotrypsin inhibitors), polyphenols and phytates.

16.5.1 Lectins (haemagglutinins)

Very many plant species contain lectins or lectin-like substances. Indeed Allen and Brilliantine (1969) found, in a survey of 2663 plant species, that 800 of them contained appreciable quantities of lectins. It has been estimated that over 600 species and cultivars of legumes in the family Leguminosae also contain appreciable quantities of lectins (Toms and Western, 1971). Most common beans, therefore, contain appreciable quantities of lectins (Jaffé, 1980). These include kidney beans which have been known for over 70 years to be toxic to rats (Johns and Finks, 1920), due to the presence of heat-labile haemagglutinating lectins (Pusztai and Palmer, 1977). Of all common beans, the kidney bean has been most extensively studied for its lectin content (Pusztai and Palmer, 1977; Pusztai *et al.*, 1975, 1979, 1982). Of 13 kidney bean cultivars studied 11 have been shown to have lectins at 10% of total protein, levels high enough to be highly toxic, even fatal to rats. Five isolectins, isolated from kidney bean varieties, have been shown to possess mitogenic properties. These workers have shown that ingestion of common bean lectins results in abnormal development of the microvilli of the small intestine. This lectin-induced disruption of intestinal microvilli is thought to result in interference with the intermediate and final steps of nutrient hydrolysis in the gut (Pusztai *et al.*, 1982). The application of moist heat effectively destroys lectins in beans. Autoclaving beans for 1 h at 16 psi has been reported to destroy completely lectins in beans (Gabriel *et al.*, 1974). Similar results were reported by Grant *et al.* (1982) who showed that although heating presoaked beans to temperatures a little above 75°C significantly reduced toxicity and haemagglutinating properties, the only safe way to eliminate toxicity completely was to heat the fully hydrated beans to 100°C for at least 10 min.

16.5.2 Protease inhibitors

The existence of protease inhibitors in various common beans has been confirmed by several investigators. These beans include kidney bean, navy bean, pinto bean and the Great Northerm bean (Pusztai *et al.*, 1975; Wang, 1975; Whitley and Bowman, 1975; Birk, 1976; Pusztai, 1986).

In an extensive review of the chemistry of dry bean proteins, Chang and Satterlee (1982) found that trypsin inhibitors isolated from dry beans are chemically similar to one another. They are all low-molecular-weight proteins in the range 8000–13 000 daltons, except for the navy bean trypsin inhibitor which has a molecular weight of 23 000 daltons. Most of the trypsin inhibitors are able to inhibit both trypsin and chymotrypsin simultaneously. Of the trypsin inhibitors isolated from dry beans, only the navy bean trypsin inhibitor was found to be a glycoprotein (Wagner and

Riehm, 1967). Evidence has been adduced to the effect that there is little correlation between digestibility of a bean and its content of trypsin and chymotrypsin (Jaffé, 1975). It has also been suggested that trypsin inhibitors play very little role in growth depression in rats fed on kidney beans (Pusztai *et al.*, 1975). For effective destruction of protease inhibitors in beans, it has been suggested that beans be subjected to moist heat application at a minimum of 100°C for not less than 10 min (Liener, 1962; Kakade and Evans, 1966, Molina *et al.*, 1975).

16.6 NUTRITIONAL EVALUATION

The nutritive value of a food depends to a large extent on its content of proximate principles especially protein and energy, and the digestibility of these proximate principles. It also depends on its amino acid composition and the bioavailabity of these amino acids. That beans and other legumes constitute an excellent and yet inexpensive source of protein and calories which are important in meeting the nutritional needs of low-income people worldwide has been documented (Bressani *et al.*, 1963). Excellent reviews of the published literature on aspects of the nutritional quality of *P. vulgaris* have appeared in the last 10 years (Bressani *et al.*, 1982; Chang and Satterlee, 1982; Norton *et al.*, 1985 Sgarbieri, 1989). These indicate that the common bean is deficient in the sulphur amino acids and tryptophan, but contains adequate lysine (Sgarbieri, 1989). Also the heat processing which is usually required to inactivate the antinutritional factors in most beans, may render the constituent amino acids insoluble and perhaps indigestible (Bressani *et al.*, 1963; Evans and Bauer, 1978). Thus not only are the sulphur amino acids the first limiting amino acids in various beans, they may also be rendered unavailable by various heat and other processing treatments designed to improve overall protein digestibility (Evans and Bauer, 1978). It is important therefore that application of heat, preferably moist heat, be controlled in both intensity and duration to enhance overall digestibility without destroying the more heat-labile amino acids. Thirdly, the proteins of some varieties of beans are refractory to digestive enzymes, this property and the attendant poor digestibility having nothing to do with the content of trypsin inhibitor in the beans (Jaffé and Vegalette, 1968). This effect was clearly demonstrated by Siedl *et al.* (1969) who isolated a large globulin fraction (30% of total protein) from black beans. This fraction was resistant to hydrolysis by seven proteases before and after heat treatment. Other results on the growth inhibitory effects of unheated GI globulin fraction (not a protease inhibitor) were reported by Liener and Thompson (1980) who fed the GI globulin fraction isolated from navy beans to rats consuming a synthetic diet. Heat treatment of the GI globulin fraction significantly improved digestibility and growth response of the rats. This partly explains why some common bean varieties still have less than optimal digestibility coefficients even after they have been properly cooked and supplemented with limiting amino acids.

From a survey of the huge amount of scientific literature published on nutritional quality of the various cultivars of the common bean, two conclusions immediately stand out. First, as a result of the deficiency of methionine, cystine and to a lesser extent tryptophan, the chemical score of most beans varies from 40 to 60, which is much lower than the chemical scores of other foods which have a higher content of the sulphur amino acids and other essential amino acids (Pusztai *et al.*, 1979). Where bean-based rat diets have been supplemented with 0.3% DL-methionine, growth and performance has been excellent – comparable with rat growth on casein and egg albumin diets (Tobin, 1975). The second conclusion is that both undercooking and overcooking have detrimental effects on digestibility, protein efficiency ratio (PER), net protein utilization (NPU) and other measures of protein quality in beans. For example, Yadav and Liener (1977) reported the effect of various heat treatments on digestibility and PER and found the highest digestibility and PER values in beans autoclaved for 15 min. Autoclaving for longer than 15 min depressed both digestibility and PER as a result of heat-induced denaturation of protein. Similarly, Antunes and Sgarbieri (1979) found a digestibility coefficient of 62–63% for Roshina G2 beans cooked at 97°C for 2.5, 10 or 30 min, whereas beans cooked at a higher temperature for the same time periods all had significantly depressed digestibility and PER.

REFERENCES

Allen, N.K. and Brilliantine, L. (1969) A survey of hemagglutinins in various seeds. *J. Immunol.*, **102**, 1295–9.

Antunes, P.L. and Sgarbieri, V.C. (1979) Influence of time and condition of storage on technological and nutritional properties of dry bean (*Phaseolus vulgaris* L.) variety Roshina G2. *J. Food Sci.*, **44**, 1703–6.

Birk, Y. (1970) Trypsin inhibitors from garden beans (*Phaseolus vulgaris*). *Methods Enzymol.*, **45** (pt B).

Bressani, R., Elias, L.G. and Valiente, A.T. (1963) Effect of cooking and amino acid supplementation on the nutritive value of black beans (*Phaseolus vulgaris*). *Br. J. Nutr.*, **17**, 69–73.

Bressani, R., Elias, L.G. and Braham, J.E. (1982) Reduction of digestibility of legume proteins by tannins. *J. Plant Foods*, **1**, 43–55.

Burkitt, D.P. and Trowell, H.C. (1975) *Refined Carbohydrate Foods and Disease: Some Implications of Dietary Fiber*. Academic Press, New York.

Cerning-Beroard, J. and Filiatre, A. (1977) A comparison of the carbohydrate composition of legume seeds: horse beans, peas, lupines. *Cereal Chem.*, **53**, 968–78.

Chang, K.C. and Satterlee, L.D. (1982) Chemistry of dry bean proteins. *J. Food Process* and *Preservation*, **6**, 203–25.

Duke, J.A. (1981) *Handbook of Legumes of World Economic Importance.* Plenum Press, New York.

Evans, R.J. and Bauer, D.H. (1978) Studies of the poor utilization by the rat of the methionine and cystine in heated dry bean seed (*Phaseolus vulgaris*). *J. Agric. Food Chem.*, **26**, 779–88.

FAO (1990) *Production Yearbook.* Food and Agriculture Organization of the United Nations, Rome.

Gabriel, G.N., Hussein, L. and Morcos, S.R. (1974) Some nutritional studies on kidney bean proteins. *Qual. Plant. Plant Foods Hum. Nutr.*, **24**, 61–70.

Geervani, P. and Theophilus, F. (1981) Influence of legume starches on protein utilization and availability of lysine and methionine to albino rats. *J. Food Sci.*, **46**, 817–18.

Grant, G., Moore, L.J., McKenzie, N.H. and Pusztai, A. (1982) The effect of heating on the hemagglutinating activity and nutritional properties of bean (*Phaseolus vulgaris*) seed. *J. Sci. Food Agric.*, **33**, 1324–6.

Jaffé, W.G. (1975) Factors affecting the nutritional value of beans, in *Nutritional Improvement of Food Legumes by Breeding*, (ed. M. Milner), Wiley-Interscience, New York.

Jaffé, W.G. (1980) Hemagglutinating (lectins), in *Toxic Constituents of Plant Foodstuff*, (ed. I.E. Liener), Academic Press, New York.

Jaffé, W.G. and Vegalette, C.L. (1968) Heat-labile growth inhibiting factors in bean (*Phaseolus vulgaris*). *J. Nutr.*, **94**, 203–10.

Johns, C.O. and Finks, A.J. (1920) Studies in nutrition. II. The role of cysteine in nutrition as exemplified by nutrition experiments with the proteins of the navy bean, *Phaseolus vulgaris J. Biol. Chem.*, **41**, 379–90.

Kakade, M.L. and Evans, R.J. (1966) Effect of soaking and germination on the nutritive value of navy beans. *J. Food Sci.*, **31**, 781–6.

Kamath, M.V. and Belavady, B. (1980) Unavailable carbohydrates of commonly consumed Indian foods. *J. Sci. Food Agric.*, **31**, 194–202.

Kaplan, L. (1965) Archeology and domestication in American *Phaseolus* (beans). *Econ. Bot.*, **19**, 358–68.

Kosson, R. and Bakowski, J. (1986) The effect of processing on the nutritional quality of bean seeds (*Phaseolus vulgaris* L.). *Acta Aliment. Poloni.* **12**, 197–203.

Liener, I.E. (1962) Toxic factors in edible legumes and their elimination. *Am. J. Clin. Nutr.* **11**, 281–6.

Liener, I.E. and Thompson, R.M. (1980) In-vitro and in-vivo studies on the digestibility of the major storage proteins of the navy bean (*Phaseolus vulgaris*). *Qual. Plant. Plant Foods Hum. Nutr.*, **30**, 13–25.

Mathur, K.S. Singhal, S.S. and Sharma, R.D. (1964) Effects of bengal gram on experimentally induced high levels of cholesterol in tissues and serum in albino rats. *J. Nutr.*, **84**, 201–4.

Meiners, C.R., Denise, N.L., Lau, H.V. *et al.* (1976) Proximate composition and yield of raw and cooked mature dry legumes. *J. Agric. Food Chem.*, **24**, 112–26.

Molina, M.R., de La Fuente, G. and Bressani, R. (1975) Interrelationships between storage, soaking time, cooking time, nutritive value and other characteristics of the black bean (*Phaseolus vulgaris*). *J. Food Sci.*, **40**, 587–91.

Moraes, R.M. and Angelucci, E. (1971) Chemical composition and amino acid contents of Brazilian beans (*Phaseolus vulgaris*). *J. Food Sci.*, **36**, 493–4.

Norton, G., Bliss, F.A. and Bressani, R. (1985) Biochemical and nutritional attributes of grain legumes, in *Grain Legume Crops* (eds R.J. Summerfield and E.H. Roberts), Collins Professional and Technical Books, London.

Pusztai, A. (1966) The isolation of two proteins, a glycoprotein and trypsin inhibitor from the seeds of kidney bean (*Phaseolus vulgaris*) *Biochem. J.*, **101**, 379–86.

Pusztai, A. (1986) The role in food poisoning of toxins and allergens from higher plants. *Dev Food Microbiol.*, **2**, 179–94.

Pusztai, A. and Palmer, R. (1977) The nutritional evaluation of kidney beans (*Phaseolus vulgaris*). The toxic principle. *J. Sci. Food Agric.*, **28**, 620–3.

Pusztai, A. and Watt, W.B. (1970) Glycoprotein II. The isolation and characterization of a major antigenic and non-hemagglutinating glycoprotein from *Phaseolus vulgaris. Biochem. Biophys. Acta*, **207**, 413–31.

Pusztai, A., Grant, G. and Palmer, R. (1975) Nutritional evaluation of kidney beans (*Phaseolus vulgaris*). The isolation and partial characterization of toxic constituents. *J. Sci. Food Agric.*, **26**, 149–56.

Pusztai, A., Clarke, E.M.W., King, T.P. and Stewart, J.C. (1979) Nutritional evaluation of kidney bean (*Phaseolus vulgaris*): chemical composition, lectin content and nutritional value of selected cultivars. *J. Sci. Food Agric.*, **30**, 843–8.

Pusztai, A., King, T.P. and Clarke, E.M.W. (1982) Recent advances in the study of the nutritional toxicity of kidney bean (*Phaseolus vulgaris*) lectins in rats. *Toxicon*, **20**, 195–7.

Rao, C.N. and Rao, B.S.N. (1978) Influence of starches from various sources on protein utilization in rats. *Br. J. Nutr.*, **40**, 1–8.

Sgarbieri, V.C. (1989) Composition and nutritive value of beans. *World Rev. Nutr. Diet.*, **60**, 132–98.

Shurpelakar, K.S., Sundaravalli, O.E. and Rao, M.N. (1979) Effects of legume carbohydrates on protein utilization and lipid levels in rats. *Nutr. Rep. Int.*, **19**, 119–24.

Siedl, D., Jaffé, M. and Jaffé, W.G. (1969) Digestibility and proteinase inhibitory action of kidney bean globulin. *J. Agric. Food Chem.*, **17**, 1318–21.

Tobin, G. (1975) Nutritive value of *Phaseolus vulgaris*. Bean Improvement Cooperative Annual Reports, **18**, 80–1.

Tobin, G. and Carpenter, K.J. (1978) The nutritional value of the dry bean (*Phaseolus vulgaris*). A literature review. *Nutr. Abstr. Rev.* (series A), **48**, 919–36.

Toms, G.C. and Western, A. (1971) Phytohaemagglutinins, in *Chemotaxonomy of Leguminosae*, (eds J.B. Harbone, D. Boulter and B.L. Turner), Academic Press, New York.

USDA (United States Department of Agriculture) (1986) Composition of Foods: Legumes and Legume Products. USDA Human Nutrition Information Service. *Agriculture Handbook no. 8–16.* United States Department of Agriculture, Washington DC.

Wagner, L.P. and Riehm, J.P. (1967) Purification and partial characterization of a trypsin inhibitor isolated from the navy bean. *Arch. Biochem. Biophys.*, **121**, 672–8.

Walker, A. (1982) Physiological effects of legumes in the human diet. *J. Plant Foods*, **4**, 5–14.

Wang, D. (1975) A crystalline protein-proteinase inhibitor from pinto bean seed. *Biochim. Biophys. Acta*, **393**, 383–596.

Whitley, E.J. and Bowman, D.E. (1975) Isolation and properties of navy bean proteinase inhibitor component I. *Arch. Biochem. Biophys.*, **169**, 42–50.

Yadav, N.R. and Liener, I.E. (1977) Optimizing the nutritive value of the protein of navy beans (*Phaseolus vulgaris*) by complementation with cereal protein. *Legume Res.*, **1**, 17–18.

CHAPTER SEVENTEEN

Winged bean (*Psophocarpus tetragonolobus* (L.) DC.)

E. Nwokolo

17.1 INTRODUCTION

The winged bean (*Psophocarpus tetragonolobus*) is one of six species of the genus *Psophocarpus* and the family Leguminosae (Fabaceae). *Psophocarpus tetragonolobus* is a twining, perennial herbaceous plant that is characterized by its tuberous roots and its winged pod. All species in the genus have tuberous roots and winged pods and are therefore technically all winged beans. Because *P. tetragonolobus* has been most extensively studied, the term winged bean is currently used almost exclusively for this species. Although the winged bean is perennial, it is most usually cultivated as an annual, because for pod and seed production (but not for root yield), productivity is higher in the first year than in the second year. Banerjee *et al.* (1984) found that fresh tuber weights may be 80–230 g in the first year of growth, and as high as 369–392 g per plant in the second year. Length of the winged pod averages 30 cm (Pospisil *et al.*, 1971). The number of pods per plant in the first year ranges from 60 to 77 (Banerjee *et al.*, 1984) or more depending on level of fertilization and availability of moisture.

Even though this crop is grown all over West Africa, and widely distributed in Asia, and the Pacific, Hymowitz and Boyd (1977) suggest that it probably originated in Papua New Guinea. The other species which comprise this genus include the following.

Legumes and Oilseeds in Nutrition
Edited by E. Nwokolo and J. Smartt
Published in 1996 by Chapman & Hall. ISBN 0 412 45930 2

1. *Psophocarpus palustris* Desv. is a twining, perennial herb with tuberous roots and winged pods (12–20 cm long). It is considered by Hutchinson and Dalziel (1958) as being native to Africa, very probably West Africa.
2. *Psophocarpus scandens* Endl. is a twining perennial herb with winged pods and tuberous roots. It is considered to be native to Africa, very probably East Africa, although it is reported to be cultivated to quite a significant extent in Indonesia, Brazil, Jamaica, Sri Lanka, India and Myanmar (Hymowitz and Boyd, 1977).
3. *Psophocarpus lancifolius* Harms. is a twining, perennial herb which is characterized by sharp, narrow leaves, winged pods and tuberous roots. It thrives in forest fringes and the wet grasslands of East, Central and West Africa (Verdcourt, 1971).
4. *Psophocarpus grandiflorus*, a twining perennial with tuberous roots and winged pods, is found in forests and grasslands of East and Central Africa, especially Uganda, Kenya and Ethiopia as well as in Zaire and the Democratic Republic of the Congo.
5. *Psophocarpus monophyllus*, like the others, a twining plant with winged pods and tubers, is found all over West Africa and as far north as the Sudan (Chevalier, 1920).

17.2 PRODUCTION

The winged bean may be planted as a multipurpose legume cover crop and a food crop, permitting it to be utilized as a green vegetable, as a source of edible fresh green pods, as well as achieving soil coverage. As a plant which exhibits vigorous nodulation, the winged bean is very desirable both as a cover crop and for soil improvement. Data from experimental plots in India (Banerjee *et al.*, 1984) indicate that between 1.9 and 2.2 kg fresh edible vegetable material can be harvested per plant in the first year of growth, whereas in the second year, up to 3 kg/plant may be harvested. Fresh edible tuber weight is low (84–92 g) in early harvests but increases dramatically, the later the harvest is delayed, up to 230 g in late harvests. If the plant is allowed to grow for 2 years, fresh edible tuber weight is 400 g or more. Pod production is optimum within the first year (average 60–78/plant) and decreases in the second year (average 40–55/plant). Similarly, total seed weight is highest during late harvest in the first year (average of 327/plant) and declines considerably in the second year. The seeds are relatively small. At the ripe stage, 100 seeds weigh 38.5 g. Banerjee *et al.* (1984) suggest that when the winged bean is grown for both vegetable and seed, it is better cultivated as an annual herb than as a perennial. They provide evidence to show that when the soil is properly fertilized, vegetative growth and seed yield from unstaked plants are as good as yield from staked plants not adequately fertilized.

Since staking is highly labour intensive and thus expensive to practise on a large scale, allowing the plant to trail on the ground is likely to be the only economic way to produce the plant on a large scale (Karikari, 1980).

17.3 CULTIVATION

Winged bean seeds are usually planted in and around homesteads in small plots, sometimes of less than 0.5 ha. In small peasant farms the average distance between plants is 26–35 cm and crops are planted in rows, 30 cm apart. It may be cultivated in pure stands but more usually, it is intercropped with corn, other beans, soybean, kidney beans and vegetables. Wherever winged bean is cultivated, it is grown at the beginning of the rains, when the weather is cool, a period which also corresponds to the planting season for most other crops. Most peasant farmers plant winged bean in a rotation in which a mixed crop of winged bean, corn, yams or cocoyam is followed by a 2–3 year fallow, followed by the same mixed crop. The winged bean takes 4–5 months to mature and dry, although harvesting of green pods, if desired, takes place much earlier. If tubers are being harvested in the first year of cultivation, they should be harvested at the same time as the dry pods. Yield measurements conducted on farmers' plots in Papua New Guinea indicate that under conditions of minimum management, seed yield may range from 800 to 1000 kg dry seed/ha (Khan *et al.*, 1977). Tuber yields have been estimated to vary from 5533 to 11 754 kg/ha also under conditions of minimum management (Khan *et al.*, 1977). In Ghana, seed yields of about 1400 kg/ha have been reported by Pospisil *et al.* (1971) in plots not optimally managed. The use of fertilizers, pesticides and irrigation is expected to double seed productivity.

17.4 CURRENT AND POTENTIAL UTILIZATION AS FOOD

In their study of the origin, ethnobiology and agricultural potential of the winged bean, Hymowitz and Boyd (1977) indicate that the primary use of the winged bean is for human consumption. According to them, the young, tender pods are prepared in much the same manner as French beans, although in some areas of the world, fresh immature pods are eaten raw as a salad vegetable. The unripe seeds may also be split or ground and used in soups and curries. In Papua New Guinea and Polynesia, the mature seeds are roasted and eaten as snacks in the same manner as peanuts (Jaffé and Korte, 1976). Among the traditional food crops of Papua New Guinea, the winged bean is second in importance only to the sweet potato in the valleys of the Central and Western Highlands, and is cultivated avidly throughout the Eastern Highlands (Khan *et al.*, 1977).

Khan *et al.* (1977) claim that the tuber is a flavourful delicacy and because the leaves, fresh pods and mature seeds are also eaten, they recommend the winged bean as an ideal crop for subsistence gardens.

In Africa as elsewhere, acceptability of the immature pod is very high and it is used in a variety of dishes. When pods are too mature and fibrous to be eaten, the seeds may be consumed in much the same way as green peas. Mature seeds of the winged bean are cooked and eaten in much the same way as lima bean or kidney bean. The bean is soaked prior to cooking to reduce cooking time as well as reduce flatulence. Sometimes, it is cooked overnight on a slow fire and is ready for eating in the morning. Conventional wisdom dictates that the bean be completely cooked before consumption because improperly cooked winged bean is reported to cause stomach cramps.

17.5 NUTRIENT CONTENT AND NUTRITIVE VALUE

As far back as the 1970s, the National Academy of Sciences of the United States (NAS) commissioned a study of the agricultural and nutritional potentials of the winged bean. A subsequent report (NAS, 1975) recognized this as an underutilized crop with a truly great potential to contribute immensely to the food needs of people living in the tropics. Khan *et al.* (1977), assuming a tuber protein content of 10% (fresh weight), calculated the total yield of tuber protein as 553–1175 kg/ha. This compares with only 202 kg protein/ha for a sweet potato crop. When tuber protein yield is combined with protein yield from fresh vegetables, fresh pods and mature seeds, it is apparent why the winged bean has such a huge potential advantage to many tropical countries where protein deficiency and malnutrition are the rule rather than the exception.

The chemical composition and nutritional quality of the winged bean has been extensively studied in the last two decades (NAS, 1975; Jaffé and Korte, 1976; Hymowitz and Boyd, 1977; Claydon, 1980; King and Puwastein, 1987). The protein content of different parts of the winged bean at early-stage, mid-stage and late-stage harvesting during the first year was studied by Banerjee *et al.* (1984) in the same manner as peanuts (Jaffé and Korte, 1976). For the whole plant, protein content ranged from 21.8 to 24.4%. Protein content of fresh leaves was highest at early harvest (29.8%) and decreased as the leaves matured (27.6–28.1%). Protein content of tubers increased with maturity from 17.3 to 18.5%. Protein in seeds rose from 31.7% at the early stages of seed maturity to 37.4% at full maturity. By all comparisons, these are immense quantities of protein, especially when it is remembered that potential competitors of the winged bean (the common bean, pigeon pea, etc.), are not as well endowed with protein as the winged bean. Table 17.1 presents a comparison of the nutrient content of winged bean samples analysed by Jaffé and Korte (1976), winged bean

Table 17.1 Composition of winged bean in comparison with soybean

	Composition (per 100 g edible portion)		
Nutrient	*Goroka sample*	*NAS*	*Soybean*
Water (g)	11.6	6.7 – 24.6	10.2
Protein (g)	30.6	29.8 – 37.4	35.1
Fat (g)	18.3	15.0 – 20.4	17.7
Carbohydrate (g)	26.4	31.6 – 28.0	32.0
Fibre (g)	9.4	5.0 – 12.5	4.2
Ash (g)	3.7	3.6 – 4.0	5.0
Calcium (mg)	215	204 – 370	226
Phosphorus (mg)	500	276 – 320	546
Iron (mg)	18	9.6 – 11.8	8.5
Thiamine (mg)	1.03	1.4	0.66
Riboflavin (mg)	0.33	0.20	0.22
Niacin (mg)	3.09	Not shown	2.20

Source: Jaffé and Korte (1976).

Table 17.2 Amino acid composition of the winged bean in comparison with soybean.

	Composition (mg/g N)		
Amino acid	*Goroka sample*	*NAS*	*Soybean*
Lysine	9.6	74 – 80	6.38
Methionine	1.40	1.2	1.26
Cystine	1.17	1.6 – 2.6	1.33
Total sulphur amino acids	2.57	2.8 – 3.8	2.59
Tryptophan	0.75	1.0	1.28

Source: Jaffé and Korte (1976).

samples as reported by the NAS (1975) and of soybean. As expected, there is wide variation in values among the winged bean samples reported by NAS. However, in general the winged bean samples correspond in nutrient content with the soybean, a positive indication that the winged bean is potentially a great source of nutrients. Protein content is very high and according to NAS (1975) ranges from 29.8% to 37.4%.

Amino acid content of winged bean has been reported by King and Puwastein (1987) as follows (g/16g N): lysine 7.2, histidine 2.7, arginine 6.1, aspartic acid 10.6, threonine 4.3, serine 5.6, glutamic acid 13.3, proline 6.1, glycine 4.1, alanine 4.2, cystine 1.4, valine 4.6, methionine 1.1, isoleucine 4.2, leucine 8.2, tyrosine 4.4 and phenylalanine 4.6. This profile of amino acids is quite similar to the profile of soybean reported in the literature.

This is also confirmed in Table 17.2 where Jaffé and Korte (1976) have compared the amino acid profile of winged bean samples with the profile of soybean. There is considerable similarity between the profile of the two

Table 17.3 Characteristics of oil from winged bean, peanut and soybean

	Winged bean	*Peanut*	*Soybean*
% oil in seed	15 – 20.4	43 – 48	16 – 18
Solidifying point (°C)	8 – 15	0 – 3	-7 to -12
Saponification value	176	188 – 196	188 – 195
Iodine value	82 – 95	82 – 106	125 – 138
% free fatty acids	0.5	2.7	0.9
% unsaponifiable	2.4 – 2.9	0.4 – 0.8	0.7 – 1.6

Source: Claydon (1980).

Table 17.4 Fatty acid composition of oil from winged bean, peanut and soybean (% of total fatty acids)

Fatty acid	*Winged bean*	*Peanut*	*Soybean*
Myristic, 14:0	0.1–0.4	0.1–0.5	0.1–0.3
Palmitic, 16:0	8.9–9.8	7.3–12.9	6.8–11.5
Palmitoleic, 16:1	0.3–0.8	0.2–2.4	0.1–1.0
Stearic, 18:0	5.7–6.9	2.6–6.3	2.4–5.5
Oleic, 18:1	32.3–41.6	42.3–65.7	22–34
Linoleic, 18:2	27.2–31.3	16.8–38.2	49.8–60
Linolenic, 18:3	1.0–2.0	–	2–10
Arachidic, 20:0	1.5–2.2	0.6–1.5	0.3–0.4
Gadoleic, 20:1	2.5–3.0	1.1–1.4	–
Behenic, 22:0	6.1–15.5	1.8–3.1	0.1–0.3
Lignoceric, 24:0	1.0–1.7	0.8–1.1	–

Source: Claydon (1980).

legumes although lysine content is slightly higher and tryptophan significantly lower in winged bean than in soybean. Like soybean and other edible beans, winged bean is deficient in sulphur amino acids and would benefit from supplementation with these amino acids (Smith *et al.* 1984).

17.5.1 Fat and fatty acids

Oil content of winged bean has been reported (NAS, 1975) to be in the range 15.0–20.4%. A value of 18.3% was reported by Jaffé and Korte (1976) and Claydon (1980) obtained a value of 17.0%. Khor and Cahn (1988) note that oil from the winged bean contains about 30–40% saturated and 60–70% unsaturated fatty acids. They observe that about 10–20% of the saturated fatty acids is the C22:0 fatty acid behenic acid and about 50% of the unsaturated fatty acids consist of oleic and linoleic acid. According to these workers, the high concentration of behenic acid in the winged bean lipid is of concern because an even lower level (1–3%) in peanut oil may be responsible for the increased atherogenicity of peanut oil in comparison with other edible vegetable oils.

The physical and chemical properties of oil from the winged bean compared with soybean and peanut oils are shown in Table 17.3 and the fatty acid composition of these oils is shown in Table 17.4. Winged bean oil has more saturated fatty acids and more unsaponifiable matter than peanut and soybean oils, indicating that perhaps as an edible oil, winged bean oil will be less preferable than peanut oil and a lot less preferable than soybean oil.

Claydon (1980) found that the content of essential minerals in winged bean is very high in comparison with other edible legumes. This is especially true for iron (11 mg/100 g) and calcium (230 mg/100 g) but not for phosphorus which is only moderately high (450 mg/100 g). Very similar values for calcium, phosphorus and iron have been reported by Jaffé and Korte (1976) and by NAS (1975) for various winged bean samples. In general, there is broad similarity between the mineral content of winged bean and that of soybean and other edible beans. The same can be said of vitamin content in soybean and in other varieties of edible beans. According to Jaffé and Korte (1976), the content of thiamine, riboflavin and niacin in winged bean exceeds the content of these vitamins in soybean as well as in a variety of other edible beans.

17.6 ANTINUTRITIONAL FACTORS IN WINGED BEANS

These have been extensively studied and include trypsin inhibitors (Jaffé and Korte, 1976; Kortt, 1979, 1983; de Lumen and Salamat, 1980; Ekpenyong and Borchers, 1981) and chymotrypsin inhibitors (Kortt, 1980, 1983). They also include haemagglutinins (Jaffé and Korte, 1976; Pueppke, 1979; Ekpenyong and Borchers, 1981; Gillespie *et al.*, 1981). Kortt (1983) observed considerable differences in trypsin and chymotrypsin inhibitor activities depending on the source of the winged bean. Thus trypsin inhibitor activity (mg trypsin inhibited per g of bean) was lower in Malaysian varieties (average 23.4, range 22–24) and Indonesian varieties (average 28, range 27–29) than in Papua New Guinea varieties (average 36, range 31–42.5) or Myanmar varieties (average 33, range 28–40). Kortt found that the level of chymotrypsin inhibitor activity was higher than trypsin inhibitor activity, the level ranging from 30 to 48 mg chymotrypsin inhibited per g of bean. It was further noted that trypsin inhibitor activity was similar for soybean and winged bean. In similar studies, Jaffé and Korte (1976) reported trypsin inhibitor activity in winged bean at 11 300 IU/g of bean sample, and King and Puwastien (1987) found inhibitor activity of 74 700 IU/g of sample. It is clear, therefore, that there is wide variation in published trypsin inhibitor activity of winged bean, depending most probably, on the variety analysed. That some of the reported toxicity of winged bean is due to its content of haemagglutinins has been demonstrated by Jaffé and Korte (1976). They detected haemagglutinin activity of +3 with normal rabbit erythrocytes and +5 with pronase-treated erythrocytes from a 1:10 sodium chloride extract of winged bean. They demonstrated that feeding raw beans to rats caused considerable mortality in 2 weeks; there was no mortality when the bean was autoclaved. In a similar study, Kortt (1983) reported the existence of haemagglutinins in all 27 varieties of winged bean analysed. Haemagglutinating activity measured with rabbit erythrocytes was about 154 haemagglutinin units/mg, although a few varieties had very low activity of about 77 haemagglutinin units/g. The presence of amylase inhibitors was also reported by Jaffé and Korte (1976) the inhibitor activity in raw beans being 13.4 IU and 5.2 IU in autoclaved beans (118°C, 20 min). Most trypsin and chymotrypsin inhibitors as well as haemaglutinating agents are heat-labile and are consequently destroyed rapidly by application of moist heat. When milled, defatted wing beans were autoclaved for 20 min at 1.5 kg/cm^2, about 30–40% of the trypsin inhibitor activity and only 15% of the chymotrypsin activity was inactivated by heat treatment. Haemagglutinin activity was markedly diminished to between 75–96% of the original value. However, protein was rendered insoluble to a much higher extent, 50–66%. Autoclaving for 30 min completely destroyed trypsin and chymotrypsin inhibitors and haemagglutinating agents and almost completely rendered the protein insoluble.

REFERENCES

Banerjee, A., Bagchi, D.K. and Si, L.K. (1984) Studies on the potential of winged bean as a multi-purpose legume cover crop in tropical regions. *Exp. Agric.*, **20**, 297–301.

Chevalier, N.A. (1920) *Exploration Botanique de L'Afrique Occidental Francaise*, Paris.

Claydon, A. (1980). The use of legumes as sources of edible oil with special reference to winged bean. *Proceedings of a Conference on Legumes in the Tropics*. Faculty of Agriculture, University of Pertanian, Malaysia.

de Lumen, B.O. and Salamat, L.A. (1980) Trypsin inhibitor activity in winged bean (*Psophocarpus tetragonolobus*) and the possible role of tannin. *J. Agric.*, **28**, 533–6.

Ekpengong, T.E. and Borchers, R.L. (1981) Some toxic factors in winged bean seeds I. Antitrypsin, hemagglutinating and urease activities. *Nutr. Rep. Int.*, **23**, 865–70.

Gillespie, J.M., Blagrove, R.J. and Kortt, A.A. (1981) *Characterization of Winged Bean Seed Proteins*. Proceedings of 2nd International Symposium on Developing the Potential of the Winged Bean. 19–23 Jan 1981, Colombo, Sri Lanka.

Hutchinson, J. and Dalziel, J.M. (1958) (Revised by R.W.J. Keay). *Flora of West Tropical Africa*. Published by Crown Agents for Overseas Governments and Administrations, London.

Hymowitz, T. and Boyd, J. (1977) Origin, ethnobotany and agricultural potential of the winged bean. *Econ. Bot.*, **31**, 180–8.

Jaffé, W.G. and Korte, R. (1976) Nutritional characteristics of the winged bean in rats. *Nutr. Rep. Int.*, **14**, 449–5.

Karikari, S.K. (1980) An integrated approach towards agronomic and research needs on the winged bean: a case study of winged bean collaborative progress in Ghana, in *The Winged Bean;* Proceeding of the First International Symposium in Developing the Potentials of the Winged Bean. Las Banos, Philippines, 1978, pp 150–9.

Khan, T.N., Bohn, J.C. and Stephenson, R.A. (1977) Winged bean cultivars in Papua New Guinea. *World Crops and Livestock* Sept/Oct., 208–14.

Khor, H.T. and Chan, S.L. (1988) Changes in lipid classes and fatty acid composition in developing seeds. *Phytochemistry,* **27** 2041–4.

King, R. and Puwastein, P. (1987) Effects of germination on the proximate composition and nutritional quality of winged bean seeds. *J. Food Sci.,* **52,** 106–8.

Kortt, A. (1980) Isolation and characterization of the trypsin inhibitors from winged bean seed (*Psophocarpus tetragonolobus* (L) DC. *Biochim. Biophys. Acta.,* **577,** 371–82.

Kortt, A. (1983) Isolation and properties of a chymotrypsin inhibitor from winged bean (*Psophocarpus tetragonolobus* (L) DC). *Biochim. Biophys. Acta,* **624,** 237–48.

NAS (1975) *The Winged Bean: A High Protein Crop For The Tropics.* National Academy of Sciences, Washington DC.

Pospisil, E., Karikari, S.K. and Boamati-Mensah, E. (1971) Investigations in winged bean in Ghana. *World Crops,* **23,** 260.

Pueppeke, S. (1979) Purification and characterization of a lectin from seeds of the winged bean, *Psophocarpus tetragonolobus* (L) DC. *Biochim. Biophys. Acta,* **581,** 63–70.

Smith, O.B., Ilori, J.O. and Onesirosan, (1984) The proximate composition and nutritive value of the winged bean *Psophocarpus tetragonolobus* (L) DC for broilers. *Anim. Feed Sci. Technol.,* **11,** 231–7.

Verdcourt, B. (1971) *Psophocarpus,* in *Flora of Tropical East Africa.* Published by Crown Agents for Overseas Governments and Admistrations, London.

CHAPTER EIGHTEEN

African yam bean (*Sphenostylis stenocarpa* (Hoechst ex. A. Rich.) Harms.)

E. Nwokolo

18.1 INTRODUCTION

The African yam bean belongs to the family Leguminosae (Fabaceae), an annual, climbing or prostrate vine. The botany of the crop has been described (Okigbo, 1973; Duke, 1981). The vine produces linear pods, 20–30 cm long in which 20–30 seeds are borne. The seeds are usually brown, cream, orange–brown or mottled and ovoid in shape. The yam bean is usually interplanted with yams and other vegetables in the traditional village farm setting. According to Ezueh (1984) flowering occurs 80–130 days after planting and seed maturity after 150–300 days. Seed yield is very poor in Nigeria, about 300–500 kg/ha (Ezueh, 1984) although much higher estimates of yield (2000–3000 kg/ha) have been reported for fertile soils (NAS, 1979). Tuber yields of 100–300 g/plant are common, with hectare yields estimated at about 2000 kg (NAS, 1979), also in fertile soils. The African yam bean flourishes all through tropical Africa, where both seed and tuber are important components of the diet. However, it is not well known as a food source outside Africa and is grown as an ornamental plant in Europe and elsewhere. It is distinct from the Mexican yam bean or simply yam bean, which grows extensively in Central and South America, primarily Mexico and Brazil. The Mexican yam bean is a tuberous vine, grown mainly for the delicious tuber, but not for the seeds, which are toxic.

Legumes and Oilseeds in Nutrition
Edited by E. Nwokolo and J. Smartt
Published in 1996 by Chapman & Hall. ISBN 0 412 45930 2

18.2 CURRENT AND POTENTIAL FOOD USES

The African yam bean is a popular grain legume in West Africa and other areas of tropical Africa. In southern Nigeria, it is considered superior to other pulses due to its excellent flavour, and is priced accordingly. Dried seeds are usually soaked overnight prior to cooking, a process which lasts many hours. To shorten cooking time, the bean is sometimes cooked with baking soda, a process which is reputed to cut cooking time by about 2 h. Prolonged cooking is required to soften the seeds, improve nutrient digestibility (Nwokolo, 1987), and inactivate toxic components which have recently been identified as alkaloids, flavonoids and saponins (Asuzu and Undie, 1986). Water extracts of African yam bean have been found to stimulate muscarine receptors in the intestine, thus leading to diarrhoea. A minimum cooking time of 12–14 h is suggested for complete destruction of toxic components. In addition to the seed, the plant produces small tubers which look like elongated sweet potatoes and taste like Irish potatoes. Immature tubers are watery, mature ones are starchy. Oven-dry tubers are rich in starch (65–70%) with a moderate protein content (10–11%). Fresh tubers are usually cooked like yams or potatoes and are eaten as vegetables, either alone or in combination with other foods.

In Nigeria and other parts of West Africa, the seeds are more popular as food than the tubers, probably due to an abundance of other types of edible tubers, and due to the peculiar flavour of the seeds. In East and Central Africa, however, the tubers are highly relished as food, being preferred to the seed.

18.3 NUTRIENT COMPOSITION AND NUTRITIVE VALUE

Proximate composition of the African yam bean (Table 18.1) indicates that the yam bean has a high content of starch and a moderately high content of protein, which is similar in composition to bambara groundnut. The fibre content is, however, slightly higher than in other edible beans, whereas the lipid content is as low as in most edible pulses. Like other grain legumes, the African yam bean is a poor source of most minerals (Table 18.2), being especially low in calcium and magnesium. Potassium content is high, similar in value to that observed in most legumes. Content of phosphorus is low and the microelements very low, similar to values reported in many other pulses (USDA, 1984).

The availability of most minerals in the yam bean is at best moderate, ranging from a low of 44–50% in potassium and magnesium, to a high of 71% in calcium and copper. The availability of zinc and phosphorus is moderate, estimated at 59% and 62%, respectively. The relatively low

Table 18.1 Proximate composition of Africa yam bean (air-dry basis)

	%
Dry matter	90.2
Protein	19.6
Ether extract	3.0
Crude fibre	8.5
Ash	4.6
Nitrogen-free extract	54.6

From Nwokolo (1987).

Table 18.2 Content and availability of minerals, (air-dry basis)

Mineral	*Mineral content*	*Mineral availability (%)*
Phosphorus (%)	0.30	62.3
Calcium (%)	0.05	71.2
Magnesium (%)	0.16	50.6
Potassium (%)	1.06	44.4
Copper (mg/g)	14.00	70.9
Zinc (mg/g)	24.00	58.8
Iron (mg/g)	30.00	69.9
Average mineral availability		**61.0**

From Nwokolo (1987).

content and only moderate availability of minerals in the African yam bean, indicates that this bean will only contribute to a small extent to the mineral requirements and nutrition of individuals consuming large quantities of the bean in their daily diets.

The gross energy and metabolizable energy content of the African yam bean are typical of values observed in most edible beans, the high starch content of these beans being responsible for the moderately high gross and metabolizable energy values of the bean. The metabolizable energy of the raw seed (2.864 kcal/g), is considerably lower than the energy of the cooked seed (3.150 kcal/g). Similarly, protein digestibility is significantly lower in raw (63%) than in cooked yam bean (81%). It is obvious, therefore, that heat treatment improves the nutritional value of the African yam bean, an improvement which may be due to heat-inactivation of some antinutritional factors that may be present in the bean. African yam bean, like many other pulses, is not a significant source of lipids. However, the fatty acid profile of the yam bean indicates that about 28% of the lipids are composed of C12:0–C18:0 saturated fatty acids and 66% are composed of C18:1–C18:3 unsaturated fatty acids (Table

Table 18.3 Fatty acid profile of lipids of Africa yam bean

Fatty acid	*% of lipids*
12:0	1.4
14:0	2.5
16:0	19.3
18:0	4.5
18:1	40.7
18:2	22.5
18:3	2.4
20:0	1.1
20:1	1.0
22:0	2.2
24:0	2.4
Total saturates	33.4
Total polyunsaturates	24.9
Total monounsaturates	41.7
P/S ratio	0.75

From Nwokolo (1987).

18.3). The monoenoic acid, oleic acid, singularly constitutes over 40% of the fatty acids in African yam bean lipids. Overall, the African yam bean can be considered as having a good fatty acid profile, the profile being important when considerable amounts of this bean are consumed.

Table 18.4 presents the amino acid profile of African yam bean. Although the African yam bean contains a moderate level of protein, reported to be between 19 and 21% on an air-dry basis (Evans and Boulter, 1974; Duke, 1981; Nwokolo, 1987), the tubers have a low protein content estimated at about 10–11% (Evans *et al.*, 1977). Seed proteins have a moderate content of glutamic acid, aspartic acid, arginine, leucine and lysine, a low content of many other amino acids, and an extremely low content of the sulphur amino acids. Very little nutritional evaluation of the yam bean tuber has been conducted, an amino acid analysis of the product (Evans *et al.*, 1977) indicating that in general the content of amino acids in the dry tuber is half that in the dry bean. Protein of the yam bean tuber has only a moderate content of aspartic acid, glutamic aid, proline, leucine and lysine and a low content of most other amino acids (Evans *et al.*, 1977). As in the bean, the tubers of the yam bean contain extremely low levels of the sulphur amino acids. Average amino acid availability is about 63% and ranges from a low value of 43% for glycine, to moderate values of 60–72%, for all other amino acids (Nwokolo, 1987). In comparison with other oilseeds consumed by man (melon seed, pumpkin seed, soybean), which have amino acid availabilities in excess of 80%, raw African yam bean has at best only moderate amino acid availability (63%). Cooking improves not only protein digestibility, but also amino acid availability to a mean value of 83% (Nwokolo, unpublished data).

Table 18.4 Content and availability of amino acid (air-dry basis)

Amino acid	*Content (mg/g)*	*Availability (%)*
Aspartic acid	17.3	64.7
Threonine	6.7	62.5
Serine	9.8	62.2
Glutamic acid	23.1	66.0
Proline	8.1	59.9
Glycine	7.9	43.1
Alanine	7.7	60.7
Cystine	1.1	60.6
Valine	8.5	61.9
Methionine	1.9	57.2
Isoleucine	7.2	61.5
Leucine	13.8	64.7
Tyrosine	7.7	69.0
Phenylalanine	9.1	67.8
Histidine	8.1	57.8
Lysine	14.6	71.9
Arginine	10.0	69.1
Average amino acid availability		**62.6**

From Nwokolo (1987).

18.4 ANTINUTRITIONAL FACTORS

The presence of antinutritional factors has always been suspected but not proved prior to the work of Asuzu and Undie (1986), who reported the existence of alkaloids, flavonoids and saponins in the bean. It has long been known among the locals in southern Nigeria, that consumption of improperly cooked yam bean is associated with diarrhoea, a fact that was confirmed by the experiments of Asuzu and Undie. Cooking time of at least 12 h seems to be enough to inactivate or destroy these antinutritional factors. There is also a likelihood of the existence of enzyme-inhibitory and other factors. More research is needed to elucidate the exact nature of other antinutrients which may exist in the African yam bean.

18.5 HUMAN AND ANIMAL NUTRITION STUDIES

It has been only in the last 5–10 years that nutritional scientists have turned their attention to the African yam bean. As a result of the previous benign neglect of this bean, nutritional evaluation studies with experimental animals are very few and those with humans are still lacking. The bean has also been processed to expand its contribution to the food system. For example, Abbey and Berezi (1988) evaluated African yam bean flours prepared from beans processed in various ways. Processing procedures included dry heating at 100°C for 3 h, moist heating at 121°C and 15 psi for 30 min, long cooking at 100°C for 5 h and quick cooking at 100°C for 30 min. Cooking for 5 h and moist cooking at 15 psi were associated with significantly improved protein digestibility, protein efficiency ratio and weight gain. Trypsin inhibitor concentrations were greatly diminished by all heat treatments.

Using albino rats, Obizoba and Souzey (1989) estimated the nutritive value of whole, dehusked and water-soaked African yam bean. These workers observed that dehusking reduced the cooking time of the bean. It was also noted that the advantages of dehusking outweighed those of soaking in respect of such parameters as food and nitrogen intake, weight gain, protein efficiency ratio, and nitrogen retention.

In another study Aletor and Aladetimi (1989) compared the composition of some lesser-known legumes with several cultivars of the better known cowpea. Levels of various antinutritional factors were lower in yam bean than in cowpea, and although no feeding experiments were performed, these workers extolled the nutritional quality of the bean. The African yam bean has had an enduring reputation as a good source of dietary protein and was used quite extensively in feeding the displaced and the severely malnourished in the war-affected areas, during the Nigerian civil war of 1967–1970. The bean was cooked with substantial quantities of fresh green vegetables (amaranthus, telfaria or cassava leaves) and given to the severely malnourished.

Seeds of the African yam bean which have been infected with weevils and other storage pests, are used in feeding of poultry and swine. These seeds are usually incorporated into animal diets at up to 10% of the diet, in partial replacement for other protein and energy sources, without any adverse effects on animal growth or productivity.

REFERENCES

Abbey, B.W. and Berezi, P.E. (1988) The influence of processing on the digestibility of the African yam bean (*Sphenostylis stenocarpa*) flour. *Nutr. Rep. Int.*, **37**, 819–27.

Aletor, V.A. and Aladetimi, O.D. (1989) Compositional evaluation of some cowpea varieties and some underutilized edible legumes in Nigeria. *Nahrung*, **33**, 99–107.

Asuzu, I.U. and Undie, A. (1986) Some observations on the toxic effects of the seed extract of *Sphenostylis stenocarpa* (Hochst ex A. Rich.) Harms. on intestinal muscle. *Qual. Plant. Plant Foods Hum. Nutr.*, **36**, 3–9.

Duke, J.A. (1981) *Sphenostylis stenocarpa* (Hochst. ex A. Rich) Harms., in *Handbook of Legumes of World Economic Importance*, Plenum Press, New York.

Evans, M. and Boulter, D. (1974) Amino acid composition of seed meals of yam bean (*Sphenostylis stenocarpa*) and lima bean (*Phaseolus lunatus*). *J. Sci. Food Agric.*, **25**, 919–22.

Evans, M., Boulter, D., Eaglesham, A.R.J. and Dart, P.J. (1977) Protein content and protein quality of tuberous roots of some legumes determined by chemical methods. *Qual. Plant. Plant. Food Hum. Nutr.*, **27**, 275–85.

Ezueh, M. (1984) African yam bean as a crop in Nigeria. *World Crops*, **36**, 199–200.

National Academy of Sciences (1979) African yam bean, in *Tropical Legumes: Resources for the Future*. National Academy of Science. Washington, DC, pp. 27–32.

Nwokolo, E. (1987) A nutritional assessment of African yam bean; *Sphenostylis stenocarpa* (Hochst ex A. Rich) Harms and Bambara groundnut; *Voandzeia subterranea* L. *J. Sci. Food Agric.*, **41**, 123–9.

Obizoba, I.C. and Souzey, J. (1989) The nutritive value of African yam bean (*Sphenostylis stenocarpa*): nitrogen and mineral utilization. *Ecol. Food Nutr.*, **22**, 297–305.

Okigbo, B.N. (1973) Introducing the yam bean *Sphenostylis stenocarpa* (Hochst ex A. Rich) Harms. *Proceedings of the first IITA Grain Legumes Improvement Workshop*, Oct. 29–Nov. 2, 1973, London, pp. 224–37.

USDA (1984) Composition of foods: nut and seed products (raw–processed–prepared). *Human Nutrition Information Service, Agriculture Handbook* 8–11, United States Department of Agriculture.

CHAPTER NINETEEN

Velvet bean (*Mucuna* spp.)

R.D. Vieira and N.M. de Carvalho

19.1 INTRODUCTION

The velvet bean originated in South Africa and was subsequently introduced throughout the tropics, subtropical and temperate areas of the world, being commonly grown in Brazil, Colombia, southern states of the USA, Zimbabwe, and many other countries (Bogdan, 1977). The bean has been cultivated for centuries as an ornamental crop, its value for soil improvement and as a forage crop was not recognized until the end of the nineteenth century when a variety of '*Stizolobium deeringianum*' was used for green manure in Florida (Piper and Morse, 1922; Duke, 1981).

Velvet bean is a name that refers to various species belonging to the genus *Mucuna* or *Stizolobium*. Thus, even though the name has been in use for a very long time, there is still considerable controversy over what constitutes velvet bean. In this discussion, the term velvet bean is used to designate all species and/or forms belonging to the genus called *Mucuna* or *Stizolobium*. Since a large group of plants are involved, all with various common names, only a general description will be attempted.

The velvet bean is a vigorous annual or short-lived perennial, with trailing vines of 3.0–7.6 m (with the exception of bush varieties). Usually, wherever velvet bean is grown, a number of cultivars are cultivated, some bushy, others vining, some early maturing, others late maturing. Cultivars also vary in size, seed colour and number of seeds per pod, and flower colour (Coe, 1918; Duke, 1981). In the US where an organized effort was made to introduce species of velvet bean from other countries, about 20 species were introduced. Introduced species in the US include the

Legumes and Oilseeds in Nutrition
Edited by E. Nwokolo and J. Smartt
Published in 1996 by Chapman & Hall. ISBN 0 412 45930 2

Chinese, the Lyon and the Yokohama velvet beans. Depending on the species and/or varieties, the seeds are almost spherical in shape and mottled brown, or completely black, brown, grey or white. *M. deeringianum* and *M. niveum* have mottled seeds, whereas *M. atterrimum* has black seeds. Seed size is also variable, seeds being 1.0–2.0 cm long and 0.5–1.1 cm wide, with a raised white hilum half as long as the seeds (Coe, 1918; Piper and Morse, 1922; Bogdan, 1977; Duke, 1981).

19.2 USE FOR HUMAN CONSUMPTION

Velvet bean has been used to some extent by man for human consumption. In India, people consider the Lyon velvet bean an excellent vegetable. Almost 70 years ago, it was reported that the Japanese ate cooked Yokohama velvet bean, although this was done in moderation due to the content of a toxic principle in this bean (Piper and Morse, 1922). These workers also reported the tentative use in Florida in the 1920s of velvet bean as food. However, the velvet bean had to be thoroughly cooked and the cooking water changed, to prevent purging and vomiting. When velvet bean was preprocessed in this way, it was described as very palatable when used in everyday cooking. It has also been reported that the velvet bean has been eaten in times of starvation, following boiling, repeated changes of cooking water and removal of the seed coat (Purseglove, 1968 as cited by Sarmezey, 1979). In Brazil, the starch found in these seeds has been used as a food thickener and an adhesive component (Duke, 1981). It is also reported to have pharmaceutical use, due to the amino acid L-dopa (3,4-dihydroxyphenylalanine) extracted from the seeds (Sarmezey, 1979; Duke, 1981). L-Dopa has been detected in various species of the velvet bean (Rehr *et al.*, 1973).

19.3 USE FOR ANIMAL CONSUMPTION

There is no doubt that use of velvet bean as an animal feed is important, either as a forage, a seed or as green manure. In the US, velvet bean has been used as animal feed in Florida and the Gulf states for over 70 years (Piper and Morse, 1922). It can be used as winter pasture, either to carry cattle through the winter or to fatten them. The crop is also used as hay or silage or eaten as pod plus seed, or seed alone. For grazing, the animals should be let in when the pods have attained maturity. In general, velvet bean produces a densely matted growth, which makes it difficult to harvest and cure as hay. It is also difficult to dry quickly, resulting in an off-colour dark product. In Brazil and other warm countries, there is tremendous potential for utilization of velvet bean as animal feed. Due to its excellent chemical composition, ease of cultivation, high yield, and

Table 19.1 Chemical composition of different parts of velvet bean

Components	*Green material (%)*	*Whole pod (%)*	*Seed (%)*
Moisture	–	10.0	10.0
Protein	15.1	18.1	23.4
Digestible protein	10.7	13.4	19.0
N-free extract	48.5	50.3	51.5
Crude fibre	19.3	–	–
Fibre	–	13.0	6.4
Ash	14.9	4.2	3.0
Dig. carbohydrates	49.6	–	–
Total carbohydrates	–	–	59.5
Tot. Dig. Nutrients	63.4	73.8	81.7
Fat	–	4.4	5.7
K	0.37	–	1.36
P	0.06	–	0.99
Ca	–	–	0.18

From Duke (1981).

Table 19.2 Amino acid content of the velvet bean seed

Amino acid	*Content (mg/gN)*	*Amino acid*	*Content (mg/gN)*
Isoleucine	300	Valine	344
Leucine	475	Arginine	494
Lysine	388	Histidine	131
Methionine	75	Alanine	219
Cystine	56	Aspartic acid	794
Phenylalanine	300	Glutamic acid	763
Tyrosine	319	Proline	369
Threonine	250	Serine	306

From Duke (1981).

good acceptability and palatability, velvet bean is a good forage plant either in the green form or as hay. In addition, the seeds as well as milled seeds plus pods are good concentrates, with their high content of protein and carbohydrates. Velvet bean can also be used as silage either alone or mixed with corn and/or sorghum (Otero, 1961). *Mucuna pruriens* and *M. atterrimum* are well accepted by rabbits (Vandoni, 1952). Rocha and Raimo (1954) had suggested that velvet bean meal could not be successfully used as chick feed at a level of 35%. This is too high a level of inclusion, but, at a lower level of inclusion, velvet bean may be an excellent source of dietary nutrients for poultry.

19.4 CHEMICAL COMPOSITION OF VELVET BEAN

A survey of the literature indicates that there is some variation in the composition of velvet bean. This variability can be attributed both to differences among types, ecological conditions where the bean was planted, and the experimental analytical procedures used. For example, fat content varied from a high level of 12.4% in *M. deeringianum* to a low of 2.4% in *M. atterrimum*. Bressani and Elias (1980) found that protein in the seed, present in the cotyledons, varies in concentration from 17% to 40%. Variations among other components do not seem to be significant. Another substance which has been found in seeds of six species of *Mucuna* at varying concentrations is the chemical laevodopa (Rehr *et al.*, 1973). Laevodopa (L-dopa; 3,4-dihydroxyphenylalanine), is a free amino acid, well known as a drug for the treatment of Parkinson's disease in humans (NAS, 1979; Nowacki 1980; Duke, 1981, 1990).

The protein content of velvet bean seeds and seed meals is far greater than in vegetative parts. Table 19.1 compares the chemical composition of the young velvet bean plant, the whole pod and the seed, and Table 19.2 gives the amino acid composition of the dry seed of velvet bean. Protein and nitrogen-free extract are high in the entire pod and the seed, and lower in the vegetative material. Protein and nitrogen-free extract are high in the seed and the whole pod, and slightly less in the green material (Table 19.1). Velvet bean has an excellent profile of amino acids, similar to that in the more nutritious legume seeds.

19.5 NUTRIENT VALUE

Meals prepared from legume seeds are a great source of nutrients and improve protein concentration in diets based on cereal grain or starch meals. The high content of protein and carbohydrates in the vegetative parts and seed makes velvet bean products a valuable feed for growing livestock and for milk production (Piper and Morse, 1922). An interesting result was reported when feeding ground velvet seed + pod was compared to feeding ground grains alone to cattle. It was found that 2.5 kg of velvet bean meal or 1.5 kg of ground grain was equal to the feeding value of 1 kg of high grade cotton seed meal. As in almost all legume seed meals, velvet seed meal is low in the sulphur amino acids, methionine and cystine, whereas it has a high content of lysine, which is usually found in low concentrations in cereal grains (Duke, 1981; Souza, 1987, 1989). Velvet bean meal, therefore, has a complementary effect when fed in combination with cereals. It is important, however, to ensure that these complementary products are fed in the correct proportions to make use of their complementarity.

Table 19.3 Tannic acid, trypsin inhibitors and haemagglutinin contents of velvet bean seeds

Compound	*Species*		
	M. atterrimum	M. deeringianum	
	(Souza, 1987)	*(Souza, 1987)*	*(Souza, 1989)*
Tannic acid (mg/g)[a]	79.32	67.79	53.31
Trypsin inhibitor (UTI/mg protein)[b]	22.15	49.75	43.57
Haemagglutinin (mg protein/ml)[c]	5.3	9.1	6.8

[a] Equivalent of mg of tannic acid per gram of sample.
[b] Unit of trypsin inhibitors per ml of protein.
[c] Minimum concentration of protein able to cause agglutination of cattle red blood cells.

19.6 ANTINUTRITIONAL SUBSTANCES

In general, legume seeds contain several compounds which, unless destroyed by heat, may exert unfavourable physiological effects when eaten by animals and in some cases by humans (Liener, 1980). These substances may be poisonous, medicinal and nutritional compounds (Duke, 1990), and may usually be eliminated by cooking or other forms of extraction (Liener, 1980). In most cases, any or a combination of these denaturing treatments is enough to inactivate the antinutritional substances and render the legume seed utilizable as food. Ulian (1981) suggests that when a cooked legume gives rise to low PER, this may not necessarily be due to a deficiency of methionine, but could be due to insufficient heat treatment, resulting in inappropriate denaturing of antinutritional substances.

There are six groups of generally heat-stable, toxic substances, which are present in legumes. One of them, toxic amino acids, is found in velvet bean seeds. Stizolobic and stizolobinic acids found in velvet bean seeds can be toxic (Nowacki, 1980). In addition, high concentrations of tannic acid and low concentrations of trypsin inhibitors and haemagglutinin have been reported in velvet bean seed (Souza, 1987, 1989) (Table 19.3). Thus, in experiments where *Mucuna pruriens* and *M. deeringianum* leaves harvested at flowering time were fed to rabbits (Vandoni, 1952), there was no evidence of toxicity.

L-Dopa, although not a toxic substance, may cause plants which contain it to turn dark as they mature; this dark compound makes plants like *Vicia faba* and *Lathyrus niger* less nutritious (Nowacki, 1980). Also, L-dopa could be linked to a metabolic disorder, favism, which is found principally in people of the Mediterranean region, especially those who lack the enzyme glucose-6-phosphate dehydrogenase. According to Duke (1990), *M.*

Table 19.4 Non-reducing sugar of *Mucuna pruriens* in the Philippines

	Sucrose (%)		*Raffinose (%)*		*Stachyose (%)*		*Verbascose (%)*		*Total oligosaccharides (%)*
Species	*1*	*2*	*1*	*2*	*1*	*2*	*1*	*2*	
M. pruriens									
sabawel A–2	2.60	5.02	1.40	2.70	0.97	1.87	1.20	2.32	3.57
sabawel A–5	2.37	4.43	1.12	2.10	1.02	1.91	1.22	2.28	3.36
Clitoria ternatea									
sam-samping	4.34	9.78	2.11	4.76	2.42	5.45	0.26	0.59	4.79
Canavalia ensiformis									
Jack bean A–1	1.49	2.87	0.34	0.66	1.44	2.78	0.01	0.02	1.79

From Revilleza *et al.* (1990)
1.% of dry mature seeds.
2.% of carbohydrates (N-free extract).

deeringianum and *M. pruriens* also produce another active substance called bufotein.

Table 19.4 compares the oligosaccharide content of Philippine varieties of *M. pruriens, Clitoria ternatea* and *Canavalia ensiformis* (jack bean). A relatively high level of oligosaccharides has been reported in seeds of *M. pruriens,* compared to jack bean and other edible legume species. *Clitoria ternatea* has an exceptionally high content of oligosaccharides of the raffinose family which have been described as compounds responsible for flatulence (Steggerda, 1968 as cited by Revilleza *et al.*, 1990).

19.7 HUMAN AND ANIMAL STUDIES WITH VELVET BEAN

Many studies were conducted with velvet bean at the beginning of this century in southern USA, mainly in Florida and the Gulf states. Results of extensive evaluation of velvet bean in feeding experiments with beef and dairy cattle, swine, horses and poultry were reported by Piper and Morse (1922). These workers concluded that although velvet bean was a good feed for cattle and sheep even in significant quantities, it was unsatisfactory for swine. When large amounts of velvet bean were fed, there was poor production, even vomiting and diarrhoea. The positive results following feeding of rabbits with velvet bean leaves (Vandoni, 1952) has been mentioned earlier. Rocha and Raimo (1954) replaced wheat bran and middlings with velvet bean flour for chicks. The level of replacement (35% of the diet) was probably too high, and that was responsible for the poor results obtained. Souza (1989) studied the replacement of soybean meal by raw mucuna meal. No statistical differences in production parameters were observed among the diets in which velvet bean meal was fed at 0.5 and 10% of the diet. However, at 15% of the diet, there was a decrease in general performance of layers. It was recommended that velvet bean meal be used at levels not exceeding 10% of the diet for laying hens. It is possible that it may be used at higher levels in grower or broiler diets.

Much less research information is available on the use of velvet bean in human diets, but Piper and Morse (1922) indicated that in India, velvet bean was consumed as a vegetable.

REFERENCES

Bogdan, A.V. (1977) *Tropical Pasture and Fodder Plants*. Longmans, New York, 475pp.

Bressani, R. and Elias, L.G. (1980) Nutritional value of legume crops for human and animals, in *Advances in legume science* (eds R.J. Summerfield and A.H. Bunting), Royal Botanic Gardens, Kew, pp. 135–55.

Coe, H. (1918) Origin of the Georgia and Alabama varieties of velvet bean. *J. Am. Soc. Agron.*, **10**, 175–9.

Duke, J.A. (1981) *Handbook of Legumes of World Economic Importance*. Plenum Press, New York, 345pp.

Duke, J.A. (1990) Promising phytomedicinals, in *Advances in New Crops*, (eds J. Janick and J.E. Simon), Proceedings of the first national symposium on New Crops: Research, Development, Economics. Part II. Aromatic, Spice, Medicinal and Bioactive Plants. Indianapolis, IN. October 23–26, pp. 491–8.

Liener, I.E. (1980) Heat-labile antinutritional factors, in *Advances in Legume Science*, (eds R.J. Summerfield and A.H. Bunting), Royal Botanic Gardens, Kew, pp. 157–70.

NAS (National Academy of Sciences) (1979) Green manure, soil reclamation and erosion control, in *Tropical Legume Resources For The Future*, pp. 292–304.

Nowacki, E. (1980) Heat stable antinutritional factors in leguminous plants, in *Advances in Legume Science*, (eds R.J. Summerfield and A.H. Bunting), Royal Botanic Gardens, Kew, pp. 157–70.

Otero, J.R. (1961) Informacoes sobre forageiras, 2nd edn. Ministero da Agricultura, Rio de Janeiro, 334pp.

Piper, C.V. and Morse, W.J. (1922) The velvet bean. *USDA/Farmer Bulletin Washington DC*. No. 1276, pp. 1–21.

Rehr, S.S., Janzen, D.H. and Feeny, P.P. (1973) L-Dopa in legume seeds: a chemical barrier to insect attack. *Science*, **181**, 81–2.

Revilleza, M.J.R., Mendoza, E.M.T. and Raymondo, L.C. (1990) Oligosaccharides in several Philippine indigenous food legumes: Determination, localization and removal. *Plant Food Hum. Nutr.*, **40**, 83–93.

Rocha, G.L. and Raimo, H.F. (1954) Contribuicao para o estudo dos substitutos dos farelos de trigo no alimentacao das aves. *Bol. Ind. Animal*, **14**, 31–44.

Sarmezey, A.A.V. (1979) Velvet beans (*Stizolobium* sp.). Breeding and Experiments 1967–1977. Zambia Ministry of Agriculture Research Memorandum no. 21.

Souza, P.A. (1987) *Avalicao bromatologica, nutricional e tecnologica de algumas leguminosas tropicais*. FCF/USP, Sao Paulo, 84 pp.

Souza, H.B.A. (1989) Caracterizacao bromatologica e avalicao nutricional em poedeiras de graos de mucuna anao (*Stizolobium deeringianum*). FCAV/ UNESP, Jaboticabal, 62pp. (MS Thesis).

Ulian, E.C. (1981) Caracterizacao agronomica, avaliacao bromatologica e valor nutritivo de sementes de leguminosas utilizadas como adubos verdes. FCAV/UNESP, Jaboticabal, 51pp.

Vandoni, R. (1952) Contribuicao para o conhecimento de algumas leguminosas como forrageiras. *An. Esalq.*, **9**, 196–214.

CHAPTER TWENTY

Faba bean (*Vicia faba* L.)

R.N. Adsule and M. Akpapunam

20.1 INTRODUCTION

Vicia faba L. is an important food legume in China, Egypt, Italy, Brazil and Ethiopia. China is the major *V. faba* producing country in the world.

Vicia faba L. is known by a variety of common names such as faba bean, field bean, broad bean, horse bean, tick bean, windsor bean, haba, feve, *baakla* and *bakla shim*. It belongs to the family Leguminosae (Fabaceae). *Vicia faba* L. is divided into two subspecies: *paucijuga* and *eu-faba*. The later subspecies is further divided into three botanical varieties on the basis of seed size:

1. *V. faba* L. var. *minor* Beck – tick bean with smaller seeds;
2. *V. faba* L. var. *equina* Pers – horse bean with medium sized seeds;
3. *V. faba* L. var. *major* – broad beans with larger seeds.

Faba bean is an erect, coarse, leafy annual herb. Its flowers are dull white with dark markings. The pods may be greenish black, brown or black in colour and contain three to four seeds. The seeds may be oblong or oval, flattened or rounded. The colour of seeds varies from white, reddish brown, greenish brown, purple to black. The seeds vary greatly in size depending on the variety. Faba bean has been cultivated since ancient times in the Mediterranean region and the Near East. It is now grown in temperate and subtropical regions of the world. It is difficult to grow in the tropics, where the plants produce seeds sparingly even in the cooler months (Aykroyd and Doughty, 1964; Purseglove, 1977). Faba bean is widely grown as a winter season crop in subtropical and tropical climates

Legumes and Oilseeds in Nutrition
Edited by E. Nwokolo and J. Smartt
Published in 1996 by Chapman & Hall. ISBN 0 412 45930 2

Table 20.1 Chemical composition of *Faba* bean seeds

Constituent	*Content (per 100 g dry seed)*
Protein (g)	32.5
Carbohydrates (g)	61.0
Lipids (g)	1.63
Ash (g)	1.45
Calcium (mg)	180
Phosphorus (mg)	660
Magnesium (mg)	135
Potassium (mg)	120
Iron (mg)	–
Thiamine (mg)	0.52
Riboflavin (mg)	0.16
Niacin (mg)	1.8

Compiled from Cerning *et al.* (1975), Gopalan *et al.* (1982).

and as a spring season crop in temperate regions. It grows best on rich loams and heavy clay soils. It is tolerant of acid soil conditions to a degree but performs best when lime is adequate and the water supply is steady. It thrives best between latitudes 25 and 60°. Faba beans are best grown in the temperature range 10–30°C. World production for 1988 was 4 720 000 MT (1443 kg/ha). The production figure for Africa was 1 384 000 MT (1620 kg/ha), for Asia it was 2 461 000 MT (1393 kg/ha) and for Europe it was 614 000 MT (2070 kg/ha). China was the largest producer with a figure of 2 350 000 MT (1382 kg/ha) (FAO, 1989).

20.2 UTILIZATION

Faba beans form an important part of the diet of people in developing countries. They may be consumed in the form of immature tender pods, green mature seed or as dry seeds after cooking. The methods of cooking and consumption differ widely according geographical location. Some of the popular faba bean dishes in the Middle East include fool medames (stewed faba bean), taamia or falafel (bean cake) and fool nabet (germinated bean) (Askar, 1986). The fermented product tempeh prepared from horse bean is comparable to soybean tempeh.

Green fresh faba beans are used as vegetables or for canning. The dried bean has been considered as a meat extender or substitute and as a skim-milk substitute. In India, roasted seeds are eaten like peanuts.

In the developed world, the horse bean is used for animal feeding. It can be used in compound feeds in rations for cattle, sheep, pigs and poultry. The immature crop can be harvested for silage or hay. Faba bean silage makes an excellent feed for cattle and sheep.

Table 20.2 Amino acid composition of faba beans

Amino acid	*Content (g/16 g N)*
Lysine	7.1
Methionine	0.7
Cystine	0.8
Arginine	8.9
Glycine	4.1
Histidine	2.4
Isoleucine	4.0
Leucine	7.1
Phenylalanine	4.3
Tyrosine	3.2
Threonine	3.4
Valine	4.4
Alanine	4.1
Aspartic acid	11.2
Glutamic acid	15.0
Hydroxyproline	–
Proline	4.0
Serine	4.5
Protein (N × 6.25)	26.2

Source Duke (1983).

20.3 NUTRITIONAL COMPOSITION

The average chemical composition of faba bean seeds is given in Table 20.1. The composition varies significantly due to genetic, environmental and cultural factors (Sjodin, 1982). Faba bean seeds have a relatively high protein content. The major part of this protein is characterized as globulin which is further differentiated into legumin (11S) and vicilin (7S) fractions.

The nutritional quality of faba bean proteins is low. Palmer and Thompson (1975) reported a chemical score of 48, nitrogen digestibility 83%, net protein utilization (NPU) 40% and biological value 48% for the seed proteins. Methionine is the first limiting amino acid in bean proteins (Table 20.2). Supplementation of diets of faba bean with sulphur amino acids can improve nutritional quality (Marquardt and Campbell, 1975; Onayemi and Potter, 1976).

Starch is the major component of seed carbohydrates. Cerning *et al.* (1975) reported 41% starch, 10% sugars, 8% crude fibre and 1.5% lignin in bean seeds. Linolenic, linoleic, oleic and palmitic acids are the major fatty acids in faba bean lipids.

Contents of minerals and vitamins are presented in Table 20.3.

Table 20.3 Mineral and vitamin contents of faba beans

	Content (mg/100 g)
Calcium	104
Phosphorus	301
Iron	6.7
Sodium	8.0
Potassium	1123.0
Thiamine	0.4
Riboflavin	0.2
Niacin	2.1
Ascorbic acid	–

Source: Duke (1983).

20.4 ANTINUTRITIONAL FACTORS

Raw faba bean seeds contain antinutritional factors such as protease inhibitors, tannins, favism-inducing factors, haemagglutinins, flatus producing factors and phytates.

Field bean protease inhibitors have been isolated and characterized. The level of protease inhibitors in faba beans is much lower than in soybeans. These inhibitors are almost completely inactivated by cooking or autoclaving. Heat treatment, by autoclaving at 121°C for 15 min has been reported to improve the nutritional quality of faba bean (Marquardt and Campbell, 1973). Chicks fed on diets containing heated faba bean show decreased pancreas size and improved efficiency of utilization as compared to those fed on raw beans.

Favism is a clinical syndrome in man consisting of acute haemolytic anaemia and related symptoms resulting from the ingestion of faba beans or the inhalation of pollen grains of the plant. Favism is due to the presence of alkaloid glucosides (vicine and convincine) and L-dopa in faba bean. Hydrolysis of the alkaloid glucosides by beta-glucosidase releases the aglycones in the digestive tract resulting in serious damage (favism) in people with a hereditary deficiency of red cell glucose 6-phosphate dehydrogenase. Cooking in water may eliminate the alkaloid glucosides by extracting them from the seeds since they are soluble. Dry heat treatment does not have any effect on them. Vicine and convicine can be removed from faba beans almost completely by extraction with 1% glacial acetic acid (Hegazy and Marquardt, 1983).

The level of haemagglutinins in faba beans is about one sixteenth of the level found in kidney beans (Palmer and Thompson, 1975). Verbascose is the major oligosaccharide present. Cerning *et al.* (1975) reported 0.45%

raffinose, 0.83% stachyose and 2.57% verbascose in beans. The oligosaccharide content is significantly reduced by soaking, cooking and germination of the beans. About 50% of the total phosphorus in seeds is present in the form of phytate. Soaking and germination significantly reduces the proportion of phytate phosphorus. Tannins are present in high concentration in the seed coats of coloured flower varieties. These tannins have been characterized as proanthocyanidins (Martin-Tanguy *et al.*, 1977). Dehulling removes most of the tannins in faba beans.

20.5 HUMAN AND ANIMAL STUDIES

The animal studies with faba beans mostly deal with the evaluation of nutritional quality in rats and feeding trials with poultry, cattle, swine and sheep. Hypocholesterolaemic effects of faba bean protein and the nutritional significance of field bean protease inhibitors have been studied in rats. The effects of faba bean on the growth of chicks and egg production in poultry has been studied by Martin-Tanguy *et al.* (1977).

REFERENCES

Askar, A. (1986) Faba beans (*Vicia faba* (L.)) and their role in the human diet. *Food Nutr. Bull.*, **8**, 15–24.

Aykroyd, W.R. and Doughty, J. (1964) *Legumes in Human Nutrition*. FAO Nutritional Studies No. 19. Food and Agriculture Organization of the United Nations, Rome.

Cerning, J., Saposnik, A. and Guilbot, A. (1975) Carbohydrate composition of horse beans (*Vicia faba*) of different origins. *Cereal Chem.*, **52**, 125–38.

Duke, J.A. (1983) *Hancbook of Legumes of World Economic Importance*. Plenum Press, New York, pp. 102–6, 275–8, 293–6.

FAO (1989) Year Book on Production, Vol. 42, Food and Agriculture Organization of United Nations, Rome, 1988.

Gopalan, C., Sastri, B.V.R. and Balasubramanian, S.C. (1982) *Nutritive Value of Indian Foods*, National Institute of Nutrition, Hyderabad, India.

Hegazy, M.I. and Marquardt, R.R. (1983) Development of a simple procedure for the complete extraction of vicine and convicine from faba beans (*Vicia faba* (L.)). *J. Sci. Food Agric.*, **34**, 100–8.

Marquardt, R.R. and Campbell, L.D. (1973) Raw and autoclaved faba beans in chick diets. *Can. J. Anim. Sci.* **53**, 741–6.

Marquardt, R.R. and Campbell, L.D. (1975) Performance of chicks fed faba bean (*Vicia faba*) diets supplemented with methionine, sulphate and cystine. *Can. J. Anim. Sci.*, **55**, 213–18.

Martin-Tanguy, J., Gullame, J. and Kossa, A. (1977) Condensed tannins in horse bean seeds. Chemical structure and apparent effect on poultry. *J. Sci. Food Agric.*, **28**, 757–65.

Onayemi, O. and Potter, N.N. (1976) Cowpea powders dried with methionine. Preparation, storage stability, organoleptic properties, nutritional quality. *J. Food Sci.* **42**, 48–53.

Palmer, R. and Thompson, R. (1975) A comparison of the protein nutritive value and composition of four cultivars of faba bean (*Vicia faba* L.) grown and harvested under controlled conditions. *J. Sci. Food Agric.*, **26**, 1577–83.
Purseglove, J.W. (1977) *Tropical Crops: Dicotyledons*, Vols 1 and 2. The English Language Book Society and Longman Publishers, London, pp. 273–6, 290–4, 318–21.
Sjodin, J. (1982) Protein quantity and quality in *Vicia faba*, in *Faba Bean Improvement*, (eds G. Hawtin and C. Webb), Martinus Nijhoff, The Hague, The Netherlands.

CHAPTER TWENTY-ONE

Moth bean (*Vigna aconitifolia* (Jacq.) Maréchal)

R.N. Adsule

21.1 INTRODUCTION

Moth bean (*Vigna aconitifolia* (Jacq.) Maréchal) is a minor legume crop. It is widely grown in India and the Far East and the National Academy of Sciences has identified moth bean as a possibly more significant food source for the future (NAS, 1978). Moth bean is known by common names such as kidney bean, dew bean, dew gram, moth, mat, mat bean and matki. It is considered to be native of India and Pakistan.

Moth bean (*Vigna aconitifolia* (Jacq.) Maréchal, synonym *Phaseolus aconitifolius* (Jacq.)) belongs to the family Fabaceae, (Leguminosae subfamily Papilionoideae).

Moth bean is a short semi-erect hairy annual bushy herb, with a prostrate creeping habit. The flowers are bright yellow in colour. Pods are linear, brown in colour, with a short curved beak. They contain four to ten seeds which are small, rectangular and have a light brown, yellow–brown, whitish-green or a mottled seed coat (Janoria *et al.*, 1984).

The moth bean can be grown on a wide range of soil types. The crop has a very high level of drought resistance and general hardiness. It grows best under high constant temperatures.

21.2 UTILIZATION

Moth bean seeds are mostly consumed as cooked whole seeds, split peas (dhal) or sprouts. The green immature seeds can be consumed as such or

Legumes and Oilseeds in Nutrition
Edited by E. Nwokolo and J. Smartt
Published in 1996 by Chapman & Hall. ISBN 0 412 45930 2

Table 21.1 Chemical composition of moth bean seeds

Constituent	*Content (per 100 g dry seed)*
Protein (g)	24.3
Carbohydrates (g)	68.0
Lipids (g)	3.9
Ash (g)	3.8
Calcium (mg)	133
Phosphorus (mg)	356
Magnesium (mg)	183
Potassium (mg)	–
Iron (mg)	11
Thiamine (mg)	0.50
Riboflavin (mg)	0.10
Niacin (mg)	1.7

Compiled from: Cerning *et al.* (1975), Gopalan *et al.* (1982), Borhade *et al.* (1984) and Salunkhe *et al.* (1985).

Table 21.2 Essential amino acid composition of moth bean proteins

Amino acid	*Moth bean (g/16 g N)*	*FAO reference protein (g/16 g N)*
Lysine	5.6	5.5
Leucine	7.0	7.0
Isoleucine	5.1	4.0
Valine	3.3	5.0
Methionine	1.0	} 3.5
Cystine	0.5	
Phenylalanine	4.7	} 6.0
Tyrosine	–	
Threonine	–	4.0
Tryptophan	0.70	1.0

Compiled from: Palmer and Thompson (1975), Gopalan *et al.* (1982) and FAO/WHO (1973).

after cooking. Moth bean seeds have medicinal value and are used in the diets of patients suffering from fevers. In India, moth bean dhal is used for preparation of traditional products like papad and sandge.

Moth beans are also grown for fodder. Dried moth bean plants form an excellent hay. Green plants are a good fodder for sheep, the straw is highly palatable and nutritious. It is a fairly good source of digestible protein (Ranjhan, 1980).

21.3 NUTRITIONAL COMPOSITION

The chemical composition of moth bean seeds (Table 21.1) varies significantly due to genetic and environmental factors (Salunkhe *et al.*,

1985). Seeds contain a significant amount of non-protein nitrogen. Globulin is the major protein fraction in moth bean seeds. Methionine is the first limiting amino acid in seed proteins (Table 21.2). Tryptophan is considered as the second limiting essential amino acid. The vitamin levels in seeds are relatively lower than those in soybean (Gopalan *et al.*, 1982).

21.4 ANTINUTRITIONAL FACTORS

Moth bean seeds contain antinutritional factors such as trypsin inhibitors, polyphenols, phytic acid, saponins, oxalic acid and amylase inhibitor. Trypsin inhibitor from moth bean has been characterized as a glycoprotein with molecular weight of about 9000 daltons (Mehta and Simlot, 1982). Autoclaving of moth bean meal for 10 min (120°C at 15 psi pressure) completely destroys the trypsin inhibitor activity (Kadam *et al.*, 1986).

Cooking causes significant reduction in tannins and haemagglutinins of the seed. The proportion of phytate phosphorus is decreased significantly after germination and cooking (Borhade *et al.*, 1984).

REFERENCES

Borhade, V.P., Kadam, S.S. and Salunkhe, D.K. (1984) Changes in phytate phosphorus and minerals during germination and cooking of horse gram and moth bean. *Qual. Plant. Plant Foods Hum. Nutr.*, **34**, 151–5.

Cerning, J., Saposnik, A. and Guilbot, A. (1975) Carbohydrate composition of horse bean (*Vicia faba*) of different origins. *Cereal Chem.*, **52**, 125–38.

FAO/WHO (1973) *Energy and Protein Requirements, FAO Nutrition Meeting Rep. Ser. No.52/WHO Technical Rep. Ser. No. 522,* Food and Agriculture Organization, Rome, Italy.

Gopalan, C., Sastri, B.V.R. and Balasubramanian, S.C. (1982) *Nutritive Value of Indian Foods*, National Institute of Nutrition, Hyderabad, India.

Janoria, M.P., Gour, V.K. and Singh, C.B. (1984) *Perspectives in Grain Legumes*, Jawaharlal Nehru Krishi Vishwa Vidyalaya, Jabalpur, India.

Kadam, S.S., Ghorpade, V.M., Adsule, R.N. and Salunkhe, D.K. (1986) Trypsin inhibitor in moth bean: thermal stability and changes during germination and cooking. *Qual. Plant. Plant Foods. Hum. Nutr.*, **36**, 43–6.

Mehta, S.L. and Simlot, M.M. (1982) Purification and characterization of trypsin inhibitor from moth bean (*Phaseolus aconitifolium*). *Indian J. Biochem. Biophys.*, **19**, 403–7.

NAS (1978) *Tropical Food Legumes: Resources for the Future*, National Academy of Sciences, Washington, DC.

Palmer, R. and Thompson, R. (1975) A comparison of the protein nutritive value and composition of four cultivars of faba beans (*Vicia faba* L.) grown and harvested under controlled conditions. *J. Sci. Food Agric.*, **26**, 1577–83.

Ranjhan, S.K. (1980) *Animal Nutrition and Feeding Practices in India*, 2nd edn, Vikas Publishing House, New Delhi, India.

Salunkhe, D.K., Kadam, S.S. and Chavan, J.K. (1985) *Postharvest Biotechnology of Food Legumes*, CRC Press, Boca Raton, FL.

CHAPTER TWENTY-TWO

Adzuki bean (*Vigna angularis* (Willd.) Ohwi et Ohashi)

A.C. Uwaegbute

22.1 GENERAL BOTANICAL AND PRODUCTION INFORMATION

The crop is a bushy annual that is variously referred to as adzuki bean, atsuki bean or azuki bean. Several varieties have been cultivated for centuries in Japan, Korea, India and China (Doughty and Walker, 1982). It is believed that the bean originated in the Far East probably in India, China or Japan. Unlike other beans in the genus *Vigna*, it is cultivated in the cooler regions of the subtropics, although it grows well at high altitudes in tropical regions. It has been successfully introduced into the southern United States, New Zealand, Latin America and some parts of Africa. Japan is probably the world's greatest producer and consumer of adzuki beans, even importing quantities from China, Korea, Taiwan, Columbia, Thailand and the US. It is Japan's second most important legume food crop, after soybean.

Pods of the adzuki bean are mainly straw coloured, although black and brown pods are also found. The pods are cylindrical, non-shattering and measure 6–12 cm in length and about 5 mm in diameter. Pods contain 5–12 seeds, each pod being slightly constricted between each seed. Seeds constitute about 50% of the weight of the pods, each seed being flattened and oblong in shape, usually 8×4 mm and truncated at both ends. Seed colour varies from maroon to straw, brown or black, with various combinations of these.

Two main types of adzuki bean are recognized. One is early maturing,

Legumes and Oilseeds in Nutrition
Edited by E. Nwokolo and J. Smartt
Published in 1996 by Chapman & Hall. ISBN 0 412 45930 2

Table 22.1 Energy and nutrient composition of adzuki bean

Energy/nutrient	*Content per 100 g edible portion dried seed*
Water (%)	15
Energy (kcal)	324
Protein (g)	21.1
Fat (g)	1.0
Carbohydrate (g)	59.5
Fibre (g)	3.9
Ash (g)	3.4
Calcium (g)	82
Iron (mg)	6.4
Thiamine (mg)	0.45
Riboflavin (mg)	0.15
Niacin (mg)	2.2

Source: Doughty and Walker (1982).

Table 22.2 Essential amino acid content of adzuki bean

Amino acid	*Content (g/16 g N)*
Isoleucine	4.5
Leucine	7.8
Lysine	7.0
Methionine	1.8
Cystine	1.1
Phenylalanine	5.4
Tyrosine	3.4
Threonine	3.8
Valine	5.4

Source: Doughty and Walker (1982).

has a short stem, light brown pods and small seeds. It is, however, low yielding. The other is high yielding, has a long stem and large seeds. Yield varies depending on intensity of cultivation, soil fertility, availability of ground moisture and other factors.

The main use of adzuki bean is as food for human and animal consumption. Adzuki bean may be cooked in water or milled and fried in oil as a bean cake. Dehulled and split beans may be used in soups. Adzuki bean may also be germinated for bean sprouts. In Japan adzuki beans are also milled into a paste for use as a confectionery. The paste may be prepared by first soaking the beans in water followed by cooking until soft. The soft beans are ground and fibrous matter separated by centrifugation. The residue is washed with water and pressed and the resulting paste may be used directly or dried to a powder.

22.2 NUTRITIONAL QUALITY

The main nutrients (protein, fat, carbohydrates, minerals and vitamins) in adzuki bean (Tables 22.1 and 22.2) are comparable to those in other common legumes. Carbohydrates include oligosaccharides of which stachyose is the major one. Sucrose and raffinose have also been identified. As in other legumes, the limiting amino acids are methionine and cystine. The fats contain 25% of saturated and 75% of unsaturated fatty acids. Stearic acid comprises 50%, palmitic acid 30%, and carnaubic acid 20% of the saturated fatty acids. The unsaturated fatty acids are predominantly linoleic, with small amounts of oleic and even smaller quantities of linolenic acid.

REFERENCE

Doughty, J. and Walker, A. (1982) *Legumes in Human Nutrition*. Food and Agriculture Organization of the United Nations (FAO), Rome.

CHAPTER TWENTY-THREE

Mung bean (*Vigna radiata* (L.) Wilczek)

M. Akpapunam

23.1 INTRODUCTION

Mung bean (*Vigna radiata* (L.) Wilczek) is also known as green gram, golden gram, Oregon pea, chickasano pea, chiroko or simply mung (Purseglove, 1977; Sinha, 1977; Duke, 1983). It is synonymous with *Phaseolus aureus* Roxb. The crop is said to have originated from India and must have been derived from var. *sublobata* which occurs wild throughout India and Burma (Aykroyd and Doughty, 1964; Purseglove, 1977). From there it has spread to South and East Asia, East and Central Africa, the West Indies and the United States. Mung bean is a low altitude crop grown from sea level to approximately 2000 m, usually as a dry land crop. It thrives best on a good loam soil with well distributed rainfall of 70–90 cm year. It is drought resistant and is susceptible to water-logging. Both short-day and day-neutral cultivars are found in India. It grows well between latitudes 0° and 30° north or south. The temperature range for optimum performance is between 20° and 45°C. In India, yields of mung bean ranging from 100 to 200 kg/ha have been reported (Duke, 1983), whereas averages of 1000 kg/ha and 1125 kg/ha has been reported in Sri Lanka and the United States, respectively.

In general appearance it is very similar to black gram with which it is sometimes confused. Some consider mung bean and black gram to be conspecific but there are good reasons to consider the two as distinct although closely related species (Smartt, 1990). The plant itself may be short and erect in growth (ca 30 mm in height) or suberect with a slight tendency to twine and a vine length of 1.2–1.5 m.

Legumes and Oilseeds in Nutrition
Edited by E. Nwokolo and J. Smartt
Published in 1996 by Chapman & Hall. ISBN 0 412 45930 2

23.2 CURRENT AND POTENTIAL USAGE AS HUMAN FOOD

Mung bean, like other food legumes, is a good source of protein, vitamins, minerals and calories. The ways in which it is used differ from country to country and from one region of the world to another. A brief description of the various uses of the bean for human food is given here.

Mung bean is an important food crop. In India the dried beans are boiled and are eaten whole or after splitting into dhal (seed coat removed). The beans are also parched and ground into flour after removal of the testa. The flour is used in various Indian and Chinese dishes. The green pods are eaten as a vegetable. In China and the United States mung bean is used for bean sprouts. The dried beans are soaked overnight, drained and placed in containers in a fairly warm, dark place. The beans are sprinkled with warm water every few hours and the sprouts are ready in a week.

23.3 CURRENT AND POTENTIAL USAGE AS ANIMAL FEED

The stem, leaves, roots, pods and seeds of mung beans like those of most other legumes, contain high amounts of protein even in vegetative parts harvested at an advanced state of maturity. They also contain high amounts of calcium and phosphorus and are good sources of vitamins A and D. They therefore play important roles as supplements of animal feeds. Mung bean is grown as green manure, hay, a cover crop, and for forage. Husks are soaked and used for cattle feed. Although mung beans have been used for livestock feed over the years, the potential in this direction has yet to be fully utilized. Research is needed to develop strains which are most suitable for specialized use for exploitation as human food as a pulse or as a sprouted vegetable and for animal feeding and use as a soil ameliorant. Where vegetative bulk production is important such efforts will improve the yield and availability of the legumes for livestock feed and thus conserve the usual concentrates for non-ruminant animals (pigs and poultry).

23.4 NUTRIENT COMPOSITION AND NUTRITIVE VALUE

Mung beans like most food legumes are high in protein (Table 23.1). Mung bean protein content varies from 22.9% to 24.0%. The amino acid composition is given in Table 23.2. The beans are adequate in most essential amino acids with the exception of the sulphur amino acids methionine and cystine which may be nutritionally limiting. The seeds are

Table 23.1 Proximate composition of mung beans

	%
Moisture	9.7
Protein	23.6
Ether extract	1.2
Ash	4.0
Crude fibre	3.3
Total carbohydrate	58.2

Source: Purseglove (1977), Duke (1983).

Table 23.2 Amino acid composition of mung beans

Amino acid	*Content (g/16 gN)*
Lysine	8.1
Methionine	0.5
Cystine	0.7
Arginine	5.5
Glycine	3.4
Histidine	2.9
Isoleucine	3.7
Leucine	7.1
Phenylalanine	4.9
Tyrosine	2.5
Threonine	3.3
Valine	4.1
Alanine	3.9
Aspartic acid	11.5
Glutamic acid	13.8
Hydroxyproline	–
Proline	3.7
Serine	4.7
Protein (N × 6.25)	22.9

Source: Duke (1983).

Table 23.3 Mineral and vitamin contents of mung beans

	Content (mg/100 g)
Calcium	105
Phosphorus	330
Iron	7.1
Sodium	6.0
Potassium	1132.0
Thiamine	0.5
Riboflavin	0.3
Niacin	2.5
Ascorbic acid	4.0

Source: Duke (1983).

a good source of soluble carbohydrate which is about 58.2% in mung bean. The seed also contains very high amounts of crude fibre (Table 23.1, Duke, 1983). The seeds contain high levels of phosphorus and potassium but low amounts of calcium, iron and sodium (Table 23.3). The vitamin contents are also shown in Table 23.3. It is important to note that dried mung bean contains a considerable amount of ascorbic acid (4.0 mg/ 100 g). The nutrient composition in general is comparable with that of most food legumes (Altschul and Wilcke, 1985).

23.5 PRESENCE, CHEMICAL NATURE AND DETOXIFICATION OF ANTINUTRITIONAL SUBSTANCES

The seeds of mung beans like most other food legumes contain antinutritional factors which affect their biological value. These factors include protease inhibitors, haemagglutinins, cyanogens, phytic acid, oligosaccharides and saponins. A brief summary of the chemical nature, occurrence, toxicity and mode of detoxification is given below.

23.5.1 Protease inhibitors

Protease inhibitors are found in the mung bean; a trypsin inhibitor level of 9.96 TUI/mg has been reported (Elkowicz and Sosulski, 1982). These inhibitors are inactivated by moist heat in autoclaving or cooking.

23.5.2 Haemagglutinins

Haemagglutinins are proteins with a molecular weight range of 36 000–132 000. They have the ability to cause agglutination of red blood cells *in vitro*. This effect which is highly specific for each protein, results from binding to erythrocyte plasma membranes. They also have the ability to stimulate mitosis in cell cultures. They cause impaired growth and reduced food utilization in animals fed on food containing them. A value of 3.42 HU/mg has been reported in mung bean. This value is much lower than those reported for soybean and navy beans (Elkowicz and Sosulski, 1982). The toxicity of all haemagglutinins is destroyed by moist heat. This means that adequate heat treatment as may be expected during cooking will destroy or eliminate these factors in food legumes.

23.5.3 Cyanogens

Cyanogenetic glucosides which generate hydrogen cyanide in a number of pulse species, most notably *Phaseolus lunatus*, have not been reported from the mung bean.

23.5.4 Phytic acid

Phytic acid (inositol hexaphosphate) is found in mung bean in amounts much lower than those reported for soybean and navy bean, respectively (Elkowicz and Sosulski, 1982). Phytic acid is known to bind to calcium, magnesium, iron and/or to protein in the intestinal wall, thus making them unavailable. Phytic acid is not heat liable and is not eliminated by heat treatment, although fermentation and germination are known to reduce its content in food legumes (Akpapunam and Achinewhu, 1985).

23.5.5 Oligosaccharides

Oligosaccarides (α-galactosides) comprising mainly raffinose, stachyose and verbascose have been implicated as a cause of flatulence in human and animals consuming raw or improperly processed food legumes (Akpapunam and Markakis, 1979; Reddy *et al.*, 1980). The human digestive system lacks the enzyme, α-galactosidase, which hydrolyses the oligosaccharides hence they enter the large intestine where they are fermented anaerobically to produce gas and abdominal discomfort. They also impair nutrient absorption. They are not heat labile and so cannot be removed by heat processing. However, they are soluble in water and are eliminated by adequate presoaking treatment. Germination and fermentation of the beans are known to reduce them. Oligosaccharides have been reported in mung beans (Sosulski *et al.*, 1982).

23.5.6 Saponins

Saponins are glycosides which occur in a wide variety of plants. They are reported in small amounts in mung bean. Saponins are known to cause haemolysis of erythrocytes *in vitro*. Those found in alfalfa are known to cause bloat in animals, reduce palatability and cause growth depression and mineral binding. Saponins are heat stable and their removal involves aqueous extractions with hot water or dilute alkali.

23.6 NUTRITIONAL STUDIES

The nutritional quality of legume proteins depends in part on the amino acid profile and the presence of antinutritional factors. Most of the proteins are limiting in the sulphur amino acids methionine and cystine. Supplementation of food legumes with these amino acids has been known to improve their nutritional quality (Kon *et al.*, 1974; Marquardt and Campbell, 1975; Onayemi and Potter, 1976). Similarly processing methods aimed at destroying or inactivating heat-labile compounds and other enzyme inhibitors have been found to improve the nutritive value of food legumes (Marquardt and Campbell, 1973; Onayemi and Potter, 1976). According to Khan *et al.* (1979) the first limiting amino acids in mung bean are methionine and cystine. The biological value of cooked mung bean as determined by a rat assay is 54–56%. This value is similar to those obtained for cowpeas but lower than those reported for chickpea by the same workers. The net protein utilization is in the range 45–48 as compared to 50–57 for cowpeas, and 55–60 for chickpea. This result suggests that mung bean protein is slightly inferior to chickpea protein. Khan *et al.* (1979) also showed that there is a significant correlation (r=0.97) between the biological value and the total sulphur amino acids of mung bean. Thus the protein nutritive value of the legume may be improved by supplementing with these amino acids or by consumption of foods high in these amino acids along with mung bean.

Further research is required to identify varieties of these beans with high nutritional quality. Other studies aimed at improving the protein quality of this food legume are desirable. Research is also needed to develop new, high quality food products, with a view to increasing their uses for both humans and livestock.

REFERENCES

Akpapunan, M.A. and Achinewhu, S.C. (1985) Effects of cooking, germination and fermentation on the chemical composition of Nigerian cowpea. *Qual. Plant. Plant Foods Hum. Nutr.*, **35**, 353–8.

Akpapunam, M.A. and Markakis, P. (1979) Oligosaccharides of 13 American cultivars of cowpea (*Vigna sinesis*). *J. Food Sci.*, **44**, 1317–18.

Altschul, A.M. and Wilcke, H.L. (1985) *New Proteins*, Academic Press, New York.

Aykroyd, W.R. and Doughty, J. (1964) *Legumes in Human Nutrition*. FAO Nutritional Studies No. 19. Food and Agriculture Organization of the United Nations, Rome.

Duke, J.A. (1983) *Handbook of Legumes of World Economic Importance*. Plenum Press, New York, pp. 102–6, 275–8, 293–6.

Elkowicz, K. and Sosulski, F.W. (1982) Antinutritional factors in eleven legumes and their air-classified protein and starch fractions. *J. Food Sci.*, **47**, 1301–4.

FAO (1989) *Year Book on Production*, vol. 42. Food and Agriculture Organization of the United Nations, Rome, 1988.

Khan, M.A. Jacobsen, I. and Eggum, B.O. (1979) Nutritive value of some improved varieties of legume. *J. Sci. Food Agric.*, **30**, 395–400.

Kon, S., Wagner, J.R. and Both, A.N. (1974) Legume powders and some nutritional and physiochemical properties. *J. Food Sci.*, **39**, 897–9.

Marquardt, R.R. and Campbell, L.D. (1973) Raw and autoclaved faba beans in chick diets. *Can. J. Anim. Sci.*, **53**, 741–6.

Marquardt, R.R. and Campbell, L.D. (1975) Performance of chicks fed faba bean (*Vicia faba*) diets supplemented with methionine, sulphate and cystine. *Can. J. Anim. Sci.*, **55**, 213–18.

Onayemi, O. and Potter, N.N. (1976) Cowpea powders dried with methionine. Preparation, storage stability, organoleptic properties, nutritional quality.

J. Food Sci., **41**, 48–53.

Purseglove, J.W. (1977) *Tropical Crops: Dicotyledons*, Vols 1 and 2. The English Language Book Society and Longman Publishers, London, pp. 273–6, 290–4, 318–21.

Reddy, N.R., Salunkhe, D.K. and Sharma, R.P. (1980) Flatulence in rats following ingestion of cooked and germinated black gram and rice bean. *J. Food Sci.*, **45**, 1161–4.

Sinha, S.K. (1977) Food Legumes: Distribution, Adaptability and Biology of Yield. *FAO Plant Production and Protection Paper no. 3*. Food and Agriculture Organization of the United Nations, Rome.

Smartt, J. (1990) *Grain Legumes – Evolution and Genetic Resources*. Cambridge University Press, Cambridge.

Sosulski, F.W., Elkowicz, L. and Reichert, R.D. (1982) Oligosaccharides in eleven legumes and their air-classified protein and starch fractions. *J. Food Sci.*, **47**, 498–502.

CHAPTER TWENTY-FOUR

Bambara groundnut (*Vigna subterranea* (L.) Verdc.)

E. Nwokolo

24.1 INTRODUCTION

The Bambara groundnut, which belongs to the family Leguminosae (Fabaceae), is an annual short-creeping herbaceous plant that is either bunched or semibunched. The botany of the crop has been described by Duke (1981) and by Hepper (1963). Pods are buried beneath the ground, each pod bearing one or two seeds. The seeds are large (1.5 cm diameter) and smooth, rounded or unilaterally flattened and hard. Seeds are variously coloured, some being white, cream or brown, others being red or mottled; a few are black. The Bambara groundnut is grown extensively in the tropics, especially throughout Africa. It grows satisfactorily on many types of soil and in most types of tropical and subtropical climates, doing as well in hot, humid, sub-Saharan regions as in cool, dry, mountainous areas. In Africa south of the Sahara, it constitutes an important food crop for villagers farming soils that are poor in nutrients (Doku and Karikari, 1971). Bambara groundnuts are also cultivated in India and Sri Lanka, Malaysia and Indonesia, the Philippines and the South Pacific, Australia and Papua New Guinea as well as in tropical Central and South America (Duke, 1981; NAS, 1979). There are many cultivars of the Bambara groundnut, seed yield varying among these various cultivars, and although yields from unfertilized soils are as low as 500 kg/ha (Karikari and Lavoe, 1971), yield of the crop is estimated at 2–3 tonnes of seed/ha, in fertilized soils (Duke, 1981). Harvesting occurs about 4 months after seed planting, the crop being in many cases planted

Legumes and Oilseeds in Nutrition
Edited by E. Nwokolo and J. Smartt
Published in 1996 by Chapman & Hall. ISBN 0 412 45930 2

together with other crops in the same field. Ezedinma and Maneke (1985) in an investigation of the flowering, agronomic and yield characteristics of 22 varieties of Bambara groundnuts cultivated in Nigeria, found significant differences among these varieties. In a similar study, Elia (1985) has also investigated seed yield, seed yield components and protein content among eight varieties of Tanzanian bambara groundnuts, reporting that the speckled-seeded variety had the highest protein content (24.6%).

24.2 PRESENT AND POTENTIAL FOOD USES

Bambara groundnuts may be eaten as immature or as mature dry seeds. Immature seeds are usually harvested after 2 months and before 3 months. Although a farmer may harvest his crop as immature seed for his own use, Bambara groundnuts of commerce are available only as mature, dry, hard seeds. Immature seeds are slightly sweeter and easier to cook than mature seeds which usually need 2–3 hours of cooking. In West Africa, seeds of Bambara groundnuts are cooked or roasted. When cooked, the seeds usually constitute the main meal and are flavoured with spices and garnished with fish or meat and vegetables.

Bambara groundnuts may be wet-milled to a fine powder, flavoured with spices, onions and tomatoes, and steamed, usually in small plastic or metal containers. These steamed bean cakes may be eaten with or without additional protein supplements. The nuts may be roasted in hot ashes and used either as snacks or as thickeners in soups and gravies. Bambara groundnuts are however consumed primarily by the poor peasant farmers and the urban poor. They are not as popular as cowpeas and, like pigeon peas, they are consumed in most households only when the price of cowpeas is prohibitive.

24.3 NUTRIENT COMPOSITION AND NUTRITIVE VALUE

The proximate composition of Bambara groundnut shown in Table 24.1 is quite similar to that of most commonly consumed pulses. Duke (1981) reports the nutrient content of dried Bambara groundnuts, per 100 g of air-dry seed as follows: 367 kcal, 10.3% moisture, 18.8% protein, 6.2% fat, 61.3% total carbohydrate, 4.8% fibre and 3.4% ash, values similar to those reported by Nwokolo (1987). The protein content of the bean, particularly, is similar to the value in most varieties of beans consumed in the tropics. Bambara groundnut is primarily an energy and protein source in human diets, containing about 22% protein, and 56% nitrogen-free extract, mainly starch. Gross and metabolizable energy values (3.913 kcal/g and 2.988 kcal/g, respectively), are moderately high, and typical of values

Table 24.1 Proximate composition (air-dry basis)

Nutrient	*Dry seed*[a]	*Immature seed*[b]
Dry matter (%)	90.4	42.7
Protein (%)	22.5	7.8
Ether extract (%)	2.9	3.1
Crude fibre (%)	6.2	3.0
Ash (%)	5.1	1.8
Nitrogen-free extract	55.8	30.0

[a]From Nwokolo (1987).
[b]From Duke (1983).

Table 24.2 Fatty acid profile of lipids of bambara groundnut

Fatty acids	*% of fatty acids*
C12:0	1.05
C14:0	2.03
C16:0	23.81
C18:0	20.54
C18:1	23.72
C18:2	22.47
C18:3	2.09
C20:0	1.46
C20:1	0.44
C22:0	2.39
Lipid content	2.85%
Saturated fatty acids	51.28%
Polyunsaturated fatty acids	24.56%

From Nwokolo (1987).

Table 24.3 Content and availability of minerals (air-dry basis)

Mineral	*Content*	*Availability (%)*
Phosphorus	0.31%	62.9
Calcium	0.07%	74.9
Magnesium	0.17%	55.6
Potassium	1.19%	45.5
Copper	11.00 mg/g	75.1
Manganese	9.00 mg/g	76.4
Zinc	25.00 mg/g	60.4
Iron	24.00 mg/g	70.3
Average availability		65.1

From Nwokolo (1987).

Table 24.4 Content and availability of amino acids (air-dry basis)

Amino acids	*Content (mg/g)*	*Availability (%)*
Aspartic acid	6.8	67.6
Threonine	11.2	90.4
Serine	25.1	93.8
Glutamic acid	29.3	85.8
Proline	8.2	82.7
Glycine	8.3	74.2
Alanine	8.1	82.2
Cystine	1.0	88.9
Valine	8.5	83.9
Methionine	1.2	71.4
Isoleucine	7.2	84.8
Leucine	14.9	86.0
Tyrosine	5.9	85.3
Phenylalanine	9.0	86.9
Histidine	8.3	78.6
Lysine	12.2	82.3
Arginine	13.9	86.2
Average availability:		83.0

From Nwokolo (1987).

observed in various edible beans and other pulses. This is primarily because most of these pulses have a very low oil content. Lipid content and fatty acid profile of Bambara groundnuts are presented in Table 24.2. Bambara groundnut contains a very low level of fat (2.9%) and is not considered as a vegetable oil source. The lipids are primarily saturated, comprising over 50% as saturated fatty acids, remaining solid at room temperature. It has a low content of polyunsaturated fatty acids and therefore a low polyunsaturated/saturated fatty acid ratio. Content and availability of minerals in Bambara groundnuts is presented in Table 24.3. Phosphorus content in Bambara groundnut at 0.31%, is similar to the content in pigeon pea (0.25%) and African yam bean (0.30%). Calcium content is low in both Bambara groundnut (0.07%) and African yam bean (0.05%), values lower than 0.15% in pigeon pea. There is considerable similarity in content of magnesium and potassium, among these three pulses. Among the trace minerals, copper, iron, manganese and zinc levels in Bambara groundnut and pigeon pea show a great deal of similarity, whereas African yam bean has significantly higher levels of these minerals. In general therefore, Bambara groundnut has a low content of dietary minerals, in comparison with other pulses. Availability of minerals is moderate (65.1%), and ranges from 45.5% (potassium) to 76.4% (copper). This is slightly higher than the average availability of 58.2% in pigeon pea and 61.0% in African yam bean. In pigeon pea, African yam bean, cowpea and other oilseeds and legumes, the lowest availability has always been in potassium, although this mineral is also the most abundant in many oilseeds and legumes. There is high availability of calcium but not of phosphorus. Probably as a result of their low concentration, the trace minerals also show high availability. Table 24.4 shows the amino acid content of Bambara groundnuts, expressed in mg of amino acids/g of air-dry seed. Bambara groundnut is a good source of amino acids, in comparison with pigeon pea, cowpea and other pulses. It has a low content of essential amino acids, especially the sulphur amino acids, cystine and methionine. Availability of amino acids from Bambara groundnut is good, average availability being about 83.0%, and ranges from 67.7% for aspartic acid, to 93.8% for serine. Amino acid availability for lysine, cystine and methionine is 82.3%, 88.9% and 71.4% respectively.

24.4 ANTINUTRITIONAL FACTORS

Bambara groundnuts are suspected to contain trypsin and chymotrypsin inhibitors. Since many enzyme inhibitors are in general heat-labile, application of heat to the seeds during the cooking process may usually be enough to inactivate the enzyme inhibitors. Bambara groundnuts are very hard seeds which require up to two hours of cooking to make them soft. Cooking of the intact seed for two hours or steaming of the milled seed for about one hour will usually be enough to destroy these antinutritional factors.

24.5 HUMAN AND ANIMAL STUDIES

In the first of two experiments, Obizoba (1984) investigated the protein supplementary value of Bambara groundnut in comparison with brown cowpeas and white cowpeas. Legumes were whole or dehulled and were cooked to a soft consistency prior to sun drying and milling. Protein quality parameters were higher in cowpeas than in Bambara groundnut. In a second study, Obizoba (1989) showed that either dehusking or soaking improved the nutritive value of Bambara groundnut which was demonstrated to be a good source of protein, minerals and other nutrients.

Because of the high cost of Bambara groundnut, it is not used in animal nutrition. However, seeds which have been infected by weevils or storage pests, and seeds not suitable for human use are incorporated into animal diets, to enhance productivity. Bambara groundnut is a good source of dietary energy and like most pulses, is a moderately high source of protein. If Bambara groundnut is consumed as a whole meal, it needs no energy supplementation, but because of its poor content of sulphur amino acids, it may need substantial amino acid supplementation which may be provided in the form of meat, fish or fresh green vegetables.

REFERENCES

Doku, E.V. and Karikari, S.K. (1971) Bambara groundnut. *Econ. Bot.*, **25**, 255–62.

Duke, J.A. (1981) *Voandzeia subterranea* (L). Thouars, in *Handbook of Legumes of World Economic Importance*. Plenum Press, New York, pp. 307–10.

Elia, F.M. (1985) Variability of yield and seed yield components in Bambara groundnut (*Voandzeia subterranea* Thouars). *Trop. Grain Legume Bull.*, **31**, 45–7.

Ezedinma, F.O.C. and Maneke, F.O. (1985) Preliminary studies on Bambara groundnut (*Voandzeia subterranea* Thouars) in the derived savanna belt of Nigeria. *Trop. Grain Legume Bull.*, **31**, 39–44.

Hepper, F.N. (1963) The Bambara groundnut (*Voandzeia subterranea*) in West Africa. *Kew Bull.*, **16**, 398–407.

Karikari, S.K. and Lavoe, S.K. (1977) Preliminary evaluation and utilization of fourteen cultivars of Bambara groundnut (*Voandzeia subterranea*, Thouars). *Acta Hortic.*, **53**, 195–9.

NAS (National Academy of Sciences) (1979) Bambara groundnuts, in *Tropical Legumes, Resources for the Future*. National Academy of Science. Washington DC, pp. 47–53.

Nwokolo, E.N. (1987) A nutritional assessment of African yam bean (*Sphenostylis stenocarpa*) and Bambara groundnut (*Voandzeia subterranea*). *J. Sci. Food Agric.*, **41**, 123–9.

Obizoba, I.C. (1984) The nutritive value of legume proteins supplemented with rice fed rats at equal intake. *Ecol. Food Nutr.*, **14**, 71–5.

Obizoba, I.C. (1989) The nutritive value of mixtures of two varities of Bambara groundnut and sorghum. *Cereal Chem.*, **66**, 249–51.

CHAPTER TWENTY-FIVE

Rice bean (*Vigna umbellata* (Thunb.) Ohwi et Ohashi)

N.M. de Carvalho and R.D. Vieira

25.1 INTRODUCTION

The rice bean (*Vigna umbellata* (Thunb.) Ohwi et Ohashi), a member of the leguminosae (Fabaceae) family, is an annual legume belonging to a group of little known and limitedly exploited plants. At the present time, the names *Azukia umbellata* (Thunb.) Ohwi, *Phaseolus calcaratus* Roxb., and *Phaseolus pubescens* B1, are considered synonyms (Duke, 1981) and this variability in nomenclature contributes towards making the rice bean less well known among people who might otherwise have some interest in it. These include students, researchers, farmers, and so on. A series of common names also increases the uncertainties about it. It has been variously referred to in English as: rice bean, red bean, climbing mountain bean, Mambi bean and oriental bean (Duke, 1981; Mejia, 1984). It has also been variously designated in Spanish as frijol arroz, frijol de arroz, judia arroz, kanime, poroto arroz (Mejia, 1984). In addition has many regional names, like tapilan, in the Philippines (Manguiat *et al.*, 1985), and, in some countries like Brazil, there is a somewhat generalized tendency to name as adzuki bean, plants belonging to both *Vigna umbellata* and *Vigna angularis*. The confusion between rice bean and adzuki bean is quite understandable as these two species are very similar. According to Sacks (1977) analyses performed by Nguyen-Thanh-Guing showed the two species to be closely related, the only difference between them being in the cell arrangement and seed coat. Part of the following general botanical description of the

Legumes and Oilseeds in Nutrition
Edited by E. Nwokolo and J. Smartt
Published in 1996 by Chapman & Hall. ISBN 0 412 45930 2

plant is an attempt to provide more precise information about the rice bean.

25.2 THE SEED AND PLANT

Rice beans are small, although variations exist in consequence of regions, climates and cultivars. Arora *et al.* (1980), in an examination of a collection of 272 entries from five regions of India, found values ranging from 2.8 to 23.0 g/100 seeds. Bulisani and Leitao-Filho (1975) found that seeds of cultivars grown in Brazil have an average length of 6 mm and an average width of 4 mm. In shape, the seeds have been described as varying from round to cylindrical (Bulisani and Leitao-Filho, 1975), or as Duke (1981) put it, from oblong to a strongly elongate, subtrapezoidal. Seed surface is smooth and seed colour is very variable. Arora *et al.* (1980) describe them as varying from uniform to speckled, the uniform colours are either black or creamy whereas among the speckled ones, the dark greenish are the more common.

The adult plant is herbaceous, commonly erect and of determinate growth habit. Arora *et al.* (1980) found that, among 300 collections from the tribal-dominated mountainous tracts of eastern and northeastern India, the majority are of the vine type, and only a few are of the semi-bushy type. Plant height is between 40 and 50 cm, although variation has been reported (Sastrapradja and Sutarno, 1977). According to a description by the National Academy of Sciences (NAS, 1979), the known cultivars in the world vary from short stemmed to twining ones that may grow up to 3 m long and need stakes for maximum yields. This report also notes that the whole plant life cycle is usually very short, being completed in 90 days but under equatorial conditions, as in Angola, a 60-day cycle has been reported. However, considering that rice bean is a short-day plant, its life cycle, under certain conditions can be very long, one example being the observation made by Vieira (1971) in Brazil where one of the cultivars had a cycle of 6 months. Mukherjee *et al.* (1980) consider this an advantage, since long vegetative phases are more convenient when the objective is forage production.

The rice bean plant seems to be one of the best adapted to thrive under tropical climatic conditions (NAS, 1979; Chatterjee and Mukherjee, 1979) which makes it appropriate to the low altitude tropics, even though it grows well at altitudes up to 1500 m. It has also been reported as ranging from moderately tolerant (Chatterjee and Mukherjee, 1979; NAS, 1979) to very tolerant of drought (Mukherjee *et al.*, 1980). On the other hand, the plant is highly susceptibile to water logging, mainly at the seedling stage. Duke (1981) also points to its high susceptibility to frosts. The plant seems to thrive well even under heavy soil conditions, but rather prefers light, sandy, well-aerated soil.

Probably due to its hirsute nature, the rice bean plant is very resistant to pests and diseases. A test conducted by the Asian Vegetable Research and Development Centre (AVRDC) showed that, of the pulses, the rice bean is least attacked by the bean fly. AVRDC has also found this species to be very resistant to powdery mildew, damping-off and leaf spot (NAS, 1979). Arora *et al.* (1980) examining 272 rice bean collections from all over India observed no sign of yellow mosaic and no *Cercospora* or bacterial blight. On the other hand susceptibility to root knot nematode has been reported (NAS, 1979).

25.3 PRODUCTION

Rice bean used to be a more intensively cultivated crop in the past, mainly in South and Southeast Asia, in a system in which it is rotated chiefly with rice. But modern techniques of multiple cropping of rice recently introduced in those areas have resulted in an increasing reduction of acreage dedicated to rice bean (NAS, 1979). More recently, some developments in the use and better knowledge of the plant's characteristics are resulting in a more promising future for the rice bean. Among those characteristics are the seed's chemical composition which makes it very appropriate for human consumption. It is rich in protein and contains high amounts of Ca, Fe, P and the vitamins thiamine, niacin and riboflavin. The nutritional quality of its protein is so good that it has been ranked as one of the best of all pulses and this has led it to being included in school children's nutritional programmes in the Philippines (NAS, 1979). It is also very suitable for rotation with corn which has recently led the government of Thailand to launch special programmes in which rice bean will be intercropped with corn (Rerkasem and Rekasem, 1988). In areas with acute shortage of land, it fits in well with forage production programmes due to its being highly productive, in addition to having a short life cycle, i.e. it is possible to get high yields of forage without hampering the production of grain cereals (Das and Chatterjee, 1977). The management of forage production, according to data by Mukherjee and Chatterjee (1979a, b) should avoid direct grazing on the rice bean plants as it does not tolerate frequent defoliation. This is particularly important to observe, when cutting rice bean plants for silage preparation. In this case, the ideal procedure is to cut the plants at a height of 25 cm above the soil and make a maximum of two cuttings per year.

25.4 HARVESTING AND USE

Bulisani (1990) recommends that harvesting is carried out when the plants have dropped their leaves and the pods are almost dry. This should take

Table 25.1 Chemical composition rice bean seed and forage

Constituent	Seed (per 100g dry matter	% of forage at various stages	
		Vegetative	Flowering
Moisture	13.3%	–	–
Dry matter	–	16.0	24.0
Calories	327.0	–	–
Protein	20.9 g	–	–
Crude protien	–	18.0	14.5
Fat	0.9 g	1.1	1.0
Total carbohydrate	60.7 g	–	–
Crude fibre	4.8 g	31.5	32.1
Ash	4.2 g	9.5	10.8
Nitrogen free extract	–	39.9	41.6
Calcium	200.0 mg	1.4	1.25
Phosphorus	390.0 mg	0.35	0.4
Iron	10.9 mg	–	–

From Duke (1981).

place about 90 days after sowing. The leaves, pods and seeds are the plant parts used for food. The part traditionally consumed is the seed. Usually the dry seeds are boiled and used in soups, or are consumed following sprouting, the young sprouts being used to prepare salads. Young pods and green leaves are also used to prepare salads. In an indirect manner, rice bean serves the purpose of food production indirectly when it is used as a cover crop and a green manure. An additional use for the rice bean has been developed in Indonesia where it is grown in gardens and sometimes in households thus providing privacy in addition to food. The entire plant can also be used as forage for livestock, especially pigs (Duke, 1981). Its palatability seems to be very good, and according to the National Academy of Sciences (1979) 'rice bean plants make a forage that is eagerly consumed by livestock'.

25.5 NUTRIENT COMPOSITION AND NUTRITIONAL VALUE OF THE PLANT AND SEED

Analyses of chemical composition of rice bean seed and plant parts are fairly well detailed (Duke, 1981) (Table 25.1). A 100 g sample of dried seeds contain 0.49 mg of thiamine, 0.21 mg of riboflavin and 2.4 mg of niacin. The composition of amino acids (mg/g N) is as follows: arginine 462, histidine 380, leucine 606, isoleucine 387, lysine 769, methionine 769, cystine 44, phenylalanine 325, tyrosine 262, threonine 294, and valine 394. The fatty acid composition of the lipids (%) is as follows: myristic acid 6.3–7.3, palmitic acid 5.6–6.0, stearic acid 2.1–4.4, behenic acid 4.6–5.8,

Table 25.2 Crude chemical composition of food and forage species: comparative values of some legume seeds

Species	*Calories (kcal)*	*Protein (g)*	*Fat (g)*	*Total CHO (g)*	*Fibre (g)*	*Ash (g)*
Arachis hypogaea (peanut)	587	24.8	47.9	24.6	3.1	2.7
Glycine max (soybean)	444	39.0	19.6	35.5	4.7	5.5
Phaseolus vulgaris (common bean)	–	22.3	1.5	61.2	4.4	3.8
Psophocarpus tetragonolobus (winged bean)	450	36.4	18.8	40.5	–	4.4
Vigna umbellata (rice bean)	390	21.5	1.2	71.6	3.4	2.3

Data in National Academy of Sciences (1979) compiled by J.A. Duke, of the US Department of Agriculture.

arachidic acid 3.0–3.9, lignoceric acid 2.9–3.2, linoleic acid 7.5–12 and oleic acid, 61.3–68.0. Seeds contain B-sitosterol. Duke also observed that rice bean is free of cyanogenic glycosides but contains trypsin and chymotrypsin inhibitors.

Analyses performed by other authors also indicates that rice bean is particularly rich in methionine and tryptophan as well as in the vitamins thiamine, niacin and riboflavin (Chatterjee and Mukherjee, 1979). Rice bean seeds are also poor in phytin content, a positive factor in view of the fact that phytic acid inhibits mineral availability and protein digestibility. Other data (NAS, 1979) comparing the chemical composition of rice bean with that of some other legumes show that rice bean has a good nutritive value either as food or feed (Table 25.2).

25.6 ANTINUTRITIONAL SUBSTANCES IN THE SEED OR VEGETATIVE PARTS

Although known to contain trypsin or chymotrypsin inhibitors (Duke, 1981), rice bean is considered to be relatively free of antinutritional substances. Del Rosario *et al.* (1980) classify rice bean as a legume with one of the lowest levels of antinutritional compounds. Among the legumes analysed (winged bean, lima bean, cowpea, chick pea, rice bean and mung bean) only mung bean had a lower trypsin inhibitor content than rice bean. Nowacki (1980) emphasizes that of the six main chemical groups which are considered toxic or antinutritional compounds and which interfere with protein digestibility (alkaloids, toxic amino acids, cyanogenic glucosides, saponins, flavones and isoflavones, and the pyrimidine glucosides) none were reported as occurring in significant levels in rice bean. Revilleza *et al.* (1990) also reveal an interesting facet of rice bean nutritional value. The authors investigated the occurrence of oligosaccharides in the seeds of seven indigenous food legumes in the Philippines. The oligosaccharides of the raffinose family were shown to be responsible for flatulence manifested after consumption of the seeds of many species of the legume family. The species with the least content of oligosaccharides was the jack bean, followed by the rice bean. Considering that the gases caused by bean seed consumption can be very distressing, rice bean nutritional characteristics become of higher value when data presented by these authors are examined. They also investigated some procedures to detoxify the seeds and chose the hyacinth bean for this study due to its high content of oligosaccharides in addition to its high acceptance by the consumers. The procedures examined consisted of soaking, dry roasting, boiling and germinating the seeds. These procedures resulted in a total (dry roasting) to almost total (seed germination) elimination of the oligosaccharides. Germination was capable of eliminating raffinose, stachyose and verbascose but not sucrose. Boiling, which is the most common procedure mainly in developing countries, in some cases, raised the level of oligosaccharides.

25.7 HUMAN STUDIES WITH RICE BEAN SEED

The National Academy of Sciences (1979) points to the fact that although rice bean can be considered as one of the best pulses in terms of nutritional value, its acceptance in certain areas of the world will be difficult due to its taste being 'weak' compared to what people in those areas are used to. This becomes apparent when rice bean is compared with the common bean (*Phaseolus vulgaris*) in Latin America. When compared to the taste of boiled common bean, the taste displayed by rice bean is considerably 'weaker'. An experiment was conducted in Brazil (Vieira *et al.*, 1989) in which rice bean was compared with common bean, cowpea, and lima bean, a total of 31 people tasted a soup prepared by boiling the seeds. The results showed that rice bean was placed last due to this so-called 'weak' taste. In areas of the world where different cultural practices prevail, results might be more favourable to a rice bean. On the other hand, Vieira *et al.* (1989) showed that rice bean is easy to cook. It took 63 min to cook 100 g of rice bean seeds whereas the same amount of common bean seeds required an average of 110 min and lima beans required 143 min. Mashing the seed in a liquefier improved the taste of the rice bean soup but even so it had the lowest grade.

REFERENCES

Arora, R.K., Chandel, K.P.S., Joshi, B.V. and Pant, K.C. (1980) Rice bean: tribal pulse of Eastern India. *Econ. Bot.*, **34**, 260–3.

Bulisani, E.A. (1990) Feijao-adzuki (*Vigna angularis* (Wild.) Ohwi & Ohashi, *Vigna umbellata* (Wild.) Ohwi & Ohashi and *Vigna umbellata* (Thumb.) Ohwi & Ohashi), in *Instituto Agronomico de Campinas*, ed. Instrucoes Agricolas para o Estado de Sao Paulo. Boletim 200, p. 96.

Bulisani, E.A., and Leitao, H.F. (1975) Contribuicao ao conhecimento botanico do feijao- adzuki e especies afins. Instituto Agronomico de Campinas, Circular no. 47, 17 pp.

Chatterjee, B.N. and Mukherjee, A.K. (1975) A new rice bean for fodder and grain pulse. Reprinted from August 1975 issue of *Indian Farming*.

Das, P.K. and Chatterjee, B.N. (1977) Competition in grass–legume mixed forage. *Indian J. Agric. Sci.*, **47**, 47–51.

Del Rosario, R.R., Lozano, Y., Pamorasamit, S. and Noel, M.G. (1980) The trypsin inhibitors activity of legume seeds. *Phil. Agr.*, **63**, 339–44.

Duke, J.A. (1981) *Handbook of Legumes of World Economic Importance*, Plenum Press, New York, 345 pp.

Manguiat, I.J., Mendoza, D.M. and Tilo, S.N. (1985) Influence of the legume–rhizobia symbiosis on the nitrogen economy of a legume based cropping system. *Philippine Agric.*, **68**, 44–55.

Mejia, M.M. (1984) Scientific and common names of tropical forage species. Cali, Colombia, CIAT, 75 pp.

Mukherjee, A.K. and Chatterjee, B.N. (1979a) Herbage growth analysis of rice bean in tropical India. I. Uninterrupted growth condition. *Forage Res.*, **5**, 117–23.

Mukherjee, A.K. and Chatterjee, B.N. (1979b) Herbage growth analysis of rice bean in tropical India. II. Under interrupted growth condition. *Forage Res.*, **5**, 125–33.

Mukherjee, A.K., Roquib, M.A. and Chatterjee, B.N. (1980) Rice-bean 'k-1' for the scarcity period. Reprinted from March 1980 issue of *Indian Farming*.

NAS (National Academy of Sciences) (1979) *Tropical Legumes. Resources for the Future*. NAS, Washington DC, 331 pp.

Nowacki, E. (1980) Heat-stable antinutritional factors in leguminous plants, in *Advances in Legume Science*, (eds R.J. Summerfield and A.H. Bunting), Royal Botanical Garden, Kew, pp. 171–7.

Rerkasem, K. and Rekasem, B. (1988) Yields and nitrogen nutrition of intercropped maize and rice bean (*Vigna umbellata* (Thumb.) Ohwi and Ohashi). *Plant Soil*, **108**, 151–62.

Revilleza, M.J.R., Mendoza, E.M.T. and Raymundo, L.C. (1990) Oligosaccharides in several Philippine indigenous food legumes: determination, localization and removal. *Plant Foods Hum. Nutr.*, **40**, 83–93.

Sacks, F.M. (1977) A literature review of *Phaseolus angularis*: the adzuki bean. *Econ. Bot.*, **31**, 9–15.

Sastrapradja, S. and Sutarno, H. (1977) *Vigna umbellata* (L.) DC in Indonesia. *Ann. Bogoriensis*, **6**, 155–67.

Vieira, C. (1971) Nota sobre o comportamento de *Phaseolus calcaratus* Roxb. em Vicosa, Minas Gerais. *Rev. Ceres*, **XVII**, 303–7.

Vieira, R.F., Romeiro, E.M.C., Souza, L.R.P. *et al.* (1989) Tempo de cacao, rendimento alimentar e aceitabilidade de feijoes secos dos generos *Vigna* e *Phaseolus*. *Rev. Ceres*, **36**, 525–33.

CHAPTER TWENTY-SIX

Cowpea (*Vigna unguiculata* (L.) Walp.)

E. Nwokolo and S.N. Ilechukwu

26.1 INTRODUCTION

The cowpea is an old world crop which is believed to have originated in Africa (Rachie and Roberts, 1974; Ng and Maréchal, 1985; Steele *et al.*, 1985). A survey of the germplasm accessions from Nigeria, Niger, Burkina Faso and Ghana showed greater diversity than accessions from other areas in Africa (Ng and Maréchal, 1985), leading to the conclusion that West Africa was the primary centre of domestication. A review of the earlier literature on the cowpea was presented by Rachie and Roberts (1974), whereas a book by staff of the International Institute for Tropical Agriculture, Ibadan, Nigeria, on cowpea research, production and utilization (Singh and Rachie, 1985) is a compendium of later information on the cowpea.

Cowpea is a crop which grows very well in hot-to-moderate temperatures and is, therefore, ideally suited to tropical lowlands, doing well both in hot dry and hot humid ecosystems. The cowpea grows well in soils which range from acidic to neutral, and, therefore, performs satisfactorily even on soils with low fertility. The plant withstands reasonable periods of drought and can be successfully cultivated in the northern savanna countries bordering the Sahara desert in Africa. It also withstands considerable shade and, therefore, does well in high rainfall and high forest areas of the tropical rain forests, especially if the soil is well drained. Additionally, growth, flowering and maturity of the cowpea are not susceptible to daylength variation. All these characteristics make the cowpea an ideal tropical crop.

Legumes and Oilseeds in Nutrition
Edited by E. Nwokolo and J. Smartt
Published in 1996 by Chapman & Hall. ISBN 0 412 45930 2

26.2 PRODUCTION

Cowpeas are cultivated throughout the tropics, more in some areas than in others. Cultivation in Africa is extensive. Current production data for most countries have been very difficult to obtain because the FAO has ceased to publish cowpea production data since the late 1970s. However, from FAO production data for total pulses, it has been possible to subtract production data for dry beans (*Phaseolus vulgaris*), broad beans, peas, chickpea and lentils, in order to obtain an estimate for cowpea production. Total cowpea production in Africa for 1990 is estimated at 3.36 million MT, of which about 1.5 million MT is produced in Nigeria. Niger produced 368 000 MT, Kenya 235 000 MT, Madagascar 177 000 MT, and Cameroun 135 000 MT. Other producers in Africa include Tanzania with 127 000 MT, Ethiopia with 122 000 MT, Morocco with 84 000 MT, Uganda with 68 000 MT and Egypt with 37 000 MT. Nigeria is by a wide margin the world's largest producer of cowpeas, although India is also a major producer of cowpeas along with many other pulses. Cultivation in Asia, especially in India and Sri Lanka, is high, primarily of cowpea and asparagus bean (*Vigna unguiculata sesquipedalis*) and to a lesser extent the catjang (*Vigna unguiculata cylindrica*). It is a minor crop in the southern United States and in California where most of the cowpea produced in the United States is cultivated. Land area devoted to cowpea in Africa is large but yield per hectare is still very poor, reflecting agricultural practices which have not changed in hundreds of years, as well as resistance to or inability to use fertilizers, pesticides and other newer technologies.

26.3 CURRENT AND POTENTIAL USAGE AS FOOD

The cowpea may be one of the world's minor legume crops but it is a crop which has been identified as possessing tremendous potential for increased contribution to human nutrition in the tropics (PAG, 1975). It is consumed extensively throughout Africa, and to a lesser extent in Asia, Latin America and the southern United States in a wide variety of dishes. In West Africa, where it is the single most important legume and contributes very significantly to daily protein and vitamin intake, cowpeas may be cooked alone to be eaten as a companion dish or may be cooked as a porridge in combination with rice, corn, yams or cocoyams, and flavoured with salt, pepper and spices. Cooked cowpeas are important as a companion dish eaten with garri (grated fried cassava). In this case the garri is soaked in water and sweetened with sugar prior to eating. Cooked cowpeas with fried plantains are a favourite dish for breakfast, lunch and dinner and are regular fare at upmarket restaurants as well as roadside kitchens because of the ease of preparation. In these and other forms, cowpea dishes are an integral component of school lunch in many residential primary, secondary and tertiary institutions of higher learning. Cooked cowpeas also feature prominently in the menus of hospitals, prisons, and other institutions where food is prepared on a large scale. Phillips and McWatters (1991) report on a 10-year collaborative project between the University of Georgia and the University of Nigeria, Nsukka, on aspects of cowpea research, processing and utilization. An initial baseline study of the nutritional status of urban and rural poor had indicated that stunting occurred in 50% of the child population and wasting in 70% of the adult population in Eastern Nigeria (Nnanyelugo, 1982). There was also considerable deficiency of proteins, energy, B vitamins and calcium in children and pregnant women (King *et al.*, 1985). A study by Ndubuaku *et al.*, (1989) had found that the average weekly consumption of cowpeas per family was 2 kg. During periods of food scarcity, cowpeas contributed 31% of the protein, 24% of the iron, 35% of the thiamin and 21% of the daily intake of niacin (Nnanyelugo *et al.*, 1985).

Two very flavourful dishes prepared from cowpeas are akara and moin-moin. Akara balls are prepared by deep-frying seasoned cowpea paste whereas moin-moin is prepared in a manner similar to akara, except that steam cooking replaces deep frying. In the traditional procedure, cowpeas for akara and moin-moin preparation are soaked in water for a few hours or overnight to facilitate removal of the skin and the black eye of the cowpea. The cowpea cotyledons are milled with bell-pepper, salt, onions and other seasonings. For akara preparation, the paste is scooped with a medium-sized spoon into boiling oil for deep frying. Frying takes 2–3 min depending on the size of the ball and the heat of the oil, the akara being considered cooked when it changes to a golden brown colour. Akara is served hot and makes an excellent breakfast with corn gruel (ogi, pap). For moin-moin preparation, the seasoned paste is steam cooked, rather than deep fried. A little oil is added to the paste and small portions are wrapped in aluminium foil or wilted plantain leaves or scooped into oiled metal/plastic cups. Whatever containers the paste is put in, they are loaded into a large steaming pot to cook, cooking lasting for about 1 h. The moin-moin is eaten out of the wrapper or container. Moin-moin and akara are used as components of meals but are increasingly used as snacks and hors d'oeuvre in well-off families. In recent years, there has been considerable interest in the commercial manufacture of both akara and moin-moin for the US and European markets. Many experiments conducted by scientists at the University of Georgia have helped to expand considerably our understanding of the physical, functional and organoleptic properties of milled cowpeas and whipped cowpea paste (McWatters, 1983; McWatters and Chinnan, 1985; McWatters *et al.*, 1988; Hung *et al.*, 1987). They studied not only the effect of various processing variables on physical and functional characteristics of cowpea paste, but the effect of frozen storage on sensory and physical attributes of akara. They are currently investigating akara as a fully cooked, frozen and reheatable product with a possible aim of introducing it to the US market (Phillips and McWatters, 1991).

Moin-moin also has an excellent potential as a fully-cooked, frozen and reheatable product. Indeed, in families where parents work outside the home in Nigeria, it is common practice to prepare moin-moin in small containers and keep them frozen for subsequent thawing and reheating in conventional ovens. Moin-moin for use at large wedding parties and other festivities is usually prepared several days in advance, frozen, and thawed just prior to the festivities.

A predominant proportion of the US crop of cowpeas is produced in the southern states, hence the name southern peas. The history of cowpea production in the southern states indicates that cowpeas were introduced together with slavery, from Africa (Steele *et al.*, 1985). Cowpeas have always featured prominently in the food of African–Americans and is consumed in a large variety of dishes, part of the famous southern cuisine and definitely an integral component of the famed soul-food. In the United States, cowpeas are produced primarily for the seeds which are harvested mature and sold as dried southern peas or black-eye peas. Cowpeas may be cooked and canned or cooked and packed in small plastic pouches for subsequent reheating and use.

26.4 BOTANY OF THE CROP

Cowpea is an annual herbaceous plant, erect, suberect or trailing (but not twining) in habit and usually about 90 cm tall. The stem has ovate to lanceolate leaflets 7–12 cm long. The flowers are borne at the end of long stalks, in inflorescences two to twelve flowers, and are of various colours usually white, yellow or purple. The pods may be curved or straight and elongated, usually up to 30 cm long. Seed size varies from very small (0.1 g) to large (0.3 g) to very large (0.5 g) and, depending on the size, there may be up to eight seeds in a pod. Seeds may be wrinkled or smooth, globular to reniform in shape, and of various colours or combination of colours. Major colours are brown, tan, white, black or red, with varying degrees of mottling. Cowpea seeds are typified by having a black or dark 'eye' around the hilum. Other varieties of cowpea which are increasingly seen in West Africa and other cowpea-producing areas are the asparagus bean (*Vigna unguiculata* var *sesquipedalis*) and the catjang (*Vigna unguiculata* var *sinensis*). Both varieties are believed to be native to Asia but have spread either by deliberate importation or accidental contamination of planting material. The asparagus bean, also called the yardlong bean, is a vigorous-growing annual or occasional

Table 26.1 Proximate composition of cowpeas

	Content (per 100 g fresh weight)	
Constituents	*Raw*	*Cooked*
Water (g)	11.95	70.04
Food energy (kcal)	336.00	116.00
Protein (g)	23.52	7.73
Total lipids (g)	1.26	0.53
Total carbohydrate (g)	60.03	20.77
Crude fibre (g)	4.58	2.31
Ash (g)	3.24	0.94

Source: USDA (1986).

perennial. It may be erect or climbing, and twining, with typical trifoliolate leaves. The pods are long to very long (30–100 cm) and usually hang down from stakes or tree branches or may simply lie on the ground. Seeds are large and widely separated in the pod. The catjang is more similar to the cowpea than the asparagus bean, bearing even shorter pods than the cowpea. Seeds are smaller too, usually less than 0.5 cm long.

Cowpeas may be planted for seed, forage or silage. For seed purposes, early planting is important because whereas early maturing cultivars like New Era may produce pods in 50 days and seed in 90 days, late maturing cultivars produce seeds in 210–240 days. Because the cowpea is indeterminate, pod production and ripening are irregular and it is common to see flowers, green pods and mature dry pods on the same vine. A practical procedure for small-scale harvesting is to harvest for edible tender pods every 2–3 days, mature green pods every 3–4 days and dry pods weekly. The usual procedure for commercial seed production is to harvest when about two-thirds of the pods are dry. According to Duke (1981) seed yield is very low in Africa (199 kg/ha) in comparison with Asia (629 kg/ha) and North America (715 kg/ha).

26.5 NUTRITIONAL QUALITY

26.5.1 Proximate constituents

The content of major nutrients in cowpea is shown in Table 26.1. In raw cowpeas, protein content ranges from 22 to 25% with a mean of about 23.5%. Food energy content is high due to a high carbohydrate content, although cowpea is a very poor source of dietary lipids. Crude fibre and dietary fibre content is moderately high, one of the great health advantages of consumption of this and other pulse legumes.

Table 26.2 Amino acid content of cowpea

Amino acid	Content (g/100 g fresh weight)	
	Raw	*Cooked*
Tryptophan	0.29	0.10
Threonine	0.90	0.29
Isoleucine	0.96	0.31
Leucine	1.80	0.59
Lysine	1.59	0.52
Methionine	0.34	0.11
Cystine	0.26	0.06
Phenylalanine	1.37	0.45
Tyrosine	0.76	0.25
Valine	1.12	0.37
Arginine	1.63	0.54
Histidine	0.73	0.24
Alanine	1.07	0.35
Aspartic acid	2.84	0.93
Glutamic acid	4.45	1.46
Glycine	0.97	0.32
Proline	1.06	0.35
Serine	1.18	0.39

Source: USDA (1986).

26.5.2 Protein and amino acids

The amino acid content of cowpeas is shown in Table 26.2. The essential amino acid content of cowpea has been the subject of several investigations (Jaffé, 1949; Orr and Watt, 1957; Bressani *et al.*, 1961), all the results indicating that the sulphur amino acids are the first limiting amino acids in cowpea protein. Other investigations highlighted the high lysine content of cowpea despite its poor sulphur amino acid content (Jaffé, 1949; Borchers and Ackerson, 1950; Chavez *et al.*, 1952; Sherwood *et al.*, 1954). Elias *et al.* (1964) found that leguminous seeds such as cowpea can supply the limiting amino acids in corn protein and be excellent supplements for cereal-based diets. When legumes provide more than 50% of the dietary protein in a predominantly vegetarian diet, sulphur amino acids are still the most deficient amino acids (Norton *et al.*, 1985). However, if cereals predominate, sulphur amino acids are no longer deficient since cereals contribute quite significantly to total sulphur amino acids, while making up their deficiency in lysine from its abundance in legumes. Norton *et al.* (1985) note that the sulphur amino acid limitation in most legumes seeds is further exacerbated by an intrinsic poor digestibility of protein. Thus they argue that although increased digestibility without supplemental sulphur amino acids does not

Table 26.3 Vitamin content of cowpea

	Content (per 100 g fresh weight)	
Vitamin	*Raw*	*Cooked*
Ascorbic acid (mg)	1.5	0.40
Thiamine (mg)	0.85	0.20
Riboflavin (mg)	0.23	0.06
Niacin (mg)	2.08	0.50
Pantothenic acid (mg)	1.50	0.41
Pyridoxine (mg)	0.36	0.10
Folacin (μg)	632.6	207.9
Vitamin A (IU)	50.0	15.0

Source: USDA (1986).

Table 26.4 Mineral content of cowpea

	Content (mg/100 g fresh weight)	
Mineral	*Raw*	*Cooked*
Calcium	110	24
Iron	8.27	2.51
Magnesium	184	53
Phosphorus	424	156
Potassium	1112	278
Sodium	16	4
Zinc	3.37	1.29
Copper	0.85	0.27
Manganese	1.53	0.48

Source: USDA (1986).

necessarily improve efficiency of protein utilization, an increase in sulphur amino acid content without any enhanced protein digestibility would improve efficiency of protein utilization.

26.5.3 Vitamins and minerals

The contents of various vitamins are shown in Table 26.3. In an investigation of the vitamins in Nigerian cowpeas, Ogunmodede and Oyenuga (1968, 1969, 1970) showed that cowpeas contain appreciable quantities of B vitamins and, given the high level of consumption of cowpeas in Nigeria, that this legume could account for a significant proportion of daily vitamin intake. However, cooking also destroys a significant proportion of thiamin (62%) and niacin (45%) (Elias *et al.*, 1964). This high loss of vitamins in cooked cowpeas has been confirmed by

Edijala (1980a,b) who estimated thiamine loss as 51% after cooking of cowpeas in tap water, noting that addition of sodium bicarbonate further enhanced vitamin destruction.

The content of minerals is cowpea is shown in Table 26.4. Fairly high contents of magnesium, sodium, calcium and iron were observed in newly developed cowpea cultivars obtained from the International Institute for Tropical Agriculture in Ibadan, Nigeria (Fashakin and Ojo, 1988b).

26.5.4 Oligosaccharides and flatulence

Although all the causative factors in flatulence formation are still not known, it has been suggested and generally accepted that low-molecular-weight oligosaccharides, primarily raffinose, stachyose and verbascose present in most legume seeds, are linked to flatulence (Akpapunam and Markakis, 1979). There is considerable variation in the oligosaccharide content of 13 cultivars of cowpeas cultivated in the United States. Flatulence is characterized by stomach cramps, nausea, diarrhoea, and intestinal and gastric discomfort resulting from the production of large amounts of gas in the gastrointestinal tract. Many researchers have suggested practical procedures for the reduction of flatulence in cooked and processed cowpeas. These include removal of seed coat prior to cooking or processing (Ndubuaku *et al.*, 1989), soaking in water and germination (Nnanna and Phillips, 1990), germination (Ologhobo and Fetuga, 1986) and cooking with trona, a mixture of sodium carbonate and bicarbonate (Ndubuaku *et al.*, 1989).

26.6 ANTINUTRITIONAL FACTORS

26.6.1 Trypsin inhibitors

Cowpeas, like other legumes contain trypsin inhibitors. Like other trypsin inhibitors they have been shown to be heat-labile and can be destroyed by heating for 60 min at 100°C. However, destroying the trypsin inhibitor in cowpea does not automatically improve the digestibility of the legume (Jaffé, 1950). Protein digestibility in beans is not primarily dependent on the presence of trypsin inhibitors, as many other factors including amino acid imbalance and refractory proteins are implicated (Norton *et al.*, 1985). The content of trypsin inhibitors in raw cowpea has been shown to be in the range 19.6–28.2 (average 24) TIU/mg protein (Ologhobo and Fetuga, 1984). The content of trypsin inhibitors in cowpea is much lower than in soybean, lima bean and faba bean. In processed cowpea, trypsin inhibitor level has been shown to decrease by up to 80–90%, depending on the severity of the processing. Heat treatment almost completely destroys trypsin inhibitor activity in cowpea.

26.6.2 Lectins (haemagglutinins)

The presence of lectins has also been demonstrated in raw cowpea. Ologhobo and Fetuga (1984) reported haemagglutinating activity in ten cultivars of cowpea in Nigeria, values which ranged from a low of 33.5 HU/mg protein in the cv. Kano 1696, to a high of 98.9 HU/mg protein in the cv. Prima. Average haemagglutinating activity was 61.4 HU/mg protein. Based on the content of lectins in these cultivars, Nigerian cowpeas are classified into low (33.5–49.6 HU/mg protein) and medium (80.4–98.9 HU/mg protein) haemagglutinating cultivars.

26.6.3 Polyphenols

Kachere *et al.* (1988) showed that seed colour was associated with content of polyphenols. Cowpea cultivars with coloured seed coats contain larger amounts of polyphenols than those with white or cream-coloured seeds. This observation was also made by Ologhobo and Fetuga (1984) who also noted that consumers preferred those with light-coloured seed coats and less polyphenol content. Moist heat application has been shown to be effective in removal of polyphenols in dark-red cowpea cultivars. Laurena *et al.* (1987) showed that simple boiling reduced polyphenol content by 61–80%, pressurized cooking reduced it by 48–83%, and application of dry heat (roasting or microwave heating) inactivated 58–71% of the tannins. Autoclaving of cowpea also causes a loss of tannins from the bean (Ekpenyong, 1985).

26.7 PROCESSING

The main aim of the collaborative project by the University of Georgia and University of Nigeria was in part, to develop suitable small-scale procedures and machinery for processing cowpea, thus facilitating usage of cowpea and processed cowpea products in Nigeria. The research that accompanied the collaborative project has been the most concentrated effort on understanding the physical, functional and sensory attributes of cowpea. In attempting to reduce the laborious stages involved in cowpea processing, McWatters and Chinnan (1985) developed a process for the production of ready-to-use cowpea paste which would simplify the in-home and commercial preparation of akara and moin-moin. According to Chinnan *et al.* (1989), a village-scale mechanical process for cowpea meal production involves the hydration of dry seeds to a moisture content of 25%, equilibration for 30 min, followed by drying in a forced draught oven to a moisture content of 10%. The seed coats are removed by rubbing against a dry abrasive surface followed by a water wash. The skins float to the surface and are aspirated. The cleaned decorticated seeds are washed thoroughly and ground to pass through a 1.0 mm screen.

A cowpea powder with no beany flavour adequate for use in bakery products had earlier been developed by Okaka and Potter (1977, 1979a). Other studies also showed that production of good quality akara depended on obtaining appropriate particle size distribution in the milled product (McWatters, 1983)

26.7.1 Weaning diets from cowpeas

Various attempts have been made at formulation of protein-enriched weaning foods containing cowpeas. Most have been tried on experimental rats, or with human test panels and pronounced acceptable. These weaning foods include cooked corn–cowpea–crayfish mixtures (Abbey and Nkanga, 1988; Abbey *et al.*, 1988), malted and roller-dried sorghum and cowpeas, and germinated rice, corn and cowpea (Malleshi *et al.*, 1989). Other foods include mixtures of corn, sorghum and cowpeas (Akinyele and Fasaye, 1988), a mixture of skim milk with processed sorghum and cowpeas (Oyeleke *et al.*, 1985) and a mixture of rice, cowpeas and skim milk powder (Roman *et al.*, 1987). In all these diets and trials, the amino acid complementarity between cereals and legumes is exploited to produce weaning foods with enhanced nutritive value, a nutritional quality which is superior to that of corn, sorghum, rice or cowpea individually.

26.8 NUTRITIONAL QUALITY DETERMINATIONS

The nutritional quality of cowpea and various cowpea blends has been estimated in numerous experiments over the last 40 years. Early experiments by research scientists at the Institute of Nutrition of Central America and Panama (INCAP) established the excellent nutritional quality of cowpea in spite of its deficiency of the sulphur amino acids and tryptophan.

Because cowpea or cowpea products are consumed by most Nigerian families once or twice a week (Ndubuaku *et al.*, 1989) and contribute 30–36% of the protein intake and 17% of the energy intake of children (Akinyele and Akinlosotu, 1987), cowpeas have justifiably attracted the attention of nutritional researchers in Nigeria and elsewhere. Reports abound in the scientific literature of nutritional evaluations of cowpeas and cowpea products. Parameters evaluated have included the influence of germination on nutrient density and digestibility, as well as direct protein quality estimates of cowpeas and cowpea blends. Nnanna and Phillips (1989) showed that germination had little impact on amino acid profile and protein digestibility whereas it significantly increased the content of niacin, thiamin and riboflavin. Germination was also reported by Abbey and Mark-Balm (1988) to eliminate completely trypsin inhibitors in cowpea, although protein efficiency ratio and net protein utilization were not effected. There is evidence that germination improves the nutritional quality of weaning foods partly through improved digestibility of carbohydrates. Marero *et al.* (1990) demonstrated that in germination of rice and cowpea, amylolytic activity produced during germination was retained in the flour despite drying. These amylases induced a partial digestion of carbohydrates into more easily utilizable malto-oligosaccharides. Akinlosotu and Akinyele (1991) observed a gradual decrease in oligosaccharides and an increase in monosaccharide content with germination. They also reported an increase in energy and protein content and in content of ascorbic acid and niacin. Content of calcium, iron and thiamine decreased with germination. Two growth studies with rats (Obizoba, 1989, 1990) demonstrated that germination and cooking were very effective processing procedures to utilize the nutrients in cowpeas and other grain legumes.

That cowpea flour and protein isolates can be successfully utilized in bakery products has been clearly demonstrated (Okaka and Potter 1977, 1979a, b; Mustafa *et al.*, 1986). In all these experiments, it seems that cowpea flours may be used at up to 10% of the composite flour without adverse effect on sensory and physical characteristics of the bakery products.

In nutritional evaluation of various cowpea–cereal blends, a 70:30 blend of germinated rice/mung bean, and a 70:30 blend of germinated corn/cowpea was shown to meet the FAO reference pattern for all amino acids except the sulphur amino acids (Marero *et al.*, 1990). Similarly, Akinyele and Fasaye (1988) studied various cereal–cowpea mixtures and noted that 30% cowpeas could be incorporated into corn for ogi preparation without any depreciation in acceptability, while significantly enhancing the protein content of the final products. In a comparison of cowpea–rice or cowpea–yam diets, Uwaegbute and Nnanyelugo (1987) found the diets to be the same in respect of protein efficiency ratio, net protein retention and biological value. Both diets were, however, inferior to a control casein diet. Williams (1985) estimated with rats, the digestible and metabolizable energy of raw cowpea meal, autoclaved cowpea paste or deep-fried akara. Metabolizable energy was 2.29–2.42 and 3.38–3.52 kcal/g for raw cowpea meal and autoclaved cowpea paste, respectively. Much higher metabolizable energy values were obtained for fried akara, in the range 5.06–5.70 kcal/g.

REFERENCES

Abbey B.W. and Mark–Balm, T. (1988) Nutritional quality of weaning food prepared from composite flours of maize, ungerminated and germinated cowpea. *Nutr. Rep. Int.*, **38**, 519–26.

Abbey, B.W. and Nkanga, U.B. (1988) Production of high quality weaning products from maize–cowpea–crayfish mixtures. *Nutr. Rep. Int.*, **37**, 951–7.

Abbey, B.W., Abbey, S.D. and Ikpe, I.P. (1988) Physiochemical, organoleptic and storage properties of maize–cowpea–crayfish mixture. *Nutr. Rep. Int.*, **38**, 1017–27.

Akinlosotu, A. and Akinyele, I.O. (1991) The effect of germination on the oligosaccharide and nutrient content of cowpeas (*Vigna unguiculata*). *Food Chem.*, **36**, 157–65.

Akinyele, I.O. and Akinlosotu, A. (1987) Contribution of cowpea (*Vigna unguiculata*) in a mixed diet to the nutrient intake of rural children in Ibadan. *Br. J. Nutr.*, **58**, 31–9.

Akinyele, I.O. and Fasaye, O.A. (1988) Nutrient quality of corn and sorghum supplementated with cowpeas (*Vigna unguiculata*) in the traditional manufacture of Ogi. *J. Food Sci.*, **53**, 1750–2, 1755.

Akpapunam, M.A. and Markakis, P. (1979) Oligosaccharides of 13 cultivars of cowpea (*Vigna sinensis*). *J. Food Sci.*, **44**, 1317–18.

Borchers, R. and Ackerson, C.W. (1950) The nutritive value of legume seeds. X. Effect of autoclaving and the trypsin inhibitor test for 17 species *J. Nutr.*, **41**, 339.

Bressani, R., Elias, L.G. and Navarrete, D.A. (1961) Nutritive value of Central American beans. The essential amino acid content of samples of black beans, red beans, rice beans, and cowpeas of Guatamala. *J. Food Sci.*, **26**, 525–30.

Chavez, N., Teodosio, N.R., Gomez de Matos, A. *et al.* (1952) The proteins of cowpea (*Vigna sinensis*) in nutrition. *Rev. Brasil. Med.*, **9**, 603–10.

Chinnan, M.S., McWatters, K.H., Ngoddy, P.O. and Nnanyelugo, D.O. (1989) Improving cowpea processing technology for enhancing cowpea utilization, in *Trends in Food Processing* II. Proc. of 7th World Congress in Food Science and Technology, Singapore Oct. 1987.

Duke, J.A. (1981) *Handbook of Legumes of World Economic Importance*, Plenum Press, New York.

Edijala, J.K. (1980a) Effects of processing on the thiamine, riboflavin and protein content of cowpeas; soaking, cooking and wet milling treatments. *J. Food Technol.*, **15**, 435–43.

Edijala, J.K. (1980b) Effects of processing on the thiamin, riboflavin and protein contents of cowpeas (*Vigna unguiculata* L. Walp.) II. Alkali (potash) treatment. *J. Food Technol.*, **15**, 445–53.

Ekpenyong, T.E. (1985) Effect of cooking on polyphenolic content of some Nigerian legumes and cereals. *Nutr. Rep. Int.*, **31**, 561–5.

Elias, L.G., Colindres, R. and Bressani, R. (1964) The nutritive value of eight varieties of cowpea (*Vigna sinensis*). *J. Food Sci.*, **29**, 118–22.

Fashakin, J.B. and Ojo, F.A. (1988a) Chemical composition and nutritive changes of some improved varieties of cowpea (*Vigna unguiculata*). 1. Some selected varieties from the IITA. Ibadan, Nigeria. *Trop. Sci.*, **28**, 111–18.

Fashakin, J.B. and Ojo, F.A. (1988b) Chemical composition and nutritive changes of some improved varieties of cowpea (*Vigna unguiculata* (L.) Walp) 2. New breeds of varieties from IITA Ibadan, Nigeria. *Trop. Sci.*, **28**, 191–9.

Gatta, C. Piergiovanni, A.R., Ng, N.Q. *et al.* (1989) Trypsin inhibitor levels in raw and cooked cowpea (*Vigna unguiculata*) seeds. *Lebensm. Wissenschaft. Technol.*, **22**(2), 78–80.

Hung, Y.C., Chinnan, M.S. and McWatters, K.H. (1987) Effect of pre-decortication drying treatment on the textural quality of cowpea products. *Am. Soc. Agric. Eng.*, Microfiche Collection no.87–6062, 17 pp.

Jaffé, W.G. (1949) Limiting essential amino acids of some legume seeds. *Proc. Soc. Exp. Biol. Med.*, **71**, 398–401.

Jaffé, W.G. (1950) Protein digestibility and trypsin inhibitor activity of legume seeds. *Proc. Soc. Exp. Biol. Med.*, **75**, 249–55.

Kachere, D.P., Chavan, J.K. and Kadam, S.S. (1988) Nutritional quality of some improved cultivars of cowpea. *Plant Foods Hum. Nutr.*, **38**, 155–62.

King, J., Nnanyelugo, D.O., Obong, H. and Ngoddy, P.O. (1985) Household consumption profile of cowpea (*Vigna unguiculata*) among low income families in Nigeria. *Ecol. Food Nutr.*, **16**, 209–15.

Laurena, A.C., Garcia, V.V. and Mendoza, E.M.T. (1987) Effects of heat on the removal of polyphenols and *in vitro* protein digestibility of cowpea (*Vigna unguiculata* (L.) Walp). *Plant Foods Hum. Nutr.*, **37**, 183–92.

Malleshi, N.G., Daodu, M.A. and Chandrasekhar, A. (1989) Development of weaning food formulations based on malting and roller drying of sorghum and cowpea. *Int. J. Food Sci. Technol.*, **24**, 511–19.

Marero, L.M., Payumo, E.M., Aguinaldo, A.I.R. and Homma, S. (1990) Malto-oligosaccharide composition of flours, weaning foods and gruels prepared from germinated rice, corn, mung bean and cowpea. *J. Nutr. Sci. Vitaminol.*, **36**, 55–64.

McWatters, K.H. (1983) Compositional, physical and sensory attributes of akara processed from cowpea paste and Nigerian cowpea flour. *Cereal Chem.*, **60**, 333–40.

McWatters, K.H. and Chinnan, M.S. (1985) Effect of hydration of cowpea meal on physical and sensory attributes of a traditional West African food. *J. Food Sci.*, **50**, 444–8.

McWatters, K.H., Chinnan, M.S., Hung, Y.C. and Branch, A.L. (1988) Effect of predecorticated drying temperature treatment on cowpea paste characteristics and functionality in preparation of akara. *Cereal Chem.*, **65**, 23–7.

Mustafa, A.I., Al-Wessali, M.S., Al-Basha, O.M. and Al-Amir (1986) Utilization of cowpea flour and protein isolate in bakery products. *Cereal Foods World*, **31**, 756, 758–9.

Ndubuaku, V.O., Uwaegbute, A.C. and Nnanyelugo, D.O. (1989) Flatulence and other discomforts associated with consumption of cowpea (*Vigna unguiculata*). *Appetite*, **13**; 171–81.

Ng, N.Q. and Maréchal, R. (1985) Cowpea taxonomy, origin and germ plasm, in *Cowpea Research, Production and Utilization* (eds S.R. Singh and K.O. Rachie), Wiley, Chichester, pp. 11–21.

Nnanna, I.A. and Phillips, R.D. (1989) Amino acid composition, protein quality and water-soluble vitamin content of germinated cowpea (*Vigna unguiculata*). *Plant Foods Hum. Nutr.*, **39**, 187–200.

Nnanyelugo, D.O. (1982) Nutritional practices and food intake measurement and their relationship to socioeconomic grouping, location and their apparent nutritional adequacy in children. *Appetite*, **3**, 229–36.

Nnanyelugo, D.O., King, J, Ene–Obong, H.N. and Ngoddy, P.O. (1985) Seasonal variations and the contribution of cowpea (*Vigna unguiculata*) and other legumes to nutrient intake in Anambra State, Nigeria. *Ecol. Food Nutr.* **17**, 271–87.

Norton, G., Bliss, F.A. and Bressani, R. (1985) Biochemical and nutritional attributes of grain legumes, in *Grain Legume Crops* (eds R.J. Summerfield and E.H. Roberts, Collins Professional and Technical Books, London.

Obizoba, I.C. (1989) Effects of germination, dehulling and cooking on the nutritive value of cowpea (*Vigna unguiculata*) flour. *J. Food Sci.* **54**, 1371–2.

Obizoba, I.C. (1990) Nutritive quality of blends of corn with germinated cowpeas (*Vigna unguiculata*), pigeon pea (*Cajanus cajan*) and Bambara groundnut (*Voandzeia subterranea*). *Cereal Chem.*, **67**, 230–2.

Ogunmodede, B.K. and Oyenuga, V.A. (1968) Estimation of vitamins A, D, and E values of varieties of cowpea (*Vigna unguiculata* Walp.) grown in Nigeria. *Nig. Agric. J.* **5**, 65–7.

Ogunmodede, B.K. and Oyenuga, V.A. (1969) Vitamin B content of cowpea (*Vigna unguiculata* Walp.). 1. Thiamine, riboflavin and niacin. *J. Sci. Food Agric.*, **20**, 101–3.

Ogunmodede, B.K. and Oyenuga, V.A. (1970) Vitamin B content of cowpea (*Vigna unguiculata* Walp). II. Pyridoxine, pantothenic acid, biotin and folic acid. *J. Sci. Food Agric.* **21**, 87–91.

Okaka, J.C. and Potter, N.N. (1977) Functional and storage properties of cowpea powder–wheat flour blends in bread-making. *J. Food Sci.* **42**, 828–33.

Okaka, J.C. and Potter, N.N. (1979b) Sensory, nutritional and storage properties of cowpea powders processed to reduce beany flavor *J. Food. Sci.* **44**, 1539–44.

Okaka, J.C. and Potter, N.N. (1979a) Physiochemical and functional properties of cowpea powders procesed to reduce beany flavor. *J. Food Sci.*, **44**, 1235–40.

Ologhobo, A.D. and Fetuga, B.L. (1984) Effect of processing on the trypsin inhibitor, hemagglutinin, tannic acid and phytic acid content of seeds of ten cowpea varieties. *Trop. Agric.* **61**, 261–4.

Ologhobo, A.D. and Fetuga B.L. (1986) Changes in carbohydrate content of germinating cowpea seeds. *Food Chem.*, **20**, 117–25

Orr, M.L. and Watt, B.K. (1957) *Amino Acid Content of Foods*. USDA Home Economics Report N. 4. Washington, DC.

Oyeleke, O.A., Morton, I.D. and Bender, A.E. (1985). The use of cowpea (*Vigna unguiculata*) in improving a popular Nigeria weaning food. *Br. J. Nutr.* **54**, 343–347.

PAG (United Nations Advisory Group) (1975) A Call For Action: Upgrading human nutrition through the improvement of food legumes, in: *Nutritional Improvement of Food Legumes by Breeding*, (ed. M. Miller), Wiley, New York.

Phillips, R.D. and McWatters, K.H. (1991) Contribution of cowpeas to nutrition and health. *Food Technol.* **45**, 127–30.

Rachie, K.O. and Roberts, L.M. (1974) Grain legumes of the lowland tropics. *Adv. Agron.* **26**, 1–132.

Roman, A.V., Bender, A.E. and Morton, I.D. (1987) Formulation and processing of a weaning food based on rice, cowpeas and skim milk powder. *Hum. Nutr. (Food Sci. Nutr.)* **41**, 15–22.

Sherwood, F.W., Weldon, V. and Peterson, W.J. (1954) Effect of cooking and of methionine supplementation on the growth-promoting property of cowpea (*Vigna sinensis*) protein. *J. Nutr.*, **52**, 199.

Singh, S.R. and Rachie, K.O. (eds) (1985) *Cowpea: Research, Production and Utilization*, Wiley, Chichester.

Steele, W.M., Allen, D.J. and Summerfield, R.J. (1985) Cowpea (*Vigna unguiculata* (L.) Walp, in *Grain Legume Crops*, (eds R.J. Summerfield and E.H. Roberts), Collins, London.

USDA (1986) Composition of Foods. Legumes and Legume Products: Raw, Processed, Prepared. *Human Nutrition Information Service. Agriculture Handbook* no. 8–16. United States Department of Agriculture, Washington, DC.

Uwaegbute, A.C. and Nnanyelugo, D.O. (1987) Nutritive and biological evaluation of processed cowpea diets compared with local weaning foods in Nigeria. *Nutr. Rep. Int.*, **36**, 119.

Williams, C.E. (1985) Digestible and metabolizable energy from cowpea (*Vigna unguiculata* (L) Walp) meal 'akara'. *Nutr. Rep. Int.*, **31**, 555–60.

PART TWO

Compositae

CHAPTER TWENTY-SEVEN

Safflower (*Carthamus tinctorius* L.)

D.C. Pavlov and N.A. Todorov

27.1 GENERAL BOTANICAL, ECOLOGICAL AND PRODUCTION INFORMATION

Safflower (*Carthamus tinctorius* L.) is an annual plant used largely as an oil and fodder source in dry regions in many countries. The plant is referred to Asteraceae, a member of the Compositae. Only three species are extensively cultivated.

Carthamus lanatus L. is a small plant-like shrub 15–45 cm high, whose seeds are small and are usually used as poultry feed. In India this species is grown in the Jammu and Kachir areas at altitudes of 1500–1800 m.

Carthamus oxyacantha Bieb. is a wild species found in the Uttar Pradesh and Punjab areas of India. Seeds contain 28% oil, about 18% of which is extractable. The rest remains in the cake. The edible oil has good quality, similar to that produced from sunflower.

Carthamus tinctorius L. is the cultivated species which is commonly found in many countries. The morphology of this species is very variable (Tar-Avanecian, 1978). The root system is extensive and penetrates to 200–300 cm in depth and extends laterally 60–80 cm. Although different cultivars attain various heights, it is usually between 60–150 cm high (Knowles, 1982). The stem of the safflower plant is naked and branched in the upper parts. Each branch terminates in a capitulum, each head containing 20–100 seeds. The seeds are shaped like small sunflower seeds and are slightly striped, oblong, with a rounded top. They are composed of a kernel surrounded by a thick fibrous hull which is very hard and difficult to remove. The seeds are usually cream coloured, but sometimes

Legumes and Oilseeds in Nutrition
Edited by E. Nwokolo and J. Smartt
Published in 1996 by Chapman & Hall. ISBN 0 412 45930 2

they are of a darker colour. A lighter coloured seed coat is preferable, because the oil takes up some of the colour of the seed coat during processing (Knowles, 1982). The seeds are small, about 20–55 g/1000 seeds.

Safflower is a widely propagated plant adapted to and cultivated in tropical and dry areas of the world. Its distribution is 0–45° latitude and 1600–2200 m altitude. It is a high temperature crop, growth being best at about 28–35°C although it tolerates 40°C. However, it does not thrive in areas of high humidity as the aerial parts of the plant are disease susceptible in such conditions. Some varieties tolerate low to very low temperatures.

Safflower is a short-day plant, requiring intense sunlight rather than prolonged day length and can therefore be cultivated successfully in northern latitudes. It grows best in neutral, fertile, well-drained and deep soils although it can be cultivated in a light soil if the soil is adequately provided with water. It is tolerant of soil acidity and can be grown at pH 5.0–8.0. Safflower is a crop which is very adaptable to latitude, altitude, salinity, alkalinity, acidic soils and variation in rainfall. In the rotation, safflower is usually sown after sorghum, millet, corn or other cereals (Moraghan *et al.*, 1984), although it has no specific need to follow or precede any particular crop in the rotation.

27.2 PRODUCTION

World safflower seed production fluctuates between 600 000 and 1 000 000 MT/year. During 1988 the world production was about 873 000 MT. It is a smaller yield compared with 1979–81, when it reached 968 000 MT seeds. The larger part (750 000 MT, 85.91%) of the world production is from the developing countries in Asia and Latin America. The developed countries produced only 123 000 MT (14%) for 1988. The largest producers of safflower seeds are India (451 000 MT, 52% of all production) and Mexico (252 000 MT, 29%).

Safflower yields vary greatly. The level of production depends on the productive ability of cropping cultivars, technology of cropping, soil and water conditions. In developing countries the yields are usually 300–400 kg/ha. In developed countries the yields are at least twice as high as in the developing ones, 1048 kg/h on the average for 1988. Yields are highest in North America at 1818 kg/ha. With normal water supply and fertilization yields may be up to 2000–3000 kg/ha (Dimova and Decov, 1984). In the USA safflower yields under dryland conditions are about 1600 kg/ha and under irrigation up to 4000 kg/ha (Robbelen, 1982).

27.3 CURRENT AND POTENTIAL USE AS HUMAN FOOD

Safflower is an important plant cultivated in many countries in tropical and warm temperature climatic areas in dry seasons. It is an ancient plant, known and used by people from ancient times. The centre of its origin is the Middle East, Ethiopia, Afghanistan, and the Mediterranean, where probably it was first domesticated. In Egypt it has been cultivated for oil production and as a source of a red dye obtained from the flowers. Safflower oil is a rapidly drying oil. The oil is used in high quality soft margarines and mayonnaise products (Robbelen, 1982). Its flavour quality is near to that of sunflower oil, when prepared from well dehulled seeds. The oil which is produced from undehulled seeds has a bitter flavour. Most of the cultivated safflower species have a high content of linoleic acid (80% of total fatty acids). Some cultivars have a low content of linoleic acid and high percentage of oleic acid, such oil produced from high oleic acid-safflower being chemically similar to olive oil. Cooking oil-high in oleic acid has proven to be excellent (Knowles, 1982; Robbelen, 1982) and has been shown to reduce low density lipoprotein (LDL) cholesterol levels (Smith, 1985; Mattson and Grundy, 1985). There are beneficial effects connected with lowering levels of cholesterol in blood when this oil is consumed, thus reducing the incidence of atherosclerosis (Knowles, 1982).

It is stable under oxidation (Purdy, 1985) and high temperature applications (Fuller *et al.*, 1967; Knowles, 1969). Oil which is high in oleic acid does not polymerize as readily under high temperature regimes as does oil high in polyunsaturated fatty acids (Irving *et al.*, 1988). It is anticipated that in the future safflower oil will play an important role in human nutrition and as an industrial oil source in developing countries. This is because in addition to its use for cooking, safflower oil is useful in the manufacture of paints, varnishes, enamels, linoleum, soap and other products.

Protein isolates extracted from defatted safflower meal are rich in essential amino acids and can be used in human consumption and because they contain a very high protein level (96%), they could be useful in meeting dietary protein needs in poor countries.

Safflower is cultivated mainly for the oil but in addition it could be used successfully as an animal feed, in the form of oil-cake, groats, or the whole plant. Safflower meal is sometimes produced from unhulled seed, because it is too difficult to remove the kernel. The undecorticated oil meal is also called whole pressed seed meal, whereas the decorticated oil meal is referred to as safflower meal. There is a good market for the meal in the USA, where it is used as a feed for poultry or livestock (Knowles, 1982). The oilcake and meal produced from undecorticated safflower seeds are very rough, with a high content of crude fibre and can be fed to ruminants only. Safflower oil meal from undecorticated seed is about equal to linseed meal as a protein supplement for dairy cows, beef cattle or sheep. When mixed with other feeds, they are eaten, without danger of changing the flavour of milk. Undecorticated oil meal has only about 18–22% protein and contains 30% fibre or more – as much as or more than hay. It requires about 2 kg safflower oil meal from unhulled seed to equal 1 kg of such high protein oil meals as soybean meal. Oil cake and groats produced from well dehulled seeds can be included in the diet of all categories of animals. Such an oil meal contains 42–44% crude protein, the crude fibre being about 8–10% which is rather high for broilers, but acceptable for laying hens and pigs. Safflower meal is rather low in methionine and lysine to be a good substitute for soybean oil meal in non-ruminant diets. Such substitution is possible following addition of synthetic lysine and methionine or by combination of safflower oil meal with a good source of methionine and lysine, like fish meal. The levels of inclusion of safflower oil meal in pigs and poultry rations depends on content of crude fibre and other nutrients. Because of the rapid oxidation of the oil, the meal has to be used immediately or be stabilized with antioxidants.

27.4 PROXIMATE COMPOSITION OF SAFFLOWER SEEDS AND MEAL

Safflower seeds comprise 60–70% kernel and 30–40% hulls. The hulls contain mainly carbohydrates (55–62% crude fibre, 30–32% NFE), 4–5% oil, 3.5–4.5% crude protein and 1–2% ash. The kernel contains mainly oil (60–65%), protein (25–30%) and some sugars. The chemical composition of the whole seed depends on the proportion of kernel and hull.

The principal constituent of seed is oil. In the dry matter of decorticated seeds the oil content ranges from 25% to 40%, and in some new varieties up to 43% (Robbelen, 1982). Percentage of lipids is not equal in different parts of the seed, cotyledons containing 49.5%, embryo 9% and integument 41.5% of the total lipids (Fedeli *et al.*, 1972).

Crude protein comprises 14.9% (Latha and Prakash, 1984) to 17% (Zazueta and Price, 1989) of dry matter. Total sugar comprises 3.2–9.2% and reducing sugar 2.1–2.8% of dry matter (Dhindsa *et al.*, 1973; Latha and Prakash, 1984).

According to Zazueta and Price (1989) whole seeds of safflower contain on a dry matter basis 17.3% crude protein, 39.8% oil, 21.7% crude fibre, 18.5% NFE and 2.7% ash, whereas according to Ensminger and Olentine (1978) safflower seeds contain 19.5% crude protein, 19.2% ether extract, 30.9% crude fibre, 15.5% NFE, 0.29% calcium (Ca), 0.24% phosphorus (P), 0.74% tannin and 3804 kcal kg metabolizable energy (ME) for ruminants (Raj *et al.*, 1983). Dhindsa *et al.* (1973) note that in the different safflower

Table 27.1 Fatty acid composition of safflower oil

	Content (%)		
Fatty acids	*1*	2	3
Linoleic-C18:2	75	68.9	72.8
Oleic-C18:1	15	16.5	14.5
Other unsaturated	–	–	0.6
Total unsaturated	93	–	88
Stearic-C18:0	2	3.9	2.7
Palmitic-C16:0	4	10.6	6.4
Others saturated	–	–	2.1
Total saturated	7	–	11.2

Sources: (1) Robbelen (1982); (2) Fuller *et al.* (1967); (3) Mattson *et al.* (1985).

Table 27.2 Chemical composition of safflower oil cake (mechanically extracted)

	(Content (% of dry matter)							
Type	*Dry matter*	*Crude protein*	*Crude fibre*	*Ash*	*Ether extract*	*NFE*	*Ca*	*P*
With hulls-India	92.7	30.5	24.9	6.1	9.0	29.5	0.12	0.59
With hulls-USA	92.0	20.7	35.9	4.3	6.5	32.6		
Without hulls-India	91.0	49.7	9.0	8.2	11.0	22.1	0.06	1.10
Without hulls-USA	92.0	39.1	19.0	8.0	8.3	25.6		
Hulls-USA	91.3	4.2	58.2	1.5	5.1	31.0		

Source: FAO (1981).

strains, phosphorus content ranges from 0.13 to 0.23% and potassium from 1.5 to 1.9%. Dehulling influences the chemical composition of safflower seed significantly. Carbohydrate and crude fibre are significantly reduced whereas protein and oil content increase. Well-hulled seeds contain 26.9% crude protein, 61.5% oil, 1.6% crude fibre, 6.8% NFE and 3.2% ash (Zazueta and Price, 1989).

27.4.1 Safflower oil

Safflower oil contains about 90–92% unsaturated and about 8–10% saturated fatty acids. The iodine count of safflower oil is 129–150 and the solidification point -20 to -13°C. Oil from commercially cultivated safflower species has a very high content of linoleic acid (Fuller *et al.*, 1967; Robbelen, 1982; Mattson and Grundy, 1985; Westphal *et al.*, 1985) with a low content of oleic acid (15–16%) (Table 27.1). The content of saturated fatty acids is low, stearic acid ranging from 2 to 4% and palmitic acid ranging from 5 to 10% (Knowles, 1982). High-yielding varieties of safflower grown in Pakistan contain 65.9%–73.4% of linoleic acid, 14.0–15.7% of oleic acid, 9.4–12.0% of palmitic acid, 2.3–5.5% of stearic acid and 0.9–3.1% of myristic acid. This ratio of linoleic to oleic fatty acids is typical for high linoleic cultivars of safflower. In high oleic cultivars, oleic acid content reaches 64–86% of total fatty acids (Knowles, 1982; Mattson and Grundy, 1985). In safflower oil, the content of sterols is 0.25g/100 g whereas the content of tocopherols is 250–500 mg/kg. Tiscornia and Bertini (1974) reported the existence of other sterol fractions in safflower oils, namely: 5-avenasterol at a concentration of 3.4–4.8% and 7-avenasterol at a concentration of 3.5–6.9% of total sterols.

27.4.2 Protein cakes

These days safflower seeds are processed to extract oil by prepressing followed by solvent extraction. The chemical composition of the residue cake or meal is quite variable depending on the level of decortication of seeds (Table 27.2). Crude protein content ranges from 20–50% (Table 27.1).

According to Ensminger and Olentine (1978) undecorticated safflower cake after oil extraction contains about 60% hulls. The presence of this hull influences the proportion of the other nutrients in the cake. Undecorticated oilcake has a low content of crude protein (18–21%) and a high percentage of crude fibres (22–36%). Partly dehulled safflower meal after extraction contains 15% fibre and 40% protein. Dehulled and defatted meal has a low content of fibre (4–5%) and significantly high protein content reaching 70% in debittered meal.

Safflower meal contains more crude protein (47.4%) than groundnut meal (43.9%) and nigerseed meal (43.1%). But the gross protein value (41.8%) and net protein value (30.3%) is lower than in groundnut and niger. The nitrogen corrected metabolizable energy of safflower (2.59 kcal/g) is equal to that of groundnut cake (2.61 kcal/g) and nigerseed cake (2.56 kcal/g). Safflower cake contains 1.27% methionine and 2.30% lysine. The lysine content is lower than in groundnut and niger. Decorticated meal contains (on a dry matter basis): 0.4% Ca, 1.3% P, 0.67% K, 0.03% Cl, 0.1% Na, 0.06% S. It also contains (per kg of dry matter: 48 mg Fe; 15.6 mg Mg; 8.5 mg Cu and 33 mg Zn. According to the content of Ca, P, Mg, K and Fe, safflower meal can be compared with soybean meal, but Na and other trace elements are not comparable, being very low. The meal also contains the following vitamins (mg/kg dry matter): biotin 1.4 mg, choline 20.67 mg, folacin 0.4 mg, niacin 22 mg, riboflavin 2.4 mg, vitamin E 0.7 mg and pantothenic acid 4.0 mg. As a food ingredient, safflower meal is a good source of niacin, thiamine and riboflavin, but it does not contain appreciable quantities of the fat-soluble vitamins and vitamin B_{12}.

27.4.3 Protein fractions and amino acids

Safflower seed protein contains a high level of glutamic acid (25.8%), aspartic acid (11.2%), arginine (9.4%) leucine (6.0%), valine (5.3%) and glycine (5%). The methionine, lysine and tryptophan content is low, methionine and lysine being considered the limiting amino acids in safflower meals and isolates (Betschart and Saunders, 1978).

Protein isolates can be produced from safflower meal. Extraction is with sodium hydroxide at pH 8, 9 or 10 followed by precipitation with HCl at pH 5 or 6. Protein isolates are obtained which contain 87–96% protein of good quality (Betschart, 1975; Betschart and Saunders, 1978).

27.5 ANTINUTRITIONAL SUBSTANCES

The protein meal from safflower is currently underutilized due to some antinutritional factors which are contained to different degrees. These antinutritional factors are the phenolic glucosides which are associated with the strong bitter flavour and cathartic activity of safflower seed meal. Matairsinol monoglucoside is associated with bitter flavour and 2-hydroxyarctin with carthartic activity. Phenolic glucosides in defatted safflower meal are highly soluble in water. There are molecular interactions between protein fractions and phenolic glucosides. Water extraction may remove 60% of total phenolics from defatted whole meal and up to 80% of total phenolics from dehulled, defatted meal.

27.6 SAFFLOWER OIL

27.6.1 Safflower oil as a source of energy and essential fatty acids

The energy value of safflower oil is 8800 kcal/kg (Lassiter and Edwards, 1982). A 10% safflower oil emulsion is used for surgical patients requiring total parenteral nutrition to provide 4% of energy given every day or every other day for 2–4 weeks. No significant changes are observed in haematological values, regulation studies, cholesterol and triglycerides in serum and no clinically significant adverse effects. A safflower oil emulsion, Lyposin-10%, is used in some cases after surgery to provide energy and to correct essential fatty acid deficiency. Taken every other or every third day, Lyposin emulsion could provide 30–50% of the energy required daily for a short time. Patients tolerate the lipid well and no adverse clinical effects are attributable to the lipid infusion. Liver function tests show that some patients develop eosinophilia (Wong and Deitel, 1981). Lyposin-10% is used in parenteral nutrition of infants after surgery for 2 weeks. No adverse side effects have been noticed and weight gain has been similar when 10% or 20% of Lyposin is used (Marchildon, 1982). Lyposin providing linoleic acid at 2% or 4% of actual energy requirement is used in preterm infants to correct essential fatty acid status (Cooke *et al.*, 1985a,b). Safflower emulsion Lyposin-10% is infused intravenously to supplement energy intake in low birth weight infants. Gains in infants have been satisfactory and resembled those of the infants given human milk. Rassias and Gibson (1986) have given safflower oil to breast feeding mothers (20 ml daily) to increase the linoleic acid content of breast milk. Essential fatty acid deficiency is observed in patients with massive bowel resection who are placed on home parenteral nutrition. Linoleic acid contained in safflower oil may improve the linoleic status of cystic fibrosis patients.

27.6.2 Safflower oil and atherosclerosis

It is known that polyunsaturated fatty acids may reduce the plasma cholesterol when exchanged for saturated fatty acids. It is self evident that the rates of coronary heart disease tend to be low in populations consuming high carbohydrate and low fat diets or when monounsaturated acids are predominant in the fatty acids composition of diets. Safflower oil included in the formula diet supplying 40% of fat calories reduces total cholesterol. High oleic safflower oil reduces total cholesterol content in man and experimental animals. High oleic safflower oil reduces cholesterol by 12.4% and high linoleic safflower oil reduces it by 15.6% compared to palm oil. Both reduce plasma LDL cholesterol and high linoleic safflower oil reduces concentrations of HDL cholesterol. High oleic acid safflower oil is as effective as high linoleic acid safflower in lowering LDL cholesterol levels in normotriglyceridaemic patients. Neither of the unsaturated safflower oils has a striking effect on lipoprotein levels of hypertriglyceridaemic patients.

27.6.3 Safflower oil and carcinogenesis

Dietary oils like safflower oil which contain very high levels of polyunsaturated fatty acids, may not be very safe for long-term consumption. They may reduce HDL, alter the composition of cell membranes and even promote carcinogenesis in experimental animals (Mattson and Grundy, 1985). Cohen *et al.* (1986a,b), demonstrated that although serum cholesterol value in rats is significantly lowered when safflower oil is given, there is a positive association between dietary linoleic acid and tumour incidence in induced-tumour animals. Rats fed with high fat safflower and maize oil diets have more mammary tumours than those fed on high olive and coconut oil diets. These workers consider that the tumour-promoting properties of high fat diets are more a function of differences in fatty acid composition than of fat content or total energy intake. Most studies have shown that diets rich in omega-6 fatty acids (safflower oil or corn oil) increase cancer induction and that there is a linear increase in mammary carcinogenesis in rats fed with increasing amounts of linoleic acid up to a level of 4%. Short-chain fatty acids, trans fatty acids and omega-3 fatty acids appear to affect carcinogenesis differently. Diets containing eicosapentaenoic acid, the principal omega-3 fatty acid in menhaden fish oil plus linoleic acid result in significant reduction in incidence and number of tumours per rat.

27.6.4 Studies with monkeys and dogs

Safflower oil is used in clinical experiments with monkeys or dogs for estimation of the influence of polysaturated fat and cholesterol on the development of induced atherosclerosis or other chemical changes in different organs. Melchior *et al.* (1978) found that mature squirrel monkeys (*Saimiri sciureus*) given diets containing safflower oil have a smaller mean bile acid pool and a higher mean lithogenic index. Of the ten monkeys given safflower oil, nine developed gallstones. In an experiment where seven monkeys were given butter and seven given coconut oil, three monkeys and one monkey, respectively, developed gallstones, whereas none of the monkeys fed without fat developed gallstones. Atherosclerotic monkeys fed on a 'saffola' diet (containing safflower oil), stock diet and low-fat low-energy diet for months show significant reduction in atherosclerogenesis on the basis of serum and aortic lipid analysis. The greatest effect found in monkeys on a diet containing

Table 27.3 Metabolizable energy of safflower products for poultry

Product	*kcal/kg*	*Reference*
Safflower oilcake undecorticated pressed	1870	FAO (1981)
Safflower oilcake undecorticated extracted	1290	FAO (1981)
Safflower oilcake decorticated extracted	1870	FAO (1981)
Safflower meal decorticated	1600	Lassiter and Edwards (1982)
Safflower hulls	1000	FAO (1981)

safflower oil where a microscopic estimation of the aorta shows definite regression of atherosclerotic lesions.

Experimentally induced atherosclerosis in the aorta of monkeys undergoes partial regression on a diet with polyunsaturated fat or one low in fat and energy. Despite the equal amount of dietary cholesterol and fat, dogs receiving safflower oil are completely protected from atherosclerosis. The level of plasma lipids is raised only slightly and the induced atherosclerotic lesions seem to disappear.

27.6.5 Studies with farm animals and poultry

Safflower meal and oil are used in the nutrition of different categories of animals as a partial or complete replacement of other meals. The results are somewhat contradictory in different investigations and categories of animals but most of the experiments have shown that safflower meal can be used with success. Safflower meal produced from decorticated seed has good (75–85%) protein digestibility. The digestibility of nitrogen is best (96–98%) in protein isolates. (Betschart and Saunders, 1978).

Halloran and Almquist (1973) found that whereas undecorticated safflower meal has a metabolizable energy (ME) value of 1.75 kcal/g dry matter, decorticated meal has an ME value of 2.24 kcal/g, estimated in experiments with turkeys. If the meal is autoclaved, pasting of the beaks in young chicks is prevented (FAO, 1981). When used for poultry feeding, ME values (kcal/kg) have been reported for the following safflower products: undecorticated pressed cake 1870, undecorticated extracted 1290, decorticated extracted 1870, hulls 1000 (FAO, 1981) (Table 27.3).

Wylie *et al.* (1972) and Rehman and Malik (1986) have established that safflower meal reduces weight gain in chicken when used to replace soybean, the reason being a deficiency of lysine and methionine compared to soybean meal. When a high level of safflower meal is used, methionine and lysine supplementation is necessary to increase growth, whereas at a low level of safflower meal methionine is needed.

REFERENCES

Betschart, A.A. (1975) Factors influencing the extractibility of safflower protein (*Carthamus tinctorius* L.). *J. Food Sci.*, **40**, 1010–13.

Betschart, A.A. and Saunders, R.M. (1978) Safflower protein isolates; influence of recovery conditions upon composition, yield and protein quality. *J. Food Sci.*, **43**, 964–8.

Cohen, L.A., Thompson, D.O., Maenra, Y. *et al.* (1986a) Dietary fat and mammary cancer. 1. Promoting effects of different dietary fats on *N*-nitrosomethylurea-induced rat mammary tumorigenesis. *J. Natl. Cancer Inst.*, **77**, 33–42.

Cohen, L.A., Thompson, D.O. Choi, K. *et al.* (1986b) Dietary fat and mammary cancer. 2. Modulation of serum and tumor lipid composition and tumor prostaglandins by different dietary fats; association with tumor incidence patterns. *J. Natl. Cancer Inst.*, **77**, 43–51.

Cooke, R.J., Zee, P. and Yeh, Y.Y. (1985a) Safflower oil emulsion administration during parenteral nutrition in the preterm infant. 1. Effect on essential fatty acid status. *J. Pedriatr. Gastroenterol. Nutr.*, **4**, 799–801.

Cooke, R.J., Buis, M., Zee, P. and Yeh, Y.Y. (1985b) Safflower oil emulsion administration during parentheral nutrition in the preterm infant. 2. Effects on triglyceride and fatty acid levels. *J. Pediatr. Gastroenterol. Nutr.*, **4**, 804–7.

Dhindsa, K.S., Mehrotra, N. and Gupta, S.K. (1973) Yield and chemical composition of some promising strains of safflower for fodder purposes. *Haryana Agric. Univ. J. Res.*, **3**, 190–2.

Dimova, R. and Decov, D. (1984) *Crop Production of Tropics and Semitropics.* Zemizdat, Sofia, Bulgaria.

Ensminger, M.E. and Olentine, C. (1978) *Feed and Nutrition.* Ensminger Publishing Company, Clovis, CA.

FAO (1981) Tropical Feeds. *Animal Production and Health Series,* no. 12, Food and Agriculture Organization of the United Nations, Rome.

Fedeli, E., Cortesi, N., Camurdti, F. and Jacini, G. (1972) Regional differences of lipid composition in morphologically distinct fatty tissues. 4. Safflower and sunflower seeds. *J. Am. Oil Chem. Soc.*, **49**, 233–8.

Fuller, G., Diamond, M.J. and Applewhite, T.H. (1967) High oleic safflower oil: stability and chemical modification. *J. Am. Oil Chem. Soc.*, **44**, 264–6.

Halloran, H.R. and Almquist, H.J. (1973) Metabolizable energy determinations of safflower meals for turkeys. *Poultry Sci.*, **52**, 1674–6.

Hanelt, P. (1963) Monographishe Ubersicht der Gattung *Carthumus L. (Compositae)* Feeds Repert. *Z. Bot. Taxon, Geobot.*, **67**, 41.

ICRISAT (1973–1974) Fertilization. *Annual Report*, pp. 59–63.

Irving, W.D., Shannon, M.C., Breda, V.A. and Mackey, B.E. (1988) Salinity effects on yield and oil quantity of high linoleic and high oleic cultivars of safflower (*Carthamus tinctorius* L). *J. Agric. Food Chem.*, **36**, 37–42.

Knowles, P.F. (1969) Modification of quantity and quality of safflower oil through plant breedig. *J. Am. Oil Chem. Soc.*, **46**, 130–2.

Knowles, P.F. (1982) Safflower. *Genetics and Breeding Improvement of Oil Seed and*

Industrial Crops by Induced Mutations. Proceedings of the Advisory Group Meeting, IAEA, Vienna.
Lassiter, J.W. and Edwards, H.M. Jr (1982) *Animal Nutrition.* Reston Publishing Company, Reston, VA.
Latha, T.S. and Prakash, V. (1984) Studies on the proteins from safflower seed (*Carthamus tinctorius* L.). *J. Agric. Food Chem.*, **32**, 1412–16.
Lloyd-Still, J.D., Johnson, S.B. and Hollman, R.T. (1981) Essential fatty acid status in cystic fibrosis and the effects of safflower oil supplementation. *Am. J. Clin. Nutr.*, **34**, 1–7.
Marchildon, M.B. (1982) Parenteral 20% safflower oil emulsion: safety and effectiveness as a caloric source in newborn infants. *J. Parenteral Enteral Nutr.*, **6**, 25–9.
Martin, F.W. (1984) *Handbook of Tropical Food Crops*, CRC Press, Boca Raton, FL.
Mattson, F.H. and Grundy, S.M. (1985) Comparison of effects of dietary saturated, monosaturated and polysaturated fatty acids on plasma lipids and lipoproteins in man *J. Lipid Res.*, **26**, 194–202.
McCullagh, K.G., Ehrahart, L.A. and Butkus, A. (1976) Experimental canine atherosclerosis and its prevention. The dietary indication of severe coronary cerebral aortic and iliac atherosclerosis and its prevention by safflower oil. *Lab. Invest.*, **34**, 394–405.
Melchior, G.W., Lofland, H.B. and St Clair, R.W. (1978) The effect of polyunsaturated fats on bile acid metabolism and cholelithiasis in squirrel monkeys. *Metabolism*, **27**, 1471–8.
Miller, D.G., Williams, S.K., Palomobo, J.D. *et al.* (1987) Cutaneous application of safflower oil in preventing essential fatty acid deficiency in patients on home parenteral nutrition. *Am. J. Clin. Nutr.*, **46**, 419–23.
Moraghan, J.T., Rego, T.J., Buresh, R.J. *et al.* (1984) Labelled nitrogen fertilizer research with urea in the semi-arid tropics. 2. Field studies on a Vertisol. *Plant Soil*, **80**, 21–33.
Morrison, F.B. (1959) *Feeds and Feeding*, The Morrison Publishing Company, Clinton, Iowa.
Purdy, R.H. (1985) Oxidate stability of high oleic sunflower and safflower oils. *J. Am. Oil Chem. Soc.*, **62**, 523–5.
Raj, A.G., Kothandaraman, P. and Kadirvel, R. (1983) Proximate composition and nutritive value of safflower (*Carthamus tinctorius* L.) seed meal. *Cheiron*, **12**, 290–3.
Rassias, G. and Gibson, R.A. (1986) The effect of dietary supplements on the fatty acid composition of breast milk. *Proc. Nutr. Soc. Aust.*, **11**, 145.
Rehman, A. and Malik, M.Y. (1986) Safflower meal as a protein source in broiler rations. *Pakistan J. Biochem.*, **19**, 39–42.
Robbelen, G. (1982) Plant breeding and management–their role in modifying the availability and composition of certain vegetable oils and fats. *Improvement of Oil Seed and Industrial Crops by Induced Mutations*. Proceedings of an Advisory Group Meeting IAEA-Vienna.
Smith J.R. (1985) Safflower: due for a rebound. *J. Am. Oil. Chem. Soc.*, **62**, 1286–91.
Tar-Avanecian, D.V. (1978) *Agriculture in India*, Colos Leningrad (in Russian).
Tiscornia, E. and Bertini, G.C. (1974) Further investigations on the composition of the sterol fractions of *Olea europea, Carthamus tinctorius* and *Helianthus annuus. Riv. Ital. Sostanze Grasse*, **51**, 50–61.
Weiss, E.A. (1971) *Castor, Sesame and Safflower*, Leonard Hill Books, London.
Wong, K.H. and Deitel, M. (1981) Studies with a safflower oil emulsion in total parenteral nutrition. *Can. Med. Assoc. J.*, **125**, 1328–34.
Wylie, P.W., Talley, S.M. and Freeman, J.N. (1972) Substitution of linseed and safflower meal in diets of growing pullets. *Poultry Sci.*, **51**, 1695–701.

Zazueta, A.J. and Price, J. (1989) Solubility and electrophoretic properties of processed safflower seed (*Carthamus tinctorius* L.) protein. *J. Agric. Food Chem.*, **37**, 308–12.

CHAPTER TWENTY-EIGHT

Sunflower (*Helianthus annuus* L.)

E. Nwokolo

28.1 INTRODUCTION

The sunflower (*Helianthus annuus* L) is a member of the family Compositae (Asteraceae). The genus *Helianthus* contains over 60 species, only a few of which are cultivated either as ornamentals or as commercial crops. The sunflower is an erect annual or perennial herb, with a rough hirsute stem 1–3 m high, crowned by a capitulum. In wild species, this head may be 9–15 cm wide, whereas in cultivated species it may be 30 cm or wider. The head contains yellow ray florets on the periphery and brown, yellow or purple disc florets; on the inside. The sunflower is cultivated mainly for the seed which is a major source of nutrients in human nutrition.

It is also cultivated for the oil obtained from the seeds; a high quality oil currently used as a salad oil and for cooking and other culinary uses. The oil is also used in manufacture of soaps and paints and as a lubricant. The flowers are a valuable source of a yellow dye important in various industrial and nutritional applications whereas the young stem and leaves are a valuable source of fodder for ruminant animals.

28.2 PRODUCTION AND PRODUCTIVITY

According to the FAO production data for 1979–1990 (FAO, 1990), sunflower is cultivated in 15 countries in Africa, 12 countries in Asia, 13 countries in Europe, seven countries in Latin America, as well as the

Legumes and Oilseeds in Nutrition
Edited by E. Nwokolo and J. Smartt
Published in 1996 by Chapman & Hall. ISBN 0 412 45930 2

former USSR, Canada, USA, Mexico and Oceania. In 1990 on a worldwide basis, 15.36 million hectares were planted with sunflower, producing about 22.07 million metric tonnes (MT) of sunflower seeds, an approximate yield of 1355 kg/ha. In 1990, the six major producers of sunflower seeds were, in order of prominence (1) the former USSR, 6.5 million MT, (2) Argentina, 3.85 million MT, (3) France, 2.31 million MT, (4) China, 1.5 million MT, (5) Spain, 1.31 million MT and (6) USA, 1.03 million MT. Only about 0.96 million MT were produced in Africa and 1.50 million MT in Asia. The major African producers of sunflower seeds were South Africa with 585 000 MT and Morocco 160 000 MT. In Asia, the major producers were China 1.5 million MT, Turkey 900 000 MT and India 550 000 MT. Yield of feed per hectare ranged from an average low value of 980 kg/ha in Africa, through a median value of 1378–1387 kg/ha in North, Central and South America, to a high average yield of 1577 kg/ha in Europe. In individual countries yield was as low as 270 kg/ha in Tanzania to as high as 3571 kg/ha in Germany.

28.3 CURRENT AND POTENTIAL USAGE AS FOOD

Current usage of sunflower seed or its by-products as human food is so low that it may as well be considered as underutilized. Its use, however, as animal feed is considerable and growing. For humans, sunflower seeds are used mainly as a snack, the seeds being roasted and eaten in much the same way as peanuts and chestnuts. Decorticated sunflower seeds feature much more in vegetarian diets and are sold primarily in health food stores as an effective alternative protein source. Vegetarians use decorticated sunflower seeds in much the same way as peanuts because they are both high sources of protein and oil, and sunflower is spared the deleterious antinutritional factors present in raw soybean. Thus the seeds may be roasted and milled for fortification of bread, pastries and cakes or as a flavour ingredient and thickener in various sauces and stews. The seeds may also be roasted and brewed as a coffee-like beverage.

The potential for utilization of sunflower seed as food is tremendous. Production data from the 1980s indicate that sunflower may be the fastest growing oilseed crop in the United States and worldwide. It has established itself as the world's fourth largest source of seed proteins after soybean, peanut and cottonseed (FAO, 1990). For the huge potential of sunflower seed to be realized, its use as food must go beyond occasional use as a bread supplement. The high protein content of the defatted meal must be made available for culinary use as a colourless, bland, protein isolate. A review of the functional and chemical characteristics, processing and utilization of sunflower oil, meal and flour (Robertson 1975) indicates that sunflower flour and protein concentrates are ideal materials for food manufacture because of their white colour, bland flour and absence of antinutritionals. However, sunflower seeds contain fairly large concentrations of two major polyphenolic compounds: chlorogenic and caffeic acids. Phenols comprise 3.0–3.5% by weight of unprocessed sunflower seed flour. Chlorogenic acid (1.94–2.08%) alone comprises over 60% of the total phenolic compounds and together with caffeic acid (0.16–0.18%) comprises nearly 66% of total phenols in sunflower flour (Sosulski and Fleming, 1977). Because chlorogenic and caffeic acids are chromophoric, they pose serious problems to the production of white protein isolates from sunflower. When the pH of sunflower flour slurry is raised to about 7.0, its colour changes from a creamy yellow through a light green to a dark green colour. It finally turns a dark brown colour on further standing to yield a highly unacceptable product. Various methods have been proposed to remove these chromogenic compounds from sunflower flour. These include extraction of the meal with aqueous ethanol (Sosulski *et al.*, 1972; Fan *et al.*, 1976) and acidic butanol (Sodini and Canella, 1977). Also investigated was treatment of the solubilized protein with sodium sulphate followed by washing of the protein precipitate with 50% isopropanol (Gheyasuddin *et al.*, 1970), and hot aqueous diffusion of cracked sunflower kernels (Sosulski *et al.*, 1973). Bau *et al.* (1983) described all these processes as not commercially or industrially feasible and proposed a simple procedure of soaking dehulled sunflower seeds in citric acid, milling, solvent extraction with hexane (or an azeotropic mixture of hexane and ethanol), and a final wash of the defatted meal in citric acid. They demonstrated this procedure to be technically feasible and easily adaptable to commercial production. Another process demonstrated to be technically feasible was aqueous extraction of defatted sunflower meal using degassed hot water under 20–23 inches of mercury in a vacuum. Both procedures have produced white protein isolates of the type acceptable for food manufacture. Additionally, these are not technically complicated processes and will be easy to adapt to conditions in most developing countries.

28.4 NUTRIENT CONTENT AND NUTRITIVE VALUE

Decorticated, dried sunflower kernels are a dense source of nutrients. With low moisture content (less than 6%), they are a concentrated source of dietary energy (570 kcal/100 g), fat (48–50%) and protein (22–25%). According to USDA (1986) dried sunflower seeds contain 22.8% protein, 49.6% fat, 18.8% total carbohydrates, 4.2% fibre and 3.5% ash. Food energy is 570 kcal/100 g (2386 kJ/100 g). Partially defatted sunflower seed flour contains 48.1% protein, 1.6% fat, 35.8% total carbohydrates, 5.2% fibre and 7.0% ash. There is a fair agreement between these values and those in Table 28.1 for sunflower kernels, defatted whole seed,

Table 28.1 Proximate analysis of sunflower kernel and meal

Sunflower products	*Protein (%)*	*Lipids (%)*	*Ash (%)*	*Ethanol solubles (%)*
Hull	2.8	1.20	2.50	0.20
Sunflower kernels	22.8	57	2.80	2.20
Defatted whole seed meal	33.0	1–3	5.87	1.26
Defatted dehulled seed meal	53.0	1–3	7.85	2.0
Defatted protein product	63.0	1–3	2.22	0.20

Source: Bau *et al.* (1983).

Table 28.2 Mineral content of sunflower products

Minerals	*Content (mg/100g fresh weight)*		
	Sunflower kernels	*Partially defatted sunflower flour*	*Sunflower seed butter*
Calcium	116	114	122
Iron	6.8	6.6	4.8
Magnesium	354	346	369
Phosphrus	705	689	736
Potassium	689	67	72
Sodium	3	3	3
Zinc	5.1	5.0	5.3
Copper	1.7	1.7	1.8
Manganese	2.00	2.0	2.1

Source: USDA (1986).

defatted dehulled seed and a protein isolate (Bau *et al.*, 1983). Sunflower seed butter, which is produced by toasting and grinding decorticated sunflower seeds, contains: moisture 0.5–1.8%, protein 22.0–25.0%, oil 43.0–49.0% and ash 3.4–4.4% (Dreher *et al.*, 1983).

Table 28.2 compares the mineral content of sunflower kernels, partially defatted flour and sunflower seed butter. There is a high content of potassium, phosphorus and magnesium, and a moderate content of calcium. Iron, zinc, sodium, manganese and copper are in very low to trace amounts. Sunflower kernels, flours and butter are low in many vitamins and completely lacking in others. They are low in thiamin and riboflavin and completely lacking in ascorbic acid, pantothenic acid, pyridoxine, folacin, vitamin B_{12} and vitamin D. Only niacin is present in any reasonable amount (4.5 mg/100 g).

The fatty acid profile of sunflower lipids (Table 28.3) indicates that the predominant fatty acid is linoleic acid, present at a concentration of

Table 28.3 Fatty acid profile of sunflower products

Fatty acids	*Content (g/100 g product)*	
	Sunflower kernels	*Sunflower butter*
Saturated	5.20	5.00
C14:0	0.05	0.05
C16:0	2.80	2.70
C18:0	2.20	2.12
Monounsaturated	9.46	9.11
C16:1	0.05	0.05
C18:1	9.36	9.00
C20:1	0.05	0.05
Polyunsaturated	32.74	31.52
C18:2	32.63	31.42
C18:3	0.07	0.07

Source: USDA (1986).

Table 28.4 A comparision of amino acid profiles of sunflower flour and sunflower concentrates

Amino acid	*Content (g/16 g N)*			
	Sunflower flour	*Sunflower concentrate*	*Soybean*	*Casein*
Tryptophan	1.3	1.1	1.5	3.1
Lysine	3.1	3.2	5.4	6.4
Histidine	2.2	2.2	2.3	2.1
Ammonia	2.2	2.2	1.9	1.5
Arginine	8.2	8.2	6.9	2.8
Aspartic acid	8.6	8.5	11.4	6.0
Threonine	3.2	3.3	3.7	3.4
Serine	3.8	3.8	4.9	4.5
Glutamic acid	21.5	21.0	19.1	9.9
Proline	3.8	3.8	4.9	9.3
Glycine	5.2	5.1	4.0	1.5
Alanine	3.9	4.0	4.1	2.5
Half cystine	1.5	1.4	1.4	0.4
Valine	4.8	4.9	4.5	5.6
Methionine	2.1	2.0	1.3	3.0
Isoleucine	3.9	3.9	4.4	4.3
Leucine	6.0	6.1	7.6	8.0
Tyrosine	2.0	2.2	3.1	4.7
Phenylalanine	4.3	4.5	4.7	4.2
Protein (%)	58.3	72.7	55.3	92.7

Source: Sosulski and Fleming (1977).

Table 28.5 Phenolic content of sunflower protein products

Sunflower sample	*Phenolic content (% of sample)*	*Free (%)*	*Bound (%)*
Hull	4.55	3.08	96.92
Defatted whole seed meal	4.61	45.35	54.65
Dehulled defatted seed meal	2.99	78.60	21.40
Defatted protein product	1.23	24.39	75.61

Source: Bau *et al.* (1983).

32.7 g/100 g of seed and representing about 68.8% of total fatty acids. The next dominant fatty acid is oleic acid, present at a concentration of 9.3 g/100 g of seed and representing 19.75% of total fatty acids. Sunflower seed oil is therefore an excellent dietary oil with a very low concentration of saturated fatty acids (5.2 g/100 g) and a preponderance of the nutritionally desirable unsaturated fatty acids (42.20 g/100 g of seed).

Table 28.4 shows the amino acid profiles of sunflower products published by Sosulski and Fleming (1977) which compares favourably with those published by USDA. Sunflower seed protein contains an excellent profile of amino acids, similar in many respects to the profile of soybean protein and only slightly inferior to casein. Soybean flour has slightly more and casein significantly more lysine, serine, proline and leucine than sunflower flour or sunflower concentrate.

Sunflower seed, flour and butter all contain varying amounts of chlorogenic acid which causes a greenish-brown or greenish grey discoloration and imparts a distinct bitter taste to sunflower products. Research on efficient and practical procedures to eliminate chlorogenic acid from sunflower products has been active, as have breeding programmes to select low chlorogenic acid cultivars (Sosulski and Fleming, 1977). According to these workers, phenolic compounds present in sunflower flours (g/100 g flour) include the following: transcinnamic acid 0.05–0.07 g, caffeic acid 0.16–0.18 g, and chlorogenic acid isomer 0.12–0.17 g. Table 28.5 shows the content of phenolic compounds in various sunflower products (Bau *et al.*, 1983). A large amount of phenolic compounds is present in the hull and in whole seed meal, the concentration of phenolic compounds decreasing as the seeds are further processed.

28.4.1 Sunflower proteins in food systems

The functional properties of sunflower flour and concentrate were compared with those of soybean flour and concentrate by Sosulski and Fleming (1977) who observed that sunflower proteins have poor water

Table 28.6 Functional properties of sunflower and soybean products

Protein product	*Water absorption (g/g flour)*	*Fat absorption (g/g flour)*	*Oil emulsification (%)*
Soybean flour	2.4	0.9	16
Sunflower flour	1.8	2.2	54
Sunflower concentrate	3.9	3.0	14

Source: Sosulski and Fleming (1977).

Table 28.7 Whippability and foam stability of sunflower and soybean products

	Volume of foam after whipping (ml)			
Protein product	*1 min*	*30 min*	*60 min*	*120 min*
Soybean flour	150	180	60	10
Soybean concentrate	360	15	10	5
Sunflower flour	605	500	490	420
Sunflower concentrate	590	445	360	120

Source: Sosulski and Fleming (1977).

solubility which will probably hinder their use in the manufacture of beverage products. The products, however, have good fat absorption and oil emulsification properties (Table 28.6), indicating that sunflower protein products could be utilized in preparation of emulsion-type meat substitutes and meat extenders. The data in Table 28.7 clearly indicate that sunflower flour and concentrate have very good foaming properties and foam stability and have considerable potential as whipping agents if low chlorogenic acid cultivars of sunflower could be developed by plant breeders.

28.4.2 Product evaluation

Ever cognisant of the great potential for utilization of sunflower oils and proteins, some scientists (Dreher *et al.*, 1983) have subjected commercially and experimentally processed sunflower butters to tests of nutritional, sensory and physical quality. Test criteria used included a 9-point hedonic test, Gardner colour, spreadability, and protein digestibility. Overall, roasting significantly improved the nutritional and organoleptic properties of sunflower butter, although taste panellists rated it lower than peanut butter. Sunflower butter has, however, been reported to have a lower protein, fat and calorie content than peanut butter, although the calcium, phosphorus and iron content are higher than in peanut butter (Falk and Holm, 1981).

Table 28.8 Pepsin and trypsin digestibility (%) of sunflower products

Sample	*Pepsin (48 h) proteolysis*	*Trypsin (16 h) proteolysis*	*Total pepsin/ trypsin*
Defatted whole seed meal	72.6	17.1	89.9
Dehulled defatted seed meal	79.1	12.4	91.5
Defatted protein product	78.2	13.6	91.8

Source: Bau *et al.* (1983).

That wheat flour can be substantially enriched through partial substitution with soybean or sunflower flour has been demonstrated by Sambucetti *et al.*, (1976). Considerable enhancement of nutritional quality of bakery products occurred in proportion to the level of substitution. Results were similar to those reported by Jain *et al.* (1975) who incorporated sunflower meal in dough for bread making. Even when sunflower flour was utilized as replacement of 10 or 15% of wheat flour, the resulting bread showed good crust, crumb structure, grain size, and improved texture. There was, however, a slight discoloration possibly due to the presence of phenolics in sunflower flour. In similar studies, Bajaj *et al.* (1991) obtained high quality nutritions cookies prepared from sunflower seeds and wheat germ. Vasquez *et al.* (1984) showed that blending precooked pea flour with 5, 10 or 15% of sunflower protein concentrate resulted in a significant increase in protein efficiency ratio (2.01, 2.09, 2.24, respectively) compared to unblended pea flour (1.60) or casein control (2.66).

28.5 NUTRITIONAL EVALUATION WITH ANIMAL AND HUMAN STUDIES

28.5.1 Animal studies

Various nutritional quality evaluations using rats have been conducted on sunflower protein products. In one such study (Bau *et al.*, 1983), sunflower products were subjected to 48 h *in vitro* pepsin and 16 h *in vitro* trypsin proteolysis (Table 28.8). Total pepsin–trypsin digestibility was similar and high on all sunflower products.

Supplementation of defatted whole seed meal, dehulled defatted meal and defatted protein product with 1% lysine significantly improved the protein efficiency ratio (PER) of these sunflower products compared with unsupplemented products (Table 28.9). Sosulski and Fleming (1977) studied the complementarity of sunflower protein concentrate with wheat flour, field peas and ground beef.

Sunflower concentrate fed alone to rats sustained a weight gain of

Table 28.9 Nutritive value of sunflower protein products

	Protein efficiency ratios	
Sample	*Diet*	*Diet and 1 % lysine*
Casein	2.50	–
Defatted whole seed meal	1.80	2.09
Dehulled defatted seed meal	1.90	2.39
Defatted protein product	1.83	2.14

Source: Bau *et al.* (1983).

52.9 g and a PER of 2.50. Supplementation with lysine improved weight gain (116.7 g) and protein efficiency ratio (3.83). Blending the sunflower concentrate with field peas caused a significant increase in weight gain (101.7 g) and protein efficiency ratio (3.31). These experiments highlighted the very good nutritional potential of sunflower concentrate because when it was supplemented with lysine or blended with field peas which contain a high level of lysine, growth and protein efficiency ratio of experimental rats far exceeded the values of these parameters in the casein control diets (72.8 g for growth, 3.13 for protein efficiency ratio, respectively).

28.5.2 Human studies

In human studies (Bruce and McDonald, 1977), substitution of sunflower oil for edible fat in diets of adult young males resulted in a sustained decrease in serum cholesterol. There was, however, only a slight decrease in serum triglycerides. Devadas *et al.* (1977) report that in India, a school lunch programme containing sunflower meal, maize, roasted chickpea flour and sesame was evaluated in 50 school children against a control of 50 other students on the regular school lunch. The supplemental lunch programme was associated with significantly enhanced nitrogen retention, digestibility and biological value of the diets, as well as improved well-being, height and weight of the children on this programme.

A recent assessment of current literature on the health effects of consumption of sunflower oil and soybean oil, dietary lipids with very high levels of polyunsaturated fatty acids, has been carried out by Meydani *et al.* (1991). They found that substitution of oils with high polyunsaturated fatty acids (PUFA), like sunflower oil, in place of cooking fat resulted in decreased total and low-density-lipoprotein cholesterol, with an associated lowering of the risk of cardiovascular disease. Additionally, thromboxane generation is slightly diminished which may reduce the risk of platelet aggregation or even vascular occlusion in the coronary circulation. They also highlight the fact that linolenic acid, a predominant component of sunflower oil, is needed for normal immune response. A deficiency of essential fatty acid impairs B and T cell-mediated responses. These workers recommended that for prevention of gross essential fatty acid deficiency, dietary calories from sunflower oil or soybean oil should be at least 3% of energy intake. For prevention of heart disease, they recommend dietary calories from sunflower oil or soybean oil at 8–10% of energy intake. In a similar study Chan *et al.* (1991) demonstrate that dietary oleic, linoleic and linolenic acids are equally capable of lowering serum cholesterol when normolipidaemic men are fed diets high in sunflower, soybean, flax, olive or canola oil.

REFERENCES

Bajaj, M., Kaur, A. and Sidhu, J.S. (1991) Studies on the development of nutritious cookies utilizing sunflower kernels and wheat germ. *Plant Foods Hum. Nutr.*, **41**, 381–7.

Bau, H.M. Mohtadi–Nia, D.J., Mejean, L. and Derby, G. (1983) Preparation of colorless sunflower protein products: effects of processing an physiological and nutritional properties. *J. Am. Oil Chem. Soc.*, **60**, 1141–8.

Bruce, V.M. and McDonald, B.E. (1977) Canadian vegetable oils and some implications for human nutrition. *J. Can. Diet. Assoc.*, **38**(2), 90–7.

Chan, J.K., Bruce, V.M. and McDonald, B.E. (1991) Dietary alpha linolenic acid is as effective as oleic acid and linoleic acid in lowering blood cholesterol in normolipidemic men. *Am. J. Clin. Nutr.*, **53**, 1230–4.

Devadas, R.P., Chandrasekhar, U., Van Santhmani, G, and Gayathri, V. (1977) Evaluation of a mixture based on sunflower meal, bengal gram flour and sesame in school children. *Ind. J. Nutr. Diet.*, **14**, 291–5.

Dreher, M.L., Schantz, R.M., Holm, E.T. and Frazier, R.A. (1983) Sunflower butter, nutritional evaluation and consumer acceptance *J. Food Sci.*, **48**, 237–42.

Falk, R.S. and Holm, E.T. (1981) Preparation and taste panel acceptance of sunflower butter. *N. Dakota Farm Res. J*, **39**, 13.

Fan, T.Y., Sosulski, F.W. and Hamon, N.W. (1976) New techniques for preparation of improved sunflower protein concentrates. *Cereal Chem.*, **53**, 118.

FAO (1990) *Production Yearbook*. Food and Agriculture Organization of the United Nations, Rome.

Gheyasuddin, S., Cater, C.M. and Mattil, K.F. (1970) Preparation of a colorless sunflower protein isolate. *Food Technol.*, **24**, 242.

Jain, N.C., Shiralkar, N.D. and Rao, B.V. (1975) Incorparation of sunflower seed meal in bread. *J. Food Sci. Technol.*, **12**, 133–4.

Meydani, S.N., Litchtenstein, A.H., White, P.J. *et al.* (1991) Food use and health effects of soybean and sunflower oils. *J. Am. Coll. Nutr.*, **10**, 406–28.

Robertson, J.A. (1975) Use of sunflower seed in food products. *Crit. Rev. Food Sci. Nutr.*, **6**, 201.

Sambucetti, M.E., Scicli, G.G. and Sanahuja, J.C. (1976) Protein enrichment of wheat flour with soya and sunflower flours in bakery products. *Arch. Latinoam. Nutr.*, **26**, 353–64.

Sodini, G. and Canella, M. (1977) Acidic butanol removal of color-forming phenols from sunflower meal *J. Agric Food Chem.*, **25**, 822.

Sosulski, F.W. and Fleming, S.E. (1977) Chemical, functional and nutritional properties of sunflower protein products *J. Am. Oil. Chem. Soc.*, **54**, 100A–104A.

Sosulski, F.W., Mcleary, F.W. and Soliman, F.S. (1972) Diffusion extraction of chlorogenic acid from sunflower kernels. *J. Food Sci.*, **37**, 253.

Sosulski, F.W., Sabir, M.A. and Fleming, S.E. (1973) Continuous diffusion of chlorogenic acid from sunflower kernels. *J. Food Sci.*, **38**, 468.

United States Department of Agriculture (1986) Composition of Foods: Legumes and Legume Products. USDA Human Nutrition Information Service. *Agriculture Handbook no. 8–16*. United States Department of Agriculture, Washington DC.

Vasquez, M., Araya, E., Videla, G. and Yanez, E. (1984) Precooked pea flour supplemented with sunflower protein concentrate. *Nutr. Rep. Int.*, **30**, 5–12

PART THREE

Cucurbitaceae

CHAPTER TWENTY-NINE

Melon
Colocynthis citrullus L.)

E. Nwokolo

29.1 INTRODUCTION

The melon, *Colocynthis citrullus* (L.), is a creeping annual of the Cucurbitaceae, many members of which grow in the tropics. The melon described here, is an unimproved form of the better known watermelon. In spite of its being less known in the western world, it is cultivated to a very great extent in tropical Africa, where the seeds constitute an important dietary source of protein and oil. The melon is a small, ovoid fruit, weighing 1–2 kg, containing a large number of seeds embedded in a bitter-tasting, watery, fibrous fruit. The seeds are dorsoventrally flattened and covered with brown-coloured, thin and brittle shells. Decorticated seeds (kernels), are white and soft in texture. Seed yield is estimated at about 2 tonnes/ha (Nwokolo and Sim, 1987). Yield of air-dry kernels is about 70–80 kg/100 kg of dry seed. In tropical countries of Africa, melon seeds are milled and usually cold-pressed or solvent-extracted for the oil. Another form of local oil extraction involves crushing the seeds, boiling them in water until the oil floats to the surface and is separated. The oil extracted using any of these procedures is clear, pale yellow in colour and has a faintly nutty odour and a bland taste.

29.2 CURRENT AND POTENTIAL FOOD USES

Decorticated, ground melon seeds are used as a thickener and flavour component of gravies. The seeds may also be cooked, ground and

Legumes and Oilseeds in Nutrition
Edited by E. Nwokolo and J. Smartt
Published in 1996 by Chapman & Hall. ISBN 0 412 45930 2

fermented for use also as a flavour enhancer in soups and gravies. About 100 g of milled melon seeds are usually used in preparation of a soup or gravy for a meal for four people, and consequently, daily consumption rarely exceeds 25 g per person per day. This level of intake may be low, but along with other protein components of the soup, melon seed contributes quite significantly to total daily protein intake. Decorticated melon seeds may be dry fried in hot pans or skillets and used as snacks.

In many areas of southern Nigeria, melon seeds are milled with dried shrimps or crayfish (crawfish), wrapped in green leaves and steamed for 30 min. This results in a very delicious and highly nutritious snack which is relished by all family members. In rural areas of Nigeria, melon seeds are milled with crayfish and an edible fungus (*Pleurotus tuber regium*). Water is added and the mixture is moulded into small boluses which are added to gravies and soups and eaten as meat substitutes.

Edible oil has always been extracted from melon seeds, the oil commanding a high commercial price, due to its general acceptance as an excellent cooking and frying oil (Oyolu and MacFarlane, 1982). Oil extracted from melon seed has been reported to show good keeping properties over a period of four months, though the solvent-extracted unrefined oil develops a slight turbidity (Girgis and Said, 1968). The residual cake from oil extraction, is a protein-rich cake which can be fried or baked and consumed as a nutritious snack. Many oil mills in Nigeria produce and market both oil and cake.

As a result of the perpetual shortage of animal protein in human diets, in the tropics, Nwokolo and Sim (1987) suggest that serious consideration should be given to commercial production of meat analogues, meat substitutes and meat extenders from various oilseeds, especially melon seed. They suggest that a starting material of high protein density (at least 40% protein, dry matter basis), like melon seed or pumpkin seed flour, is necessary for the manufacture of textured protein products. These vegetable protein products can be utilized in a variety of dishes to supplement the meat eaten in developing countries. These workers are of the opinion that other extended or simulated meat products which may be produced from melon seed or pumpkin seed include meat or poultry product extenders, extended fresh or cooked sausages, meat loaf or meat roll and simulated ground beef. These products can easily be integrated into the local people's food habits. Oyenuga and Fetuga (1975) had earlier suggested that amino acid supplementation of melon seed protein could improve its nutritional quality. Similarly, Nwokolo and Sim (1987) conclude that by judiciously supplementing with other protein sources high in lysine and methionine, melon and pumpkin seed-based meat substitutes, can be produced for malnourished people in many tropical countries.

Table 29.1 Proximate composition (air-dry basis)

	Content (%)					
Melon	*DM*	*Fat*	*CP*	*Fibre*	*Ash*	*NFE*
Raw melon seed	86.6	45.6	27.4	5.9	7.7	13.4
Fried melon seed[a]	–	53.7	36.9	2.5	4.0	2.9
Pressed melon cake	90.1	20.3	45.3	6.8	8.9	18.7
Defatted melon meal	94.7	0.9	66.2	7.2	9.1	16.6
Fried defatted meal[a]	–	2.5	69.4	5.3	9.0	13.8

From Nwokolo and Sim (1987).
[a]From Oyenuga and Fetuga (1975).

Table 29.2 Effect of toasting on proximate composition of melon seed

	Content (%)				
	Protein	*Fibre*	*Fat*	*Ash*	*NFE*
Raw undefatted	30.6	3.1	56.9	3.4	6.0
Raw defatted	68.8	5.2	2.7	8.9	14.4
Toasted undefatted	31.5	3.4	56.8	3.9	5.9
Toasted defatted	69.9	5.8	2.8	9.6	14.8
Fried undefatted[a]	30.8	3.1	56.8	3.7	5.6
Fried defatted[a]	69.4	5.2	2.5	9.0	13.8

From Nwokolo and Sim (1987).
[a]From Oyenuga and Fetuga (1975).

29.3 PROXIMATE COMPOSITION

The most striking features of melon seed are the high content of oil in the whole seed and the exceptionally high content of protein in the defatted meal (Table 29.1). Defatted melon seed meal contains 66–77% protein, much more than in soybean meal, peanut meal or sunflower meal. The very high content of oil in the whole seed (45–57%), and of protein in the defatted meal, can make commercial cultivation of melon seed for edible oil and protein, a lucrative business. Application of heat (toasting, frying or baking) slightly influences the proximate composition of melon seed as shown in Table 29.2. Oyenuga and Fetuga (1975) evaluated three cultivars of melon seed, and found protein content of 30.6–35.7% in undefatted, raw melon seed and a protein content of 68.8–76.2% in defatted raw melon seed. In fried melon seed, protein content was 30.8–36.9% in undefatted seed and 69.4–77.7% in defatted seed meal. In the samples evaluated by Oyenuga and Fetuga, fat content was 54.2–56.8% in raw undefatted samples. Fried, undefatted samples had a slightly lower fat content, due to a slight loss of oil during frying.

Table 29.3 Content and availability of minerals in defatted seed (air-dry)

Mineral	*Content*	*Availability (%)*
Phosphorus	0.80%	50.7
Calcium	0.08%	58.0
Magnesium	0.34%	54.9
Potassium	0.59%	40.8
Copper	23.00 mg/g	55.0
Manganese	30.00 mg/g	61.1
Zinc	36.00 mg/g	60.0
Iron	76.00 mg/g	48.5

From Nwokolo and Sim (1987).

29.3.1 Minerals

The content of minerals in melon seed is shown in Table 29.3. Melon seed meal has a significantly higher phosphorus, magnesium and copper content than soybean meal, whereas soybean meal has significantly more calcium, potassium, manganese and iron. The content of trace minerals is generally low in both melon seed meal and soybean meal, although Oyenuga and Fetuga reported extremely high values for the trace minerals (iron, copper, manganese and zinc) in melon seed meal.

The average availability of minerals in melon seed (54%) is in the same range as for soybean meal and other oilseeds and grains (55–65%). Availability of potassium (41%) is as low as in soybean meal (44%). Availabilities of iron, phosphorus, magnesium and copper are moderate (50–55%), whereas those of calcium, zinc and manganese are fairly high (58–61%).

29.3.2 Protein and amino acids

The true nitrogen digestibility of defatted raw and toasted melon seed is of interest. True nitrogen digestibility determined with rapidly growing broiler chicks, is significantly lower in raw melon seed meal (68.8%), than in soybean meal (82.6%). When melon seed meal is oven-toasted (85°C for 24 h), the true nitrogen digestibility rises to 82.7%, similar to that of defatted, toasted soybean meal.

True protein digestibility determined with rats (Oyenuga and Fetuga, 1975) ranged from 91 to 93% for raw and toasted, defatted and undefatted melon seed meal. True digestibility values of creeping melon and climbing melon were reported by Achinewhu (1983). Digestibility coefficients determined with weanling rats were similar (82–83%) and significantly lower than for casein (92%). The amino acid profile of melon

Table 29.4 Content and availability of amino acids in defatted meal (air-dry)

Amino acid	*Content (mg/g)*	*Availability (%)*
Aspartic acid	20.3	91.1
Threonine	31.4	97.4
Serine	81.7	98.8
Glutamic acid	95.6	97.5
Proline	22.0	95.1
Glycine	33.0	96.7
Alanine	27.3	97.4
Valine	25.0	97.3
Methionine	15.1	98.8
Isoleucine	22.3	97.6
Leucine	40.0	97.4
Tyrosine	18.9	96.8
Phenylalanine	28.8	97.5
Histidine	18.0	87.5
Lysine	9.2	93.9
Arginine	101.1	98.1
Total amino acids	599.7	Mean: 96.1

From Nwokolo and Sim (1987).

seed is shown in Table 29.4. Melon seed meal has an excellent amino acid configuration, being as good as soybean meal in its content of all essential amino acids except lysine. A comparison was made, of the amino acid content in melon seed meal with FAO/WHO recommended protein and amino acid requirements for human subjects. It indicates that this meal more than meets the protein and amino acid needs of all categories of human subjects. The high content of the essential and other amino acids in the defatted protein meal, will enable it adequately to supplement (even at low levels of inclusion), the high carbohydrate–low protein diets of people in the tropics. Amino acid availability, determined with rapidly growing broiler chicks is uniformly high for all amino acids (average 96.1%) and not significantly different for soybean meal amino acids (average 94.9%). This indicates excellent utilization of dietary amino acids in both meals.

29.3.3 Oil and fatty acids

Melon seed oil is a semi-drying oil, consisting mainly of glycerides of oleic and linoleic acids. In fatty acid composition, melon seed oil is similar to soybean oil, except that soybean oil has a higher percentage of linolenic acid than melon seed oil (Table 29.5).

This higher content of linolenic acid puts soybean oil at a distinct advantage over melon seed oil, in respect of human dietary requirements for omega-3 fatty acids. Because of the high content of unsaturated fatty acids in melon seed oil, it should prove to be a good substitute for corn oil,

Table 29.5 Fatty acid profile and characteristics of melon seed oil

Fatty acids	*% of fatty acids*
Profile[a]	
C12:0	1.60
C14:0	1.18
C16:0	11.82
C18:0	10.74
C18:1	14.47
C18:2	57.65
C18:3	2.10
Characteristics[b]	
Oil content of seed	45.6
Free fatty acid value	1.1–2.9
Saponification value	189.5–190.5
Hydroxyl value	2.6
Iodine value	114.7–121.1
Unsaponifiable matter	1.04
Refractive index	1.472
Specific gravity	0.917

[a]From Nwokolo and Sim (1987).
[b]From Girgis and Said (1968).

for use in diets intended to reduce high levels of blood cholesterol. It should also prove to be very useful in blending dietary oils in the tropics, where a preponderance of the edible oils are saturated palm oil, palm kernel oil or coconut oil.

29.4 ANTINUTRITIONAL FACTORS

The presence of antinutritional factors in melon seed is suspected. Most food legumes and oilseeds contain several antinutritional constituents which include enzyme inhibitors, goitrogens, antivitamins, haemagglutinins, saponins, phytates, cyanogenic compounds and flatulence factors.

There is evidence that all legumes and oilseeds contain trypsin inhibitors, the level of inhibitor content varying from species to species. In broiler chicks fed on raw melon seed for 28 days, feed refusal was obvious after 24 h. Weight loss was severe after 5 days and by day 7, some mortality had occurred and the feeding trial had to be discontinued. When subsequently the melon seeds were toasted, growth rate was normal. Protein digestibility was similarly very much higher in toasted than in raw melon seed meal (Nwokolo and Sim, 1987). Post-mortem examination of birds that died indicated evidence of hepatic and pancreatic hypertrophy. Trypsin and chymotrypsin inhibitors are associated with pancreatic hypertrophy, depressing growth through a massive endogenous loss of essential amino acids, particularly sulphur amino acids, secreted by a very active pancreas. The antinutritional factors in melon seed seem to be heat-labile; cooking, toasting and other heat treatments being able to enhance the nutritional quality of melon seed.

29.5 NUTRITIONAL QUALITY EVALUATIONS

Some experiments to estimate the nutritional quality of melon seed meal, have been conducted with male weanling albino rats (Wistar strain), fed on raw defatted and toasted defatted melon seed meals (Oyenuga and Fetuga, 1975). For raw defatted melon seed meal, protein efficiency ratio (PER), net protein utilization (NPU) and biological value (BV) are 0.93, 49.74 and 52.84, respectively. For toasted defatted melon seed meal, these parameters are 1.01, 50.12 and 54.18, respectively. For soybean meal, PER, NPU and BV are 1.91, 61.33 and 6.85, respectively, whereas for whole hen egg used as a control, these parameters are 4.77, 94.86 and 96.94, respectively. Oyenuga and Fetuga (1975) observed that when toasted–defatted melon seed meal is fed at high protein levels (as sole source of protein), an increase in live weight gain and PER results. A 15% protein diet consequently sustains higher live weight than a 10% protein diet, each containing toasted-defatted melon seed as sole protein. Improvement in growth rate of rats following amino acid supplementation of the 10% protein diet containing toasted–defatted melon seed meal, confirms that lysine is the most limiting amino acid in melon seed meal for rat growth.

Lysine is followed by methionine and threonine as second and third limiting amino acids. These workers observe that the levels of tryptophan and isoleucine in melon seed meal are adequate for rat nutrition. Achinewhu (1983) notes a significantly lower PER in rats fed either creeping or climbing melon (0.144–0.145), in comparison with those fed casein diets (0.258). Supplementation of cowpea diets with melon seed (Akpapunam and Markakis, 1981) resulted in significantly improved apparent digestibility and PER.

Sharma *et al.* (1986) in an investigation of some curcubit kernel protein including musk melon (*Cucumis melo*) and watermelon (*Citrullus vulgaris*), reported biological values of 59 and 64, chemical scores of 51 and 54, net protein utilization of 52 and 58 and apparent digestibility coefficients of 70 and 72%, respectively. The oil-free melon meals contained 62.1 and 76.1% protein, respectively. The profile of amino acids was good although lysine was the first limiting amino acid. These workers concluded that melon seed had a tremendous potential for utilization as a dietary protein and amino acid source. Achinewhu and Riley (1986) showed that fermentation improved the thiamine and riboflavin but not the niacin content of melon seed (*Citrullus vulgaris*). The content of thiamine, riboflavin and niacin was 0.08, 0.12 and 3.2 mg/100 g prior to fermentation and 0.23, 0.22 and

3.2 mg/100 g after fermentation. This increase in concentration was merely due to a loss of other nutrients during the fermentation process. A further study of the effects of fermentation on carbohydrate and fatty acids of melon seeds (Achinewhu, 1987) showed that fermentation was associated with an increase in soluble solids and soluble nitrogen and a decrease in sucrose content. Fatty acid content was only slightly affected by fermentation.

29.6 CONCLUSIONS

The results summarized here indicate that melon seed, more than any other tropical oilseed, has a great potential to improve the nutritional status of many malnourished people in the tropics. Unlike soybean, melon seed is already consumed in various ways by people in these tropical countries. Solvent extracted and toasted, it has a bland taste and can, therefore, be easily incorporated into food, without the beany flavour which characterizes solvent-extracted soybean. There are indications that some antinutritional factors are present in melon seed, and it is, therefore, necessary to attempt to inactivate these in order to enhance the nutritive value of melon seed meal for human use.

REFERENCES

Achinewhu, S.C. (1983) The nutritive qualities of plant foods. 2. Protein quality of bread fruit (*Artocarpus ultillis*), climbing melon (*Colocynthis vulgaris*) and creeping melon (*Citrullus vulgaris*) *Nutr. Rep. Int.*, **27**, 541–5.

Achinewhu, S.C. (1987) Carbohydrate and fatty acid composition of fermented melon seeds (*Citrullus vulgaris*). *J. Food Sci. Technol. (India)*, **24**, 16–19.

Achinewhu, S.C. and Riley, J. (1986) Effect of fermentation on the thiamin, riboflavin and niacin content of melon seed (*Citrullus vulgaris*) and African oil bean (*Pentaclethira macrophylla*). *Food Chem.*, **20**, 243–52.

Akpapunam, M.A. and Markakis, P. (1981) Protein supplementation of cowpeas with sesame and water melon seeds. *J. Food Sci.*, **26**, 960–4.

Girgis, P. and Said, F. (1968) Lesser known Nigerian edible oils and fats. 1. Characteristics of melon seed oils. *J. Sci. Food Agric.*, **19**, 615–16.

Nwokolo, E. and Sim, J.S. (1987) Nutritional assessment of defatted oil meals of melon (*Colocynthis citrullus* L.) and fluted pumpkin (*Telfaria occidentalis* Hook) by chick assay. *J. Sci. Food Agric.*, **38**, 237–6.

Oyenuga, V.A. and Fetuga, B.L. (1975) Some aspects of the biochemistry and nutritive value of the water melon (*Citrullus vulgaris*, Schrad). *J. Sci. Food Agric.*, **26**, 843–54.

Oyolu, C. and MacFarlane, N. (1982) A study of the oil and soluble protein components of five egusi (*Colocynthis citrullus* L.) cultivars. *Trop. Sci.*, **24**, 93–8.

Sharma, P.B., Lal, B.M, Madaan, T.R. and Chatterjee, S.R. (1986) Studies on the nutritional quality of some cucurbit kernel proteins *J. Sci. Food Agric.*, **37**, 418–20.

CHAPTER THIRTY

Fluted pumpkin (*Telfaria occidentalis* Hook.)

E. Nwokolo

30.1 INTRODUCTION

The fluted pumpkin, *Telfaria occidentalis* (Hook), is a tropical cucurbit, a member of a large family, the Cucurbitaceae, which includes gourds, pumpkins, squashes and melons. The botany of the crop has been described by Okoli and Mgbeogu (1983). The fluted pumpkin is a fast-growing, climbing annual which bears heavy fruits that are fluted or furrowed. The creeping stem may bear two or more fruits. Fruits take 5 months to mature and may weigh up to 10 kg. Small fruits weigh 2–5 kg. Fruits contain many seeds, small fruits containing up to 30 seeds, large fruits up to 70 seeds and very large fruits contain up to 100 seeds. Okoli and Mgbeogu (1983) report fruits measuring 105 cm in length and containing 196 seeds. They observe that the seeds are dark red in colour, non-endospermic and quite large in size, measuring 3.3–4.9 cm from end to end. Estimates of seed yield of fluted pumpkin in southern Nigeria are 2.5–3 tonnes/ha (Nwokolo and Sim, 1987). Seeds are cooked and eaten when mature. The young leaves form a very delicious vegetable when cooked, and are eaten in a variety of dishes.

30.2 CURRENT AND POTENTIAL FOOD USES

Because of their extremely high oil content, fluted pumpkin seeds are difficult to oven-dry, without considerable loss of vegetable fat. Air-drying is achieved over 3–4 days at ambient temperatures greater than

Legumes and Oilseeds in Nutrition
Edited by E. Nwokolo and J. Smartt
Published in 1996 by Chapman & Hall. ISBN 0 412 45930 2

Table 30.1 Proximate composition of pumpkin seed meal (dry matter basis)

Pumpkin seed meal	*Protein (%)*	*Fibre (%)*	*Fat (%)*	*Ash (%)*	*NFE (%)*
Raw, undefatted	26.6	3.1	53.4	2.3	14.6
Raw, defatted	69.7	8.0	6.5	9.3	6.3
Cooked, undefatted	27.1	3.1	50.6	5.2	14.0
Cooked, defatted	71.1	9.3	10.2	8.3	1.1
Mech.-pressed	46.5	6.9	19.1	8.6	9.0
Solvent extracted	71.9	8.0	0.9	9.4	10.7

From Longe *et al.* (1983), Nwokolo, unpublished data.

30°C. Oven-drying at 45°C is completed in 48 h. Because of high fat content, decorticated, dried pumpkin seeds are very difficult to store. Fresh seeds may be cooked for 1 h prior to eating, a deep purple colour developing in the cooking water. Cooked, full-fat pumpkin seeds are excellent sources of dietary fat and protein. They are used as snacks, and as vegetables in yam-based diets. They are cooked, ground and fermented to form a high protein flavour enhancer in soups, gravies and general cooking. In spite of the popular use of fluted pumpkin, there is no evidence that the oil is extracted and used as a dietary oil. Neither is there any evidence that the pressed cake is used in the same way as pressed melon seed cake. Because of the high protein content of fat-free pumpkin seed meal (71%), it is felt that there is tremendous potential to utilize the defatted meal as a protein concentrate in infant feeding. This high protein oilseed can be used in the manufacture of baby foods and protein supplements for babies. These are ways of ensuring that the most vulnerable age groups (babies and school-age children) obtain additional dietary protein to meet their daily requirements of amino acids. It can also be used in the fabrication of meat extenders and simulated meat products, to supplement the meagre animal protein intake in developing countries. High protein supplements like melon seed meal and pumpkin seed meal, can be used to supplement corn meals or sorghum meals, usually supplied as aid to refugees or victims of natural disasters. Because defatted pumpkin and melon seed powders are bland in taste, they can be used as high protein supplements to wheat flour, to be used in baking of snacks for school lunches in developing countries.

30.3 PROXIMATE COMPOSITION AND NUTRITIVE VALUE OF PUMPKIN SEED MEAL

Proximate constituents of raw, undefatted and defatted pumpkin seed meal together with cooked, undefatted and defatted pumpkin seed meal are shown in Table 30.1 which highlights the differences in nutrient content between defatted and undefatted meals. Undefatted meal has a

Table 30.2 Mineral constituents of pumpkin seed meal

	Content (mg/100 g dry matter)		
Mineral	*Raw undefatted*	*Raw defatted*	*Cooked undefatted*
Calcium	30.9	49.5	74.9
Phosphorus	162.5	359.4	148.9
Magnesium	402.5	699.0	216.9
Potassium	1111.0	2219.3	1379.0
Sodium	24.8	23.1	20.5
Manganese	2.3	3.8	1.9
Zine	10.0	13.9	2.1
Copper	2.2	4.7	21.8
Iron	37.8	31.7	30.0

From Longe *et al.* (1983).

very high fat content (51–53%), and moderate protein content (26–27%).

Table 30.3 Amino acid content and availability in defatted toasted pumpkin seed meal (air–dry)

Amino acid	*Content (mg/g)*	*Availability (%)*
Aspartic acid	53.6	97.2
Threonine	23.1	91.1
Serine	28.4	93.3
Glutamic acid	73.5	94.3
Proline	23.6	92.7
Glycine	25.1	93.1
Alanine	27.1	93.8
Valine	28.8	94.2
Methionine	8.4	93.7
Isoleucine	27.4	94.6
Leucine	50.1	94.6
Tyrosine	24.4	94.2
Phenylalanine	24.3	94.9
Histidine	19.8	87.9
Lysine	24.6	93.2
Arginine	64.8	96.5
Total amino acids	527.1	–
Average availability	–	93.8

From Nwokolo and Sim (1987).

Defatted meal has a low fat content (6.5–10%) and very high protein content (70–71%). Mechanically pressed pumpkin seed meal has very much more residual fat (19%) than solvent-extracted pumpkin seed meal (1%). The highest protein content of pumpkin seed meal (71.9%) obtained in our laboratory, was after almost complete removal of fat from pumpkin

Table 30.4 Lipid characteristics and fatty acid profile of pumpkin seed oil

Fatty acid[a]	*% of fatty acids*
C12:0	1.62
C14:0	0.63
C16:0	17.10
C18:0	15.04
C18:1	35.41
C18:22	27.09
C18:3	1.20
C22:1	1.67
Lipid characteristics[b]	
Specific gravity	0.91
Refractive index	1.47
Iodine number	120
Saponification number	190

[a]From Nwokolo and Sim (1987).
[b]From Nwokolo, unpublished data.

seed meal. The high oil content of the seed and the high protein content of the defatted meal, make pumpkin seed a potential source of commercial vegetable oil and protein in human nutrition.

Pumpkin seed meal has a high content of potassium, magnesium, and phosphorus. Calcium which is normally moderate or high in many food and feed ingredients, is extremely low in defatted pumpkin seed meal. There is only a moderate content of sodium and iron whereas there is a low content of zinc, manganese and copper (Table 30.2). In comparison, commercial soybean meal contains the following (mg/100 g): calcium, 306; phosphorus, 589; magnesium, 225; potassium, 1509; sodium, 50; manganese, 3.9; zinc, 4.0 and iron, 9.7. Availability of minerals which was determined by a chick assay is as follows: calcium 73.0%, phosphorus 65.1%, magnesium 52.2%, potassium 39.0%, sodium 58.6%, manganese 63.5%, zinc 58.2%, copper 63.1% and iron 54.8% (Nwokolo and Sim, 1987). Mineral availability is moderate for most minerals. Potassium availability is poor (39%), whereas availability of magnesium, iron, zinc and sodium is moderate (52–59%). Availability of such minerals as copper, manganese, phosphorus and copper is high (63–73%), and superior to availability of the same minerals in soybean meal. In general, average availability of all minerals in defatted pumpkin meal is 58.8%, slightly higher than average mineral availability of 54.9% in defatted pumpkin seed meal.

Amino acid content of defatted pumpkin seed meal is excellent, being comparable to that in soybean meal. For some essential amino acids, pumpkin seed meal has a greater content than soybean meal. Pumpkin seed meal has a low content of methionine but a moderate content of lysine, similar to the pattern in soybean meal (Table 30.3). Longe *et al.* (1983) compared the amino acid patterns of raw full-fat pumpkin seed and cooked full-fat pumpkin seed, with that of whole egg. There is twice as much sulphur amino acids and one and half times as much lysine in whole egg as in raw full-fat pumpkin seed meal. Pumpkin seed meal is superior to whole egg in respect of histidine and arginine, as well as in glutamic acid and proline. There is a considerable similarity in the content of tryptophan, phenylalanine and tyrosine in pumpkin seed meal and whole egg. Longe *et al.* (1983) conclude that the sulphur amino acids are the most limiting amino acids, and lysine and threonine are the second limiting amino acids, based on whole egg reference protein standard. Like most other oilseeds, pumpkin seed meal, melon seed meal, and soybean meal are deficient in methionine, cystine and lysine. Amino acid availability in defatted, toasted pumpkin seed meal is uniformly high (93.8%), as high as in soybean meal (94.9%) and in melon seed meal (96.1%).

Dried seeds of fluted pumpkin contain 51–53% fat, almost all of which is recoverable by exhaustive hexane extraction. The recovered oil has a bland taste and odour and the colour is usually a dark reddish-brown. The physical and chemical properties of the oil, summarized in Table 30.4, indicate that pumpkin seed oil closely resembles other common oils processed for food use. There is a high level of carotenoids in pumpkin seed oil (105 mg/g), similar to the level in cotton seed oil. Xanthophylls appear to be the predominant pigments present. The fatty acid composition of lipids in pumpkin seed is quite different from that in either melon seed or soybean. Fluted pumpkin lipid contains about 17% palmitic acid, 15% stearic acid, 35% oleic acid and 27% linoleic acid. This is in comparison with 12% palmitic acid, 11% stearic acid, 14% oleic acid, 58% linoleic acid and 2% linolenic acid in melon seed oil.

On the other hand, soybean oil contains 13% palmitic acid, 6% stearic acid, 28% oleic acid, 44% linoleic acid and 7% linolenic acid. Most members of the Cucurbitaceae contain oils with considerable proportions of unsaturated fatty acids, which are good salad oils and are used for cooking in various parts of the world.

Carbohydrates comprise only a small portion of the organic matter in raw pumpkin seed. Longe *et al.* (1983) found that starch accounts for only 0.5% of the pumpkin seed kernel and soluble sugars account for about 0.9% (Table 30.5). Digestible and metabolizable energy values of fluted pumpkin seeds have been determined using weanling rats. Gross energy values of undefatted pumpkin seed (raw or cooked) are extremely high (7.74–7.95 kcal/g), due to the very high content of oil in the undefatted meal. Defatted pumpkin seed meals, raw or cooked, have high gross energy content (4.82–4.97 kcal/g). This is in contrast with most cereals, peas and beans which have gross energy values of 3.9–4.2 kcal/g. In pumpkin seed meal as in many other oilseed meals and pulses, digestible

Table 30.5 Carbohydrate constituents of raw pumpkin seed

Constituent	*% of sample*
Total soluble sugars	0.9
Starch	0.5
Water soluble polysaccharide	4.6
Hemicellulose	2.1
Cellulose	2.5
Lignin	1.9

From Longe *et al.* (1983).

Table 30.6 Digestible and metabolizable energy values of pumpkin seed meal fed to weanling albino rats

Pumpkin seed meals	*GE (kcal/g)*	*DE (kcal/g)*	*ME (kcal/g)*
Raw, undefatted	7.955	6.378	4.897
Raw , defatted	4.826	4.157	3.201
Cooked, undefatted	7.740	6.832	5.614
Cooked, defatted	4.969	4.491	3.750

From Longe *et al.* (1983).
GE, DE and ME are gross energy, digetible energy and metabolizable energy values, respectively.

and metabolizable energy values are influenced by heat treatment.

Digestible energy value of raw, undefatted pumpkin seed meal is only 80% of its gross energy. For cooked pumpkin seed meal, digestible energy is 90% of its gross energy. Metabolizable energy values range from a very high 4.9 kcal/g in undefatted pumpkin seed meal, to high values of 3.2–3.7 kcal/g, in mechanically pressed meals. Solvent-extracted meals have lower metabolizable energy values, of about 2.4–2.5 kcal/g. Results reported by Longe *et al.* (1983) on the digestible and metabolizable energy content of fluted pumpkin as determined with albino rats, are shown in Table 30.6. These results are in general agreement with metabolizable energy values for other cucurbits.

30.4 TOXICOLOGICAL COMPONENTS IN PUMPKIN SEED AND MEAL

In feeding trials with rats, Longe *et al.* (1983) report 100% mortality within 8 days, among albino rats fed on diets in which raw (defatted and undefatted) pumpkin seed meal supplied 15% of the dietary protein. High mortality rates usually follow feeding of raw undefatted oilseeds to rats. The presence of naturally occurring toxins has been suggested as a causative factor in these deaths. Raw full-fat pumpkin seed meal, fed at levels which supplied 10% dietary protein, although not causing death, is associated with drastic weight loss in rats.

For pumpkin seed meal, as with melon seed meal, growth is improved with cooking and oil extraction, suggesting that growth inhibitors are removed or destroyed by these treatments. In a short duration growth trial of 28 days, using rapidly growing broiler chicks as assay animals, it has been observed that broiler chicks fed raw defatted melon and pumpkin seed meal have a high mortality rate (45%), within 7 days. Toasting eliminates the factor(s) responsible for the mortality and improves the nutritional quality of both melon and pumpkin seed meals (Nwokolo, unpublished). Protein and energy digestibility experiments with rats, have shown that toasting or other heat treatment of oilseeds and legumes, improves digestibility and utilization of the nutrients in these products. It is recognized that this improvement in nutrient utilization following toasting may be as a result of thermal destruction or denaturation of antinutritional factors, or improvement due to gelatinization of starches and polysaccharides.

The chemical nature of the antinutritional factors in fluted pumpkin have, however, not been reported, although heat-labile trypsin and chymotrypsin inhibitors are suspected to be implicated.

Jacks *et al.* (1972) report that cucurbit seeds contain considerable amounts of phytic acid, which accounts for the large amounts of phosphate reported in various cucurbit seeds and seed meals. If that is so, then availability of many minerals should be low, as phytates combine with various dietary minerals, rendering them unavailable. Cucurbit seeds have also been reported to contain several terpenoid glycosides, including the purgative colocynthin. Proper heat treatment, alone or in combination with some fermentation, however, seems to be enough to detoxify or inactivate these antinutritional factors in cucurbit seeds and meals.

30.5 ANIMAL FEEDING EXPERIMENTS

Jacks *et al.* (1972) discussed results from 27 years' research into the potential uses of oils and proteins of cucurbits, for edible purposes. Members of the family Cucurbitaceae have variable seed coat content, ranging from 18% to 60% of the weight of the seed. Melon seed has a low coat : seed ratio, whereas fluted pumpkin has a higher coat seed ratio. For many cucurbits, decorticated seeds contain, by weight, 50 ± 2.0% fat and 35 ± 2.5% protein. These workers report that a majority of the oil in many cucurbit seeds comprises unsaturated fatty acids, a smaller proportion being conjugated trienes. Most cucurbit oils are valuable edible oils, many being over 95% digestible in rats. Pumpkin seed oil is 99% digestible. *Cucurbita* seeds contain about 35% protein, of which between 80–95% is globulin (Alekseeva, 1960, 1963). Various reports on different cucurbits indicate that they are rich in aspartic acid and glutamic acid and in the essential amino acid arginine. By using the procedure of Kwolek and van Etten (1968), the nutritional value of various cucurbit seeds has been assessed, based on their content of essential amino acids. Kwolek and van Etten suggest computing the ratio between the content of each essential amino acid in a plant protein and the content of the same amino acid in egg protein. The mean value of these ratios is calculated and the closer this mean value is to unity, the more the plant protein resembles egg protein. By using these procedures, Jacks *et al.* (1972) showed that although cucurbit seed proteins are deficient in lysine and the sulphur amino acids, as are other oilseeds, cucurbit seed meals are good sources of protein. Globulins isolated from cucurbit seed meals are far superior to the meals themselves, although the globulins are inferior to casein and lactalbumin for supporting weight increases in growing rats (Hubbell *et al.*, 1953). Pumpkin seed meal has also been compared to soybean meal, Zucker *et al.* (1958) reporting that soybean meal is superior to pumpkins seed meal, for weight regains in protein-depleted rats and pigs.

Many experiments have been reported (Bressani, 1963; Bressani and Arroyave, 1963) in which various oilseeds have been compared with powdered skim milk. At marginal dietary protein levels for rats, protein efficiency ratio (PER) of pumpkin seed is approximately 80% of that of skim milk powder. At higher dietary protein levels, pumpkin seed meal is equivalent to or superior to skim milk powder in diets of growing rats. Bressani and Arroyave (1963) observed that lysine supplementation of pumpkin seed protein, drastically increases PER of pumpkin seed meal. Although supplementation with either methionine or threonine does not influence PER, supplementation with both amino acids increases not only growth rate but also feed conversion ratio in rats.

REFERENCES

Alekseeva, M.V. (1960) The seed proteins of cucurbita. *Chem. Abstr.*, **54**, 1667.

Alekseeva, M.V. (1963) A comparative study of nitrogen-containing substances in the seeds of some representatives of the gourd family. *Chem. Abstr.*, **58**, 10514d.

Bressani, R. (1963) Nutritional value of various mixtures of vegetable oilseed proteins developed to combat protein malnutrition in Latin America. *Qual. Plant.*, **10**, 73–108.

Bressani, R. and Arroyave, R. (1963) Nutritive value of pumpkin seed. Essential amino acid content and protein value of pumpkin seed (*Cucurbita farinosa*). *Agric. Food Chem.*, **11**, 29–33.

Hubbell, R.B., Vickery, H.B. and Noland, L.S. (1953) Cucurbit seed globulins. II. Use as a substitute for edestin in experimental diets. *J. Nutr.*, **25**, 99–101.

Jacks, T.J., Hensarling, T.P. and Yatsu, L.Y. (1972) Cucurbit seeds: 1. Characteristics and uses of oils and proteins, A Review. *Econ. Bot.*, **26**, 135–41.
Kwolek, W.F. and van Etten, C.H. (1968) Evaluation of amounts and patterns of essential amino acids in plant seeds. *J. Agric. Food Chem.*, **16**, 496–9.
Longe, O.G., Farinu, G.O. and Fetuga, B.L. (1983) Nutritional value of the fluted pumpkin (*Telfaria occidentalis*, Hook). *J. Agric. Food Chem.*, **31**, 989–92.
Nwokolo, E. and Sim, J.S. (1987) Nutritional assessment of defatted oil meals of melon (*Colocynthis citrullus* L.), and fluted pumpkin (*Telfaria occidentalis* Hook), by chick assay. *J. Sci. Food Agric.*, **38**, 237–46.
Okoli, B.E. and Mgbeogu, C.M. (1983) Fluted pumpkin, *Telfaria occidentalis*; West African vegetable crop. *Econ. Bot.*, **37**, 145–9.
Zucker, H., Hays, V.W., Speer, V.C. and Catron, D.V. (1958) Evaluation of pumpkin seed meal as a source of protein using a depletion-repletion technique. *J. Nutr.*, **65**, 327–34.

CHAPTER THIRTY-ONE

Bottle gourd, buffalo gourd and other gourds

E. Nwokolo

31.1 INTRODUCTION

Many varieties of gourd (*Cucurbita* spp.) are indigenous to tropical areas of the world. Most are wild, a few are cultivated for the seeds or the gourds, still used for storage of water and beverages in tropical Africa. Smaller gourds are carved into household ornaments, used as musical instruments, and as cups for drinking of water and beverages. Okoli (1984) observes that cucurbits occupy such a prominent place in the diet, culture and life of many ethnic groups in Nigeria, that virtually every farming family has at least one cucurbit species in the garden. The fruits of the bottle gourd (*Lagenaria siceraria*) show unparalleled variation in size, shape and utility. The bottle gourd is cultivated for its fruit and although the fruits are too bitter to eat, improved cultivars of this gourd are reportedly as good as the pumpkin and are used as food. Okoli reports that bottle gourds are used as receptacles for fresh and fermented milk for guinea corn gruel, drinking vessels, ladles, flutes and for storage of palm wine.

The wild buffalo gourd (*Cucurbita foetidissima*), is indigenous to western North America including Mexico. It is a very vigorous, drought-resistant annual, producing a large number of fruits containing seeds that are rich not only in protein (33%), but also in oil (33%). Bemis *et al.* (1975) describe the buffalo gourd as a potential crop for production of protein and oil in the dry areas of the world, and estimate seed yield of the buffalo gourd at about 3.3 MT/ha in desert regions of the world.

Legumes and Oilseeds in Nutrition
Edited by E. Nwokolo and J. Smartt
Published in 1996 by Chapman & Hall. ISBN 0 412 45930 2

Table 31.1 Chemical composition of some edible cucurbits

Nutrient	*Brown melon seed*	*White melon seed*	*Bottle gourd seed*
Crude protein (%)	33.8	36.3	35.4
Oil (%)	53.1	50.4	47.9
Total sugar (%)	6.0	6.6	6.1
Starch (%)	7.1	7.6	9.4
Minerals			
Calcium (%)	0.15	0.10	0.10
Magnesium (%)	0.44	0.54	0.54
Phosphorus (%)	0.95	1.10	1.05
Sulphur (%)	0.38	0.36	0.45
Potassium (%)	0.65	0.70	0.70
Manganese (ppm)	140	40	60
Iron (ppm)	290	250	250
Zinc (ppm)	58	72	78

From Okoli (1984).

31.2 CURRENT FOOD AND FEED USAGE

Seeds of most gourds are not currently used as food for human beings or as feed for animals. This is in spite of the demonstrated potential for such usage. Seeds of the bottle gourd have for example been reported (Okoli, 1975) to contain high levels of protein (35%) and oil (48%), a little higher than the content of the same nutrients in the buffalo gourd. Berry *et al.* (1976) found that the buffalo gourd has tremendous promise as a possible new source of edible oil and protein in arid regions, because the seed is high in protein and fat. They estimate that with a gross energy value of 6.17 kcal/g, and a metabolizable energy value of 4.32 kcal/g, the buffalo gourd is equivalent to other oilseeds in protein and available energy, although the fibre content is relatively high (35%). The application of existing processing technology can convert the protein and oil in the lowly gourd seed, into high quality dietary protein and oil for human culinary or industrial use. These gourd seeds can also be processed into high quality dietary supplements for domestic animals.

31.3 PROXIMATE COMPOSITION AND NUTRITIVE VALUE

There is considerable variation in proximate composition of the various gourd members of the family Cucurbitaceae (Table 31.1). Variation exists not only in protein content, but in fat, carbohydrate and mineral content.

There is also variation in size of seed and in seed coat. The weight of the seed coat as a percentage of the entire seed, ranges from 18% in *Cucurbita*

Table 31.2 Chemical composition of buffalo gourd seed

Proximate constituent	*Whole seed*	*Whole kernel*	*Defatted kernel*	*Gourd seed flour*[a]
Moisture (%)	4.9	4.5	4.1	7.6
Protein (%)	32.9	37.5	75.0	70.2
Oil (%)	33.3	48.0	1.0	0.3
ADF (%)	35.1	6.9	11.8	4.0
Lignin (%)	20.0	0.7	2.3	–
Ash (%)	3.1	4.2	7.6	8.8
Carbohydrate (%)	–	3.4	4.6	9.1
Gross energy (kcal/g)	6.17	7.03	4.70	–

From Berry *et al.* (1976).
[a]From Tu *et al.* (1978).

pepo seed, to 60% in *Lagenaria siceraria* seed. For many gourds, whole seeds have a fairly low moisture content (5%), decreasing to 4.5% in decorticated kernels, and 4.0% in defatted kernels. In whole gourd seeds, crude protein content is 32–36% (average 35%) of which 80–90% are globulins (Jacks *et al.*, 1972). These workers report that these globulins contain about 18% nitrogen, are soluble in 10% salt solution from which they crystallize on dilution, and are also soluble in both acidic and basic solutions of low ionic strength. Oil content in most gourd seeds is in the range 31–39%, with an average of 36%.

Table 31.2 compares the proximate composition of buffalo gourd seed as whole seed, whole kernel, defatted kernel and after milling as flour. After decortication, the nutritive value of the residual kernel is greatly improved. Crude protein and oil content increase (37.5%, 48.0%, respectively) whereas acid detergent fibre (ADF) and lignin decrease (6.9% and 0.7%, respectively). Defatted gourd kernels have extremely high protein content (72–75%), very low fat content (1%), and moderately low ADF (12%) and lignin (2–3%) content. The decorticated, undefatted kernels have very a high gross energy content, whereas the defatted kernels have moderate gross energy content, due to very low residual fat.

Undefatted meals of most gourd species contain extremely high levels of lipids; levels so high that in most cases, the seeds are difficult to grind. The oils usually have a high carotenoid content, giving this group of oils a deep orange coloration; a hue observed in oils from cucurbit seeds as diverse as bottle gourd, buffalo gourd, fluted pumpkin and squash seed. Vasconcellos *et al.* (1980) reported that the level of carotenoids in buffalo gourd oil was similar to that in cotton seed oil, with a mean of 110 mg/kg and a range of 51.3–231.5 mg/kg. They observed that xanthophylls were the predominant pigment in gourd seed oil, as compared to beta-carotenes in raw red palm oil, the edible oils predominantly used in many tropical countries. The refractive index, free fatty acid, specific gravity and

Table 31.3 Physicochemical properties of gourd seed oil

Free fatty acid (%)	0.5±.2
Peroxide value (mg/kg oil)	16.9±10.9
Saponification value	191.5±1.6
Iodine value	129.9±5.3
Hydroxyl value	7.3±.5
Carotenoids	109.7±55.0

From Vasconcellos *et al.* (1980).

Table 31.4 Fatty acid profile of buffalo gourd lipid

	(% of lipid)	
Major fatty acids	*Range*	*Mean ± SD*
Palmilic acid	6.6–24.4	11.8±4.8
Stearic acid	1.2–10.2	3.5±2.2
Oleic acid	10.0–31.6	22.2±6.2
Linoleic acid	39.3–77.2	60.6±11.0
Linolenic acid	0.0–0.3	0.03±0.05

From Vasconcellos *et al.* (1980).

peroxide values of gourd seed oil, are similar to values accepted as normal in many edible oils (Table 31.3). Gourd seed oils in general have a low content of saturated fatty acids and a relatively high content of linoleic acid, similar to values in corn oil and sunflower oil (Table 31.4). It is the presence of high levels of linoleic acid in corn and sunflower oils which have earned them their universal acceptance as excellent dietary oils, and as potent sources of polyunsaturated fatty acids. Because of their very good fatty acid profile, gourd seed oils have also the potential to be accepted as excellent dietary oils, when their nutritional qualities have been completely assessed.

The amino acid profile in gourd seed flour is shown in Table 31.5, which compares the profile of amino acids in gourd seed flour with that of soybean. The FAO amino acid standard profile which is designed to meet human nutrient requirements is also provided for further comparison. Amino acid profile of gourd seed flour is inferior to both the soybean amino acid profile and the FAO recommended amino acid pattern. This indicates a potential deficiency if gourd seed flour is utilized as a sole protein in human nutrition.

Tu *et al.* (1978) compared the essential amino acids in buffalo gourd seed flour with the minimum dietary amino acid requirement for rat growth (Table 31.6). These workers had computed the supply of amino acids from a diet containing buffalo seed flour at a level of 14.2%, and showed that with the exception of phenylalanine, buffalo gourd seed flour

Table 31.5 Amino acid content of gourd seed flour

	Content (g/16 g N)		
Amino acid	*Gourd flour*	*Soybean*[a]	*FAO standard*
Lysine	3.85	5.60	5.50
Leucine	6.14	7.40	7.00
Isoleucine	3.23	5.30	4.00
Methionine	1.99	1.40	3.50
Phenylalanine	4.05	4.65	6.00
Threonine	2.21	4.00	4.00
Valine	3.79	4.90	5.00
Tryptophan	0.89	1.40	1.00
Tyrosine	2.93	3.20	–
Cystine	1.95	1.50	–
Histidine	2.35	2.70	–
Arginine	15.73	4.60	–
Glutamic acid	20.65	17.60	–
Aspartic acid	8.93	11.40	–
Proline	3.19	5.10	–

From Tu *et al.* (1978).
[a]From Martin (1984).

was deficient in all other essential amino acids. DeVeaux and Shultz (1985) suggested that buffalo seed protein had an amino acid pattern similar to that of many conventional oilseeds, except soybean which was higher in lysine. These workers were of the opinion, however, that gourd seed flours should be blended with other protein sources to achieve optimal utilization of the gourd seed protein. Given the low content of most amino acids in gourd seed flour, it is not recommended that gourd seed protein be administered as the sole source of protein in human or animal diets. It will serve as a useful adjunct to other complementary protein sources, helping to meet nutrient requirements for various amino acids. Jacks *et al.* (1972) had assessed the nutritional quality of cucurbit seeds for food use, through various comparisons of amino acid composition of the seeds. They had compared the mean value of ten essential amino acids in cucurbit seeds, against the mean value of the same ten essential amino acids in chicken egg protein, and showed that cucurbit seed protein was inferior to whole egg protein. These workers also expressed each essential amino acid as a ratio of all 10 essential amino acids in both cucurbit seeds and whole egg. They showed that most cucurbit seeds were deficient in lysine and sulphur amino acids, as were most oilseeds. Further evidence of the inability of gourd seed flour to sustain growth when used as a major source of protein in rat diets, was provided by Tu *et al.* (1978) who calculated the contribution of amino acids from a rat diet containing 14% gourd flour and compared the amino acid levels with the NRC amino acid requirements for growing rats. They

Table 31.6 Essential amino acids in gourd seed flour

Essential amino acids	*Amino acid content (g/ 100 g)*	*Amino acids in 14% gourd flour (% diet)*	*Amino acid required for growth*[a]	*Deviation from growth needs (%)*
Lysine	2.70	0.38	0.75	-49
Isoleucine	2.27	0.32	0.46	-30
Leucine	4.31	0.61	0.62	-2
Methionine	1.39	0.20	0.33	-39
Phenylalanine	2.84	0.40	0.33	+39
Threonine	1.55	0.22	0.41	-46
Valine	2.66	0.38	0.49	-22
Tryptophan	0.62	0.09	0.12	-25

From Tu *et al.* (1978).
[a]NRC requirements for rats.

showed that for most amino acids, serious deficiencies would occur (Table 31.6).

In order to utilize cucurbit seed proteins effectively in animal or human nutrition, these proteins need to be supplemented with lysine and methionine, leucine and isoleucine, threonine, valine and tryptophan. For human use, small quantities of meat or fish are enough to provide the supplementation needed for optimal utilization of cucurbit seed proteins.

31.4 TOXICOLOGICAL COMPONENTS OF GOURD SEEDS

Seeds of many cucurbits contain considerable quantities of phytic acid. Phytic acid is an antinutritional factor which interferes with utilization of mineral elements in gourd seeds, gourd seed meals and diets containing gourd seed products. Bolley and McCormack (1952) report that when solvent-extracted meals of *Cucurbita foetidissima, C. digitata* and *C. palmata* are extracted with a dilute acid solution and the extract is subsequently boiled, a precipitate containing 70–80% phytin is formed. Many cucurbit seeds are also reported to contain small quantities of several tetracyclic terpenes and terpenoid glycosides, collectively called cucurbitacins (Jacks *et al.*, 1972). A larger quantity of cucurbitacins exists in the fruit pulp, giving the pulp a bitter taste. Thorough cleaning of gourd seeds before drying eliminates cucurbitacin contamination from the fruit pulp. Various feeding trials using chicks and rats, some of which are discussed in detail later, have not revealed the presence of toxic components in gourd seed meals. Growth depression observed when these gourd meals have been fed as the sole protein source, have primarily been due to amino acid imbalance. Vasconcellos *et al.* (1980) found small quantities of conjugated fatty acids which, according to them, have no important effect on the possible use of gourd seed oil as an edible oil.

31.5 FEEDING EXPERIMENTS WITH GOURD SEED PROTEIN

There are very few experiments in which seeds or seed meals of gourds have been fed to animals. This is primarily because only in recent years have any biochemical or nutritional assays of gourd seed meals been conducted. Only recently have gourd seeds been recognized as potential sources of protein and lipids in animal and human nutrition.

Amino analysis data indicate that many cucurbit seed meals, including nearly all gourd seed meals, are deficient in threonine, lysine, methionine and isoleucine. The deficiency of these amino acids in gourd seed meals implies that they are not noteworthy protein sources.

In experiments in which purified globulins from various cucurbits were compared to casein and lactalbumin in rat diets, cucurbit globulins supported significantly lower growth rates (Hubbell *et al.*, 1953).

By using a depletion–repletion technique, Zucker *et al.* (1958) evaluated the nutritional quality of pumpkin seed meal in rats and pigs. These workers reported significantly improved repletion rates in soybean-fed rats and pigs, in comparison with pumpkin-fed animals.

A number of experiments in which protein efficiency ratio (PER) was utilized to estimate the nutritional quality of various cucurbit seed meals, have been reported (Bressani, 1963; Bressani and Arroyave, 1963). These workers studied the PER values of various cucurbit seed meals developed as human food to combat malnutrition in Central America. They observed that the PER of rat diets supplemented with cucurbit seed meals was 80% of those diets supplemented with skim milk. When the protein level was increased in diets of rats fed on cucurbit seed meals, the rats grew at a rate faster than those fed skim milk. Bressani and Arroyave (1963) showed that whereas supplementation of pumpkin seed meal diets with either threonine or methionine caused no increases in growth rate, supplementation with both amino acids increased growth rate and feed efficiency. Tu *et al.* (1978) found that in rat diets containing buffalo gourd seed flour, addition of some amino acids individually (threonine, lysine, methionine, isoleucine) did not improve growth response. However, supplementation with lysine, threonine and methionine was associated with growth rate similar to that of rats fed on casein, and further supplementation with isoleucine gave growth superior to rats fed the casein diets.

These workers concluded that lysine and threonine were the most limiting essential amino acids in buffalo gourd seed flour for rat growth (Table 31.7). To ensure optimal utilization of most cucurbit seed proteins in human food, it is recommended that they are supplemented with lysine, methionine, threonine and isoleucine or with combinations of proteins that are excellent sources of these amino acids.

Table 31.7 Weight gain and protein efficiency ratio (PER) of rats fed diets containing gourd seed flour, supplemented with amino acids

Dietary treatment	*Weight gained (g)*	*PER*
Casein control	89.5±9.2	2.50
Soybean flour	65.5±4.0	1.79
Raw gourd seed flour	57.0±2.4	1.72
Heated gourd seed flour	66.1±6.8	1.84
Gourd seed flour + Thr	67.2±2.1	1.79
Gourd seed flour + Thr + Lys	92.2±5.9	2.50
Gourd seed flour + Thr + Lys + Met	97.6±3.4	2.66
Gourd seed flour + Thr + Lys + Met + Ile	117.8±9.1	2.64

From Tu *et al.* (1978).

REFERENCES

Bemis, W.P., Curtis, L.C., Weber, C.W. *et al* (1975) The buffalo gourd (*Cucurbita foetidissima* HBK). A potential crop for production of protein, oil and starch on arid lands. *Technical Series Bulletin No. 15.* Office of Agriculture, Technical Assistance Bureau, Agency for International Development, Washington, DC.

Berry, J.W., Weber, C.W., Dreher, M.L. and Bemis, P. (1976) Chemical composition of buffalo gourd, a potential food source. *J. Food Sci.*, **41**, 465–6.

Bolley, D.S. and McCormack, R.H. (1952) Separation of phytin from oil seed flours. *J. Am. Oil Chem. Soc.*, **29**, 470–2.

Bressani, R. (1963) Nutritional value of various mixtures of vegetable oilseed proteins developed to combat protein malnutrition in Latin America. *Qual. Plant.*, **10**, 73–108.

Bressani, R. and Arroyave, R. (1963) Nutritive value of pumpkin seed. Essential amino acid content and protein value of pumpkin seed (*Cucurbita farinosa*). *Agric. Food Chem.*, **11**, 29–33.

deVeaux, J.S. and Shultz, E.B. (1985) Development of buffalo gourd (*Cucurbita foetidissima*) as a semi-aridland starch and oil crop. *Econ. Bot.*, **39**, 454–72.

Hubbell, R.B., Vickery, H.B. and Nolan, L.S. (1953) Cucurbit seed globulins. II. Use as substitutes for edestin in experimental diets. *J. Nutr.*, **25**, 99–101.

Jacks, T.J., Hensarling, T.P. and Yatsu, L.Y. (1972) Cucurbit seeds. 1. Characterizations and uses of oils and proteins, a review. *Econ. Bot.*, **26**, 135–41.

Martin, F.W. (1984) *Handbook of Tropical Food Crops*. CRC Press, Boca Raton, FL.

Okoli, B. (1984) Wild and cultivated cucurbits in Nigeria. *Econ. Bot.*, **38**, 350–7.

Tu, M., Eustace, W.D. and Deyoe, C.W. (1978) Nutritive value of buffalo gourd seed protein. *Cereal Chem.*, **55**, 766–72.

Vasconcellos, J.A., Berry, J.W., Weber, C.W. *et al.* (1980) The properties of (*Cucurbita foetidissima*) seed oil. *J. Am. Oil Chem. Soc.*, **57**, 79–84.

Zucker, H., Hays, V.W., Speer, V.C. and Catron, D.V. (1958) Evaluation of pumpkin seed meal as a source of protein using a depletion–repletion technique. *J. Nutr.*, **65**, 327–34.

PART FOUR

Palmae

CHAPTER THIRTY-TWO

Babassu (*Orbigyna oleifera* B.) fruit, oil and meal

L. Lennerts and W. Lennerts

32.1 INTRODUCTION

Babassu oil and meal are by-products of the fruit of the babassu palm tree (*Orbigyna oleifera* B.). The babassu palm tree grows mainly in the South America equatorial region, where it covers an enormous area of more than 33 million acres. With a height of 15–22 m a thin, smooth sepia-coloured trunk, and a high crown, it can be regarded as one of the most beautiful palm tree species. Because the upper part of its fronds stands vertically upward, it can be recognized from a distance and easily be distinguished from other palm trees.

The fruits of the babassu palm tree form as hanging bunches each containing from 200–400 fruits. In the fully ripe condition each fruit has the shape of a goose egg, and, depending on the size, weighs 80–250 g, (average about 150 g). The colour is light brown. Each fruit consists of an outer fibrous layer, a meal-like fleshy middle layer and an inner nut. While ripening, the fruit flesh dries to a 4–8 mm thick, light brown mealy mass. Well ground, washed and pureed it forms a very tasty starchy flour. Cooked with milk, it results in a product with a chocolate-like taste, which is eaten by the local people where these palms grow. The dried fruit flesh contains about 3.3% crude protein and 10% crude fat. Apart from being used in human nutrition, it is fed to pigs and is used as a raw material for alcohol production, 2.5 kg fruit flesh usually resulting in 1 l of pure alcohol. The inner babassu nut consists of a stony material, in which five to six kernels are embedded in a longitudinal axis. The average weight of

Legumes and Oilseeds in Nutrition
Edited by E. Nwokolo and J. Smartt
Published in 1996 by Chapman & Hall. ISBN 0 412 45930 2

Table 32.1 Chemical composition of babassu kernels

	%
Water	5.0
Crude fat	67.0
Crude protein	9.0
Nitrogen-free extract	10.5
Crude Fibre	6.5
Ash	2.0

Table 32.2 Content of fatty acids in babassu fat

Fatty acids	*Content (%)*
Lauric acid	44.0–45.0
Myristic acid	15.0–16.5
Oleic acid	12.0–16.0
Capric acid	6.6–7.6
Palmitic acid	5.8–8.5
Caproic acid	4.1–4.8
Stearic acid	2.5–5.5
Linoleic acid	1.4–2.8
Indices	
Melting point (°C)	22–26
Saponification number	247–253
Iodine number	13–16

the entire fruit is approximately 150 g. The percentage of the different parts by weight being: the outer fibrous layer, 10%; fruit flesh, 24%; stored material, 58%; and kernels, 7–8%. Only the fruit flesh, and the kernels are eaten. Depending on the size of the fruit and the number of kernels, average kernel weight is 3–9 g.

32.2 BABUSSU FAT IN HUMAN NUTRITION

Table 32.1 shows the nutrient composition of babassu kernels. They contain 67% lipids and represent a potentially important source of dietary fats for human nutrition. As depicted in Table 32.2, almost half the babassu fat is made up of saturated lauric acid. Monosaturated fatty acids make up only 15% of total fatty acids. The high proportion of saturated fatty acids results in a relatively low iodine number (13–16) and a melting point which lies in the medium range (around 30°C). Because of its chemical composition, babassu fat is an excellent cooking fat. It can be used separately or in combination with other vegetable fats and oils and is best used for frying meat, mixing dough, and in potato dishes.

Table 32.3 Composition of babassu cake and meal

Feedstuff	*Content (% dry matter basis)*					
	Organic matter	*Crude protein*	*Crude fat*	*Crude fibre*	*Nitrogen-free extract*	*Ash*
Babassu cake free from stone material	94.2	25.1	7.0	17.7	44.3	5.8
Babassu oil meal with a portion of stone material	93.9	22.6	5.6	31.5	35.2	5.1
Babassu oil meal free from stone material	93.7	24.0	1.4	18.0	50.3	6.3
Babassu oil meal	94.5	22.7	2.1	27.5	42.2	5.5

Table 32.4 Content of amino acids in babassu protein

Amino acid	*% of total amino acids*
Arginine	14.0
Histidine	1.8
Isoleucine	3.9
Leucine	6.2
Lysine	4.3
Methionine	2.3
Phenylalanine	5.9
Threonine	3.1
Tryptophan	1.1
Valine	5.2
Other amino acids	52.2

Additionally, it is used in the production of chocolate and toilet soaps, in the latter case because of its good lathering properties.

32.3 BABASSU PRODUCTS AS FEEDSTUFFS FOR ANIMALS

Commercial production of babassu fat involves hydraulic/mechanical crushing and compression, followed by oil extraction. The residue is babassu cake or babassu meal depending on the extraction procedure used. Table 32.3 shows that the resulting cake or meal has a medium content of protein and fibre and has the potential for utilization as an animal feedstuff. Table 32.4 shows that the protein of babassu meal contains a relatively high proportion of essential amino acids, for example lysine (4.3%) and methionine (2.3%). With respect to its value in animal nutrition, babassu protein can be regarded as an important dietary protein for various types of domestic animals.

The content of crude fibre in babassu meal products depends on the

Table 32.5 Content of minerals and trace minerals in babassu cake

K (g/kg)	*Na* (g/kg)	*Ca* (g/kg)	*Mg* (g/kg)	*P* (g/kg)	*S* (g/kg)	*Fe* (mg/kg)	*Mn* (mg/kg)	*Cu* (mg/kg)
9.1	0.7	2.6	4.3	11.0	0.5	375	325	44.5

proportion of crushed stone materials left over in the final product. In meal products from only kernels, whereas the content of crude fibre may be regarded as significant (average 16–17%), there is considerable digestibility of cellulolytic material resulting in a positive nutritional value for animals. Thus for ruminant animals, digestibility of crude fibre may be as high as 60%, with a high content of net energy.

When babassu meal products which contain a considerable proportion of stone material are prepared, the content of crude fibre is significantly higher, almost in proportion to the quantity of stone material incorporated. Because the fibre of the stone material is indigestible, the digestibility of the meal is greatly diminished, and the metabolizable and net energy values are very much lower.

Table 32.5 shows the content of mineral and trace elements in babassu feedstuffs. The high content (average 11 g/kg) of phosphorus is noteworthy. In contrast to this, the content of calcium (average of 2.6 g/kg) is relatively low. In addition, the content of potassium (average of 9.1 g/kg) and manganese (average of 325 mg/kg) is noteworthy.

Nutrition information currently available indicates that babassu cake or meal produced using only the kernel can be regarded as an excellent concentrated feedstuff for dairy cattle, young or fattening beef cattle, or even sheep. This cannot, however, be said of meal produced from a combination of other non-kernel parts of the babassu seed.

32.4 POTENTIAL FOR INCREASED UTILIZATION

Whether the products of the babassu palm tree will have greater international economic importance in the future will depend on several factors. It will be necessary to open up vast tracts of rain forest jungle where the babassu palm grows, to allow for cultivation and transportation of the fruits. There is also need for systematic cultivation of the babassu palm possibly in plantations. Additionally, there will be a need to develop an efficient procedure to smash the hard stone material and facilitate the recovery of the babassu kernels, a procedure which is currently not available. Finally, there is the problem of the relatively high proportion of non-usable material. To get an idea about the disproportion of non-usable matters, one has to realize that the production of 1 tonne of babassu fats involves production of 10–12 tonnes of stone material. In the future, the economic importance of babassu fat will largely depend on its price in comparison to other vegetable fats and oils which are used in human nutrition.

FURTHER READING

Esdorn, J. (1961) *Die Nutzpflanzen der Tropen und Subtropen der Weltwirtschaft.* Gustav Fisher, Stuttgart.

Kling and Wohlbier (1983) *Handelsfuttermittel.* Vols 2a and 2b., Eugen Ulmer, Stuttgart.

Lennerts, L. (1984) *Olschrote, Olkuchen, Pflanzliche Ole Fette. Herkun, Gewinnung un Verwendung.* A. Strothe, Hannover.

CHAPTER THIRTY-THREE

Coconut (*Cocos nucifera* L.) kernel, oil and meal

O.C. Onwudike

33.1 GENERAL BOTANICAL, ECOLOGICAL AND PRODUCTION INFORMATION

33.1.1 Botany and ecology of the coconut

The coconut palm (*Cocos nucifera*) is a monocotyledonous stately tree about 15–30 m tall, 20–60 cm in diameter and usually unbranched. Some trees are, however, dwarf and have branches. The botany of the coconut has been fully discussed by Child (1974). The fruit is ovoid in shape and is a fibrous drupe. It consists of four parts: about 35% husk, 12% shell, 28% meat (kernel), and 15% water (Aten *et al.*, 1958). From a study of very young Sri Lankan coconuts gathered at 4 weeks, Nathanael (1958) reports average wet weights of 413 g, 177 g, 339 g and 124 g for husk, shell, kernel and coconut water. Dry weights were 330 g, 155 g, 182 g and 6 g for husk, shell, kernel and water, these fractional weights representing 49.0, 23.0, 27.1 and 0.9% of total weights.

About 90% of the total coconut hectarage in the world lies in the zone 20°N and 20°S latitudes. The major areas of production are the Philippines, India, Indonesia, Sri Lanka, South Sea Islands and Malaysia. Coconuts are now grown in all the tropical regions around the world where the temperature, humidity, soil, and elevation are suitable.

Legumes and Oilseeds in Nutrition
Edited by E. Nwokolo and J. Smartt
Published in 1996 by Chapman & Hall. ISBN 0 412 45930 2

33.1.2 Physical and chemical nature of coconut and its by-products

The important by-products of the coconut which are useful for human and animal feeding are the coconut kernel, copra (coconut cake), coconut water, coconut oil and toddy.

(a) The coconut kernel

The coconut contains about 28% inedible shell and 72% edible portion. According to Woodroof (1978), fresh coconut meat (kernel) is rich in fat and carbohydrates and contains moderate amounts of protein. It also contains 151 IU of thiamine, traces of vitamin A and 0.2 mg of tocopherol per 100 g. Coconut proteins are high in nutritive value and are fairly rich in lysine, methionine and tryptophan. Subrahmanyan and Swaninathan (1959) investigated the proximate composition of coconut. They showed that in unripe coconut kernels, moisture content is about 90.8%, protein 0.9%, fat 1.4%, carbohydrates 6.3% and ash 0.6%. In ripe coconut kernels, moisture content is only 46.3%, protein 4.08%, fat 37.3%, carbohydrates 11.3%, ash 1.03% and crude fibre 3.4%.

(b) Copra

Copra is dried coconut endosperm (kernel). The husked coconut is split into two, the kernel is removed and dried. Curing of the copra may be by sun drying, kiln drying or a combination of the two. Sun drying requires long sunny periods, with no rainfall and lasts 6–7 days on cement or bamboo floors. The copra produced amounts to about 23% of the whole nut or 32% of the weight of the husked nut. The drying is aimed at reducing moisture content from about 50% in fresh coconut, to below 7% in the cured product. Following curing, the oil content increases from 35% in the fresh endosperm to 60–65% in the well-cured copra. The dried copra is chopped into uniform pieces, sacked, weighed and shipped. Child (1937) concludes that copra if dried to 6–7% moisture and stored in well-ventilated dry rooms, will remain of good quality for a long time. At higher moisture content, damage caused by insects, bacterial and fungal attacks is usually high. Also oil from such damaged copra is likely to be rancid, coloured and of high free fatty acid content. Nathanael (1960) reports that an under-dried copra sample containing 12% moisture and stored under good conditions for 5 months suffers a loss of 10% of its weight, the oil also containing 1.5% of free fatty acids (as lauric acid). The proximate composition of copra as given by Subrahmanyan and Swaninathan (1959) consists of moisture (5.80%), crude protein (8.90%), fat (67.0%), carbohydrates (16.50%), crude fibre (4.10) and ash (1.80).

Table 33.1 Composition of coconut cake

	Expeller[ac]	*Solvent*[ac]	*X'Tralac*[ad]	*Morlac*[bd]	*Hydraulic*[d]	*Expeller*[d]
Moisture	4.84	11.14	10.95	10.82	11.0	10.0
Oil	7.74	2.76	0.53	3.26	6.0	10.0
Protein	22.89	20.30	20.38	19.38	19.80	19.10
Carbohydrate	29.15	47.46	–	–	45.3	43.8
Crude fibre	8.53	11.46	–	–	12.2	11.8
Ash	6.82	6.88	–	–	5.7	5.3

[a]Solvent extracted coconut meal from Sri Lanka.
[b]A blend of solvent-extracted with hydraulic press meal in Sri Lanka.
[c]From Fronda *et al.* (1957).
[d]From Child (1974).

Table 33.2 Amino acid composition of coconut meal

Amino acid	*% of meal*	*% of protein*
Dry matter	89.90	100.00
Alanine	0.81	3.87
Valine	0.89	4.25
Glycine	0.89	4.25
Isoleucine	0.60	2.87
Leucine	1.21	5.78
Proline	0.71	3.39
Threonine	0.66	3.15
Serine	0.96	4.59
Methionine	0.37	1.79
Hydroxyproline	0.05	0.23
Phenylalanine	0.81	3.87
Aspartic acid	1.62	7.75
Glutamic acid	3.64	17.41
Tyrosine	0.46	2.20
Lysine	0.48	2.29
Histidine	0.41	1.86
Arginine	1.96	9.37
Cystine	0.24	1.14

From Creswell and Brooks (1971).

(c) Copra cake/copra meal

Copra cake, also called coconut cake or Poonac, is produced after the extraction of coconut oil from copra and is used almost entirely for animal feeding. The yield of copra cake from copra is 33–42%. The cake is ground to produce the meal. Coconut cake of good quality is usually whitish or light brown in colour (Creswell and Brooks, 1971a). The composition varies depending on the processing methods. Table 33.1 shows the composition of coconut cake after different methods of processing. The oil

Table 33.3 Mineral composition of coconut meal

Mineral	*Content*
Dry matter (%)	89.90
Calcium (%)	0.16
Phosphorus (%)	0.55
Magnesium (%)	0.23
Potassium (%)	1.75
Zinc (mg/kg)	53.00
Copper (mg/kg)	40.00
Manganese (mg/kg)	75.00

content of solvent-extracted coconut cake is generally lower than that of coconut cake produced by other methods. The cake generally has a protein content of about 20% but it is deficient in such amino acids as tryptophan, lysine, methionine and histidine and these deficiencies coupled with its crude fibre level of about 10%, reduce its utilization at high levels in animal diets. The amino acid content of coconut meal (Table 33.2) as presented by Creswell and Brooks (1971a) indicates that coconut meal may contain less lysine, isoleucine, leucine and histidine than other commonly used plant proteins. The lysine level appears to be extremely low when compared with other protein sources. Various degrees of heat treatment especially in processing decrease lysine availability, nitrogen digestibility and NPU of coconut meal (Creswell and Brooks, 1971a).

The mineral composition of coconut cake shown in Table 33.3 (Creswell and Brooks, 1971a), indicates that the principal inorganic constituent is potassium and this may amount to more than 2% of the dry weight or 40% of the total ash. The soluble sugars in copra cake, as in coconut kernel and desiccated coconut, consist almost entirely of sucrose (Child and Nathanael, 1949). The sugar content is about 16%.

(d) Coconut oil

Coconut oil is generally produced from copra by using various procedures including pressing and solvent extraction. The oil in a tropical environment ranges from colourless to pale brown. In a temperate climate it appears as a greasy, crystalline white or yellowish solid fat which has a melting range of 20–26°C. It contains a high proportion of the lower-molecular-weight saturated fatty acids, especially lauric acid and myristic acid. It shares this feature with palm kernel oil from *Elaeis guineensis* and the oil from the Brazilian palm, *Orbigyna speciosa*. The fatty acid composition presented in Table 33.4 shows that there is about 91% saturated fatty acids, and 9% unsaturated fatty acids. The saturated fatty acids lauric and myristic acids are the most predominant fatty acids. The saponification index of coconut oil is very

Table 33.4 Constituent fatty acids of coconut oil

	% of total fatty acids		
Fatty acid	*Child (1974)*	*Kaufman (1965)*	*Woodroof (1978)*
Caproic	0.2–0.5	–	–
Capric	5.4–9.5	8.0	7.8
Lauric	44.1–51.0	7.0	7.6
Myristic	13.1–18.5	19.0	18.0
Palmitic	7.5–10.5	7.0	9.5
Stearic	1.0–3.2	4.0	5.0
Arachidic	0.2–1.5	–	–
Oleic	5.0–8.2	6.0	8.2
Linoleic	1.0–2.6	1.0	–

high (246–260) due to its high content of lauric and myristic acids. The high degree of saturation of the oil makes it very useful for confections, bakery goods and deep fat frying. It has the lowest iodine value of all the vegetable oils in common industrial use. The iodine value is usually 7–10. The glycerides are very complex, Bezard *et al.* (1971) reported 79 types of glycerides, the principal ones being caprodilaurin (20%) caprolauromyristine (12%) and dilauromyristine (11%). Coconut oil also contains 0.5% unsaponifiable constituents (i.e those constituents not reacting with caustic soda to form soap) and these are made up mainly of tocopherol (vitamin E).

(e) Coconut water

The liquid endosperm present when the endosperm is not yet hardened is referred to as coconut water. It is to be differentiated from coconut milk which is the liquid enclosed by the mature endosperm in the mature coconut fruit. Sugars are the most important constituents. It contains about 5% sugar and also various minerals, amino acids and vitamin C in the range of 2.2–3.7 mg/100 ml (Woodroof, 1978). The caloric value is about 17.4/100 g. At an early stage, the concentration of total solids is about 2.5 g/100 ml and this increases to a maximum of about 6 g/100 ml at about the seventh month and then starts to decline. The composition of coconut water is shown in Table 33.5. (Pandalai, 1958). The principal inorganic constituent seems to be potash. Vanderbelt (1945) reported the following B complex vitamins in coconut water: nicotinic acid 0.63 μg/ml, pantothenic acid 0.52 μg/ml, biotin 0.02 μg/ml, riboflavin $<$0.01 μg/ml, and folic acid 0.003 μg/ml.

Woodroof (1978) indicates that the percentages of alanine, cystine and serine in the proteins are higher than those in cow's milk, making coconut water nutritionally important. Coconut water is also reported to contain substances capable of inducing rapid proliferative growth in plant tissues.

Table 33.5 Composition of coconut water

Water	95.5%
Nitrogen	0.05%
Phosphoric acid	0.56%
Potassium	6.60%
Calcium oxide	0.69%
Magnesium oxide	0.69%
Iron	0.5 mg/100 g
Total solids	4.7 mg/100 g
Reducing sugars	0.80 mg/100 g
Total sugar	2.08 mg/100 g
Ash	0.62 mg/100 g

From Pandalai (1958).

For example, van Overbeck *et al* (1942) showed that coconut water contains factors promoting growth and development of very young embryos of *Datura stramonium*. Prevot (1968) has given a more comprehensive review of this quality of coconut water.

33.2 USE AS HUMAN FOOD

33.2.1 Coconut kernels, fresh or desiccated

Coconut kernel is eaten by man in its fresh form. In Nigeria, as in many countries of the world, it is eaten directly alone or in combination with corn, tapioca (sliced cassava) and garri (grated, fermented and fried cassava). In each case the use of coconut enhances the palatability of the corn, tapioca or garri and increases intake. Apart from being eaten in the fresh form, it is also consumed as desiccated coconut. This is the dried, shredded and sweetened or unsweetened coconut that is used in cakes, biscuits, puddings, peanut butter, marshmallows, and topping for individual bakery items where it adds nut-like chewiness and flavour. It is made from the kernel after the removal of the brown testa. After shredding it is dried at 60–75°C to a moisture content of less than 2.5%. The major feature of coconut (meat or oil) in the diets of both man and livestock is its effect on serum cholesterol levels. Prior *et al.* (1981) examined the diets of two populations of Polynesians living on atolls near the equator in relation to their serum cholesterol levels. They observed that for both groups habitual diets were rich in saturated fat and poor in cholesterol and sucrose. However, 63% and 34% of the total energy intake was from coconut for inhabitants of the Islands of Tokelau and Pukapuka, respectively. The serum cholesterol level was 35–40 mg/100 ml higher in Tokelau than in Pukapuka and the difference was ascribed to the higher intake of saturated fat in Tokelau resulting from the higher intake of coconut in Tokelau.

33.2.2 Coconut oil

Coconut oil is used in the manufacture of margarine, shortening, curries, sweets confectionery, salad and cooking oil. It is used for deep fat frying of doughnuts and french fries and as a substitute for butter fat in fluid milk substitutes. In various milk-deficient areas of the world therefore, coconut oil mixed with non-fat solids is offered as a replacement for whole milk for many purposes, including the feeding of infants (Woodroof, 1978). This is said to have better keeping qualities than milk. The special qualities of coconut oil come from the high percentage of lauric acid, myristic acid, palmitic acid, stearic acid and oleic acid. Coconut oil is also used as a liquid carrier and solvent for vitamins, hormones and antibodies for intramuscular injection. Closely related to this is the use of coconut oil as a coating for delayed action pills and tablets because of the high melting point of coconut oil. These coatings can withstand the acid environment of the stomach without disintegrating and will only breakdown and liberate the active medication when the alkaline enzymatic intestinal tract is reached (Woodroof, 1978). Related to coconut oil is coconut cream which is the thick white juice from the grated kernel. The cream is used in cooking chicken, fish, pork, rice and beans, to which it imparts a delicious flavour and taste.

The importance of coconut oil on serum cholesterol has already been mentioned. More details of this are given in section 33.3.

33.2.3 Coconut water

The water of a tender coconut (i.e the liquid endosperm) has been described as the purest and most nutritious and wholesome beverage that nature has provided for the people in the tropics (Woodroof, 1978). It is said to keep the body cool and prevent prickly heat and summer boils. The percentage of alanine, arginine, cystine and serine in the water are higher than those in cow's milk and this has made coconut water useful in infant feeding. Coconut water can also be fed intravenously. Experiments in Bangkok, Thailand, have shown the possibilities of using coconut water for feeding patients intravenously. Woodroof (1978) has therefore suggested its use in remote areas or under military emergencies when supplies of more usual sugar and salt solutions for intravenous feeding have been interrupted. Clamohoy *et al.* (1962) have reported that coconut water can be used with egg yolk as a diluent for boar semen. These workers showed that among three diluents – pure coconut water, coconut water plus 1% egg yolk, and 5% egg yolk – the use of pure coconut water alone as a diluent gave the highest breeding efficiency.

Table 33.6 Composition and digestibility of components of coconut meal

	Composition	*Digestion coefficient*
Dry matter (%)	89.9	83.7
Protein (%)	20.9	50.7
Nitrogen-free extract (%)	46.2	94.1
Crude fibre (%)	10.5	–
Ether extract (%)	5.8	100.0
Ash (%)	6.5	–
Gross energy (kcal/g)	4.2	85.5
Digestible energy (kcal/g)	3.6	–

Creswell and Brooks (1971a).

33.3 USE AS ANIMAL FEED

33.3.1 Coconut meal (copra meal)

Coconut meal has been used in the feeding of both ruminant and non-ruminant animals. In feeding coconut meal, it is important that it is not old and therefore rancid as this will cause diarrhoea. The meal swells considerably in water and should, therefore, be soaked in water before being fed to animals. Animals not used to it are at first somewhat reluctant to eat it but if it is introduced gradually, animals soon acquire a liking for it. Coconut meal has been fed to pigs and poultry with good results. It is, however, fairly rich in fibre and this restricts its inclusion in the diets of these non-ruminant animals. Earlier studies have shown that when coconut meal is incorporated at levels of up to 30% in diets of growing finishing pigs, acceptable levels of growth are indeed possible, without any detrimental side effects (Grieve *et al.*, 1966; Malynicz, 1973). Increasing the levels of coconut meal in pig diets to more than 30% has been reported to result in serious impairment of growth performance, feed intake, feed conversion efficiency and dressing percentage (Grieve *et al.*, 1966; Creswell and Brooks, 1971b; Malynicz, 1973; Thorne *et al.*, 1988). The protein of coconut meal has been shown to have poor digestibility and low quality in terms of essential amino acid balance (Butterworth and Fox, 1963; Creswell and Brooks, 1971a; Mee and Brooks, 1973; Kuan *et al.*, 1982; Thorne, 1986). This is amply illustrated by the digestibility data of coconut meal fed to pigs (Cresswell and Brooks, 1971a) (Table 33.6). While earlier researchers (Butterworth and Fox, 1963) had suggested a limit of 15–30% for the use of coconut meal in swine diets, more recent research by Thorne *et al.* (1988) has indicated that coconut meal can be used at much higher levels in diets of monogastric animals. These workers showed that the supplementation of pig diets containing 50% coconut meal with protein of superior quality or synthetic amino acids (methionine, lysine, threonine) improves performance ($P<0.001$) almost to control levels. Coconut meal is used to a very limited extent in poultry rations because of difficulties in formulating a ration balanced in amino acids, high in utilizable energy and yet sufficiently low in fibre. With poultry the amino acid of the greatest significance in feeding of coconut meal is lysine since it is usually destroyed during processing. According to Monongan *et al.* (1965), the feeding value of poultry and livestock rations containing 15–30% coconut meal is improved with methionine and lysine supplemention. Chicks receiving supplements have better growth and more efficient feed utilization and higher liver nitrogen than those receiving coconut meal without supplements. For laying hens fed different levels of coconut meal, egg production and efficiency of feed utilization are decreased as coconut meal increases. The best result are obtained when coconut meal is not more than 20% of the diets (Soldevila *et al.*, 1976).

With ruminants, coconut meal has been found to be useful for dairy and fattening cattle. It gives a firmer butter and harder body fat than do other oil cakes (Woodroof, 1978; McDonald *et al.*, 1987).

33.3.2 Coconut oil

Coconut oil has been used in the feeding of ruminants and non-ruminants with varying degrees of success. One very important feature of consumption of coconut oil by humans and livestock is its effect on serum cholesterol levels. Many studies with humans and animals have shown that coconut oil generally elevates serum cholesterol level. This is probably due to the relative proportion of saturated and polyunsaturated fatty acids which are believed to influence blood lipid levels.

With rabbits, coconut oil has been found to exert a positive influence on nutrient digestibility (Castillo *et al.*, 1962). When 0, 5, 10 or 15% of coconut oil was fed, these authors noted that the oil increased the digestibility of crude protein, crude fat and nitrogen free extract but decreased the digestibility of crude fibre. Kritchevsky *et al.* (1976) fed groups of Dutch rabbits on a semipurified diet containing 40% sucrose, 25% casein and 14% fat without added cholesterol for 10 months. The fats were peanut oil (PNO), corn oil (CO), butter oil (BO) and coconut oil (CNO). All the diets were hyperlipidaemic and hyperlipoproteinaemic. The normal serum cholesterol level in Dutch rabbits is 40–60 mg/100 ml but with the feeding of the various fats, serum cholesterol levels were elevated in all fat groups and ranged from 152 mg/100 ml in CO to 330 mg/100 ml in CNO. Serum triglyceride levels were in the normal range in PNO, CO and BO groups (28–51 mg/100 ml) but grossly elevated (286 mg/100 ml) in the CNO groups. Similar trends have been reported with rats by Kellog (1974) who compared safflower oil and CNO and also by Nair and

Gopakumar (1981) who compared CNO and PNO. Nair and Gopakumar noted that the inclusion of fish oil reduced serum cholesterol levels even in the presence of coconut oil. Fish and other marine oils are rich in polyunsaturated fatty acids and can, therefore, act as strong hypocholesterolaemic agents. Many experimental and epidemiological studies have indicated a relationship between the incidence of atherosclerosis and serum cholesterol levels. Apart from the elevation of serum cholesterol level by CNO in relation to other fats, Kritchevsky *et al.* (1976) have shown that the most severe gross atherosclerosis was in the CNO group which had 25% of the aortic surface area affected and the least was the CO group with 2% of the surface affected. Similar findings on the increased incidence of atherosclerosis with the feeding of CNO has been reported by Krichevsky *et al.* (1968, 1973) and also in experiments with pigeons in which CO and CNO were compared (Clarkson *et al.*, 1962). Labadan (1962) showed that broilers did not perform well with coconut oil fed at levels of 0, 5, 10 and 15%. Birds fed rations containing 15% coconut oil had a higher mortality as a result of greater incidence of cannibalism. The feathering of birds fed rations with no added oil was good to excellent whereas that of about 18% of the birds fed rations with added oil was poor. A greater incidence of poor fleshing over the breast and thighs was found in birds fed 15% oil.

Rohr *et al.* (1979) studied the effect of fat supplements on metabolism and performance of dairy cows. The cows were fed on diets containing coconut oil, palm oil, tallow or soybean oil. It was reported that the inclusion of high fat levels slightly increased the digestibility of crude protein and significantly enhanced ether extract digestibility. Digestibility of crude fibre was reduced by the addition of coconut oil, tallow or palm oil to the diet. It was concluded that irrespective of its nature, fat in diets of lactating cows can only be used within certain limits as an energy source.

All the studies on the feeding of coconut oil to animals, indicate that the major effect is the elevation of serum cholesterol level and a higher incidence of atherosclerosis than when other oils are fed. The high serum cholesterol resulting from the feeding of coconut oil can be greatly reduced by the feeding of polyunsaturated fatty acids from fish oil.

REFERENCES

Aten, A., Manni, M. and Cooke, F.C. (1958) Copra processing in rural industries. *FAO Agricultural Development Paper no. 63*, Rome.

Bezard, J., Bugaut, M. and Clement, G. (1971) Triglyceride composition of coconut oil. *J. Am. Oil Chem. Soc.*, **48**, 134–9.

Butterworth, M.H. and Fox, H.C. (1963) The effects of heat treatment on the nutritive value of coconut meal and the prediction of nutritive value by chemical methods. *Br. J. Nutr.*, **17**, 445–52.

Castillo, L.S., Gerpacio, A.L., Javier, T.R. *et al.* (1962) The utilization of coconut oil in livestock and poultry feeding II. In rabbit rations. *Philippine Agric.* **16**, 266–87.

Child, R. (1937) Ceylon estate copra. *Trop. Agric.* (*Ceylon*) **88**, 137–49.

Child, R. (1974) *Coconuts*, 2nd edn. Longmans Group Limited, London.

Child, R. and Nathanael, W.R.N. (1949) Annual Report. Coconut Research Scheme (Ceylon) for 1949. Ceylon Govt. Sessional Paper VIII, p. 6.

Clamohoy, L.L., Abilay, J.P., Palad, O.A. and Timu, M.W. (1962) Coconut water–egg yolk as diluter for boar semen. *Philippine Agric.* **46**, 352–68.

Clarkson, T.B., Prichard, R.W., Lofland, H.B., Jr and Goodman, H.O. (1962) Interaction among dietary fat, protein and cholesterol in atherosclerosis susceptible pigeons. *Circ. Res.* **11**, 400–4.

Creswell, D.C. and Brooks, C.C. (1971a) Composition, apparent digestibility and energy evaluation of coconut oil and coconut meal. *J. Anim. Sci.*, **33**, 366–9.

Creswell, D.C. and Brooks, C.C. (1971b) Effect of coconut meal on Coturnix quail and of coconut meal and coconut oil on performance, carcass measurement and fat composition in swine. *J. Anim. Sci.*, **33**, 370–5.

Fronda, F.M., Katigbak, T.V. and Ortigas, R.Z. (1957) Solvent process extracted copra meal versus expeller process extracted copra meal in poultry rations. *Philippine Agric.* **41**, 314–18.

Grieve, G.C., Osbourn, D.F. and Gonzales, F.O. (1966) Coconut oil meal in growing and finishing rations for swine. *Trop. Agric.* (*Trinidad*), **43**, 257–61.

Kaufman, T.G. (1965) Coconut oil derivatives in cosmetics. *Drug Cosmetic Ind.*, **97**, 172–3.

Kellog, T.F. (1974) Steroid balance and tissue cholesterol accumulation in germ-free and conventional rats fed diets containing saturated and polyunsaturated fats. *J. Lipid Res.*, **15**, 574–9.

Kritchevsky, D., Sallata, P. and Tepper, S.A. (1968) Experimental atherosclerosis in rabbits fed cholesterol-free diets. 2. Influence of various carbohydrates. *J. Atherosclerosis Res.*, **8**, 697–703.

Kritchevsky, D., Tepper, S.A. and Kitagawa, M. (1973) Experimental atherosclerosis in rabbits fed cholesterol-free diets. Comparision of fructose and lactose with other carbohydrates. *Nutr. Rep. Int.*, **7** 193–202.

Kritchevsky, D., Tepper, S.A., Kim, H.K. *et al.* (1976) Experimental atherosclerosis in rabbits fed cholesterol free diets. Comparision of peanut, corn, butter and coconut oils. *Exp. Mol. Pathol.* **24**, 375–91.

Kuan, K.K., Moli, T.F., Razak, A. and Farrell, D.J. (1982) Chemical composition and digestible energy of some feedstuffs determined with pigs in Malaysia. *Trop. Anim. Prod.*, **7**, 315–21.

Labadan, M.M. (1962) The utilization of coconut oil in livestock and poultry feeding 1. In broiler rations. *Philippine Agric.* **46**, 143–55.

Malynicz, G.L. (1973) Coconut meal for growing pigs. *Papua New Guinea Agric. J.* **24**, 142–4.

McDonald, P., Edwards, R.A. and Greenhaigh, J.F.D. (1987) *Animal Nutrition*, 4th edn. Longman Group Limited, Hong Kong.

Mee, J.M.L. and Brooks, C.C. (1973) Amino acid availibility of coconut meal protein in swine. *Nutr. Rep. Int.*, **8**, 261–9.

Monongan, V.G., Castillo, L.S. and Gerpacio, A.R. (1965) High levels of copra meal in poultry and livestock ration III. Evaluation of the feeding by depletion repletion method. *Philippine Agric.*, **48**, 399–413.

Nair, P.G.V. and Gopakumar, K. (1981) Effect of long-term feeding of coconut oil-based diet and a combined coconut oil-fish diet on serum cholesterol and fatty acid composition of different organs of albino rat. *J. Food Sci. Technol.*, **18**, 187–91.

Nathanael, W.R.N. (1958) Ceylon Coconut Research Institute Annual Report for 1958.

Nathanael, W.R.N. (1960) Economic losses to the coconut industry consequent on deterioration of under-dried copra. *Ceylon Coconut Quarterly*, **11**(3/4), 7–46 (Report submitted to the FAO group on coconut and coconut products.)

Pandalai, K.M. (1958) Coconut water and its uses. *Coconut Bull.*, **12**, 167–73.

Prevot, J. (1968) L'utilization du lait de coco comme accelerateur du croissance des vegetaux. *Oléagineux*, **23**, 177–8.

Prior, I.A., Davidson, F., Salmond, C.E. and Czochanska, Z. (1981) Cholesterol, coconuts and diet on Polynesian atolls: a natural experiment: the Pukapuka and Tokelau Island studies. *Am. J. Clin. Nutr.*, **34**, 1552–61.

Rohr, K., Daenicke, R. and Oskige, H.J. (1979) Effects of different fat supplements on metabolism and performance of dairy cows. *Nutr. Abst. Rev.*, 49.

Soldevila, M., Rojas-Deporta, M. and Carlo, I. (1976) Effect of different levels of coconut meal on egg production. *Agric. Univ. Puerto Rico*, **60**, 635–8.

Subrahmanyan, V. and Swaninathan, D. (1959) Coconut in food. *Coconut Bull.*, **13**, 153–8.

Thorne, P.J. (1986) The use of copra meal in pig diets. Ph.D. Thesis, University of Nottingham.

Thorne, P.J. Wiseman, J., Cole, D.J.A. and Machin, D.H. (1988) Use of diets containing high levels of copra meal for growing/finishing pigs and their supplementation to improve animal performance. *Trop. Agric.* (*Trinidad*), **65**, 197–201.

Vanderbelt, J.M. (1945) Nutritive value of coconut. *Nature*, **156**, 174–5.

van Overbeck, J., Conklin, M.E. and Blakeslee, A.F. (1942) Cultivation *in vitro* of small *Datura embryos*. Am. *J. Bot.*, **29**, 472–7.

Woodroof, J.G. (1978) *Coconut: Production, Processing, Products*, 2nd edn. AVI Publishing, Westport, CT.

CHAPTER THIRTY-FOUR

Oil palm (*Elaeis guineensis* Jacq.)

O. C. Onwudike

34.1 GENERAL BOTANICAL, ECOLOGICAL AND PRODUCTION INFORMATION

34.1.1 Botany of the oil palm

The oil palm *Elaeis guineenis* Jacq. is a monocotyledonous plant widely believed to be native to West Africa. It belongs to the family Palmae, order Palmales and genus *Elaeis*. It is an erect plant with a solitary columnar stem and short internodes, the stem bearing a crown of fronds (palm leaves). Other species include *Elaeis oleifera* and *Elaeis melanococca*. It is generally a monoecious plant with the male and female flowers separated but occurring on the same plant. Sometimes hermaphrodite inflorescences may be found (Hartley, 1977). The inflorescences are borne in the axils of the leaves and the female inflorescence, or spadix, gives rise to the palm bunch bearing the palm fruits. The fruit is a sessile drupe which is borne in a large compact bunch. The bunch is ovoid and may reach 50 cm in breadth. A bunch carries 500–4000 fruits with a mean of 1500 (Hartley, 1977). Each fruit may be 2–5 cm in length and about 3–20 g in weight. the fruit consists of a leathery exocarp, a fleshy mesocarp which provides the palm oil and this surrounds a nut, the shell of which (endocarp) encloses the palm kernel. The fruits are classified into three types according to their external appearance:

1. Nigrescens: the fruits are black when unripe and red when ripe with part of the top half black in colour.

Legumes and Oilseeds in Nutrition
Edited by E. Nwokolo and J. Smartt
Published in 1996 by Chapman & Hall. ISBN 0 412 45930 2

2. Virescens: the fruits are green when unripe due to a marked absence of anthocyanin pigments and light reddish orange when ripe with a greenish tip.
3. Albescens: the fruits are white when unripe due to a marked absence of carotenoids and are pale yellow when ripe.

Palm fruits are generally classified on the basis of the thickness of the shells. Classifications based on shell thickness are as follows:

1. Pisifera: these are fruits with thick mesocarp and little or no shell.
2. Dura: these are fruits with thick shells (2–8 mm) and medium mesocarp (35–55%).
3. Tenera: these are fruits with thin shells (0.5–4 mm) and medium to high mesocarp (60–96%).

34.1.2 Ecology

The natural habitat of the oil palm seems to be along forest fringes near rivers. However, it thrives well wherever a part of the forest has been cleared in soils where moisture is maintained. Geographically, the palm belt of Africa runs 7° North and South of the equator, from the Fouta D'Jallon district of Guinea, through the coastal West African countries into the equatorial region of the Congo. Areas of large-scale production of oil palm in the world include the Philippines, Malaysia, India, Sri Lanka, Indonesia and Pacific Islands.

34.2 PHYSICAL AND CHEMICAL NATURE OF OIL PALM PRODUCTS AND BY-PRODUCTS

34.2.1 Palm oil

Palm oil is produced from palm fruits following the physical breakdown of the mesocarp. A combination of heat and physical pressure is often used to rupture the cells and homogenize the fat constituents. The fibrous pulp of the fruit contains about 30–70% fat (Hill, 1952). The palm oil produced is a yellow–orange or brownish–red fluid at tropical temperatures, which is solid at ambient temperatures in temperate climates. It consists mostly of triglycerides, the composition of which determines its characteristics and uses. The non-glyceride fraction, which is less than 1%, consists of carotenes, tocopherol, sterols, and phospholipids among others (NIFOR, 1989). The carotenes impart a rich red colour to palm oil and are sources of vitamin A. The tocopherols act as antioxidants, the sterols are polymerization inhibitors, and the phospholipids are mostly emulsifiers (NIFOR, 1989). The quality of palm oil

Table 34.1 *Percentage fatty acid composition of palm oil, palm kernel oil, coconut oil and lard*

Fatty acid	*Palm oil*[a]	*Palm oil*[b]	*Palm oil*[c]	*Palm kernel oil*[d]	*Coconut oil*[e]	*Lard*[f]
Saturated fatty acids						
Caproic, C_6	–	–	–	–	0.2–0.5	–
Caprylic, C_8	–	–	–	3.0 (3–4)	5.4–9.5	–
Capric, C_{10}	–	–	–	6.0 (3–7)	4.5–9.7	–
Lauric, C_{12}	–	–	–	50 (46–52)	44.1–51.0	–
Myristic, C_{14}	2.3 (0.6–5.9)	0.7	1.0–1.5	16 (14–17)	13.1–18.5	0.7–1.1
Palmitic, C_{16}	41.2 (32.3–45.1)	37.1	41.8–45.0	6.5 (6–9)	7.5–10.5	26.32
Stearic, C_{18}	4.3 (2.2–6.4)	4.0	3.7–5.1	1.0 (1–2.5)	1.0–3.2	12–16
Unsaturated fatty acids						
Arachidonic $C_{20:4}$	–	–	–	–	0.2–1.5	–
Oleic, $C_{18:1}$	42.5 (38.6–52.4)	46.5	38.6–40.2	16 (13–19)	5.0–8.2	41–51
Linoleic, $C_{18:2}$	9.6 (5.0–11.3)	11.7	10.2–11.9	1.0 (0.5–2)	1.0–2.6	3.14

Source: [a]Eckey (1954);
[b]Devendra (1978);
[c]MOPGC (1979);
[d]Hartley (1977);
[e]Child (1974);
[f]Ministry of Commerce and Industry Malaysia (1970).

obtained from the fruit is influenced by the rapidity with which the fruits are loosened from the bunch. If the fruits are left for long, the quality of oil deteriorates since there is enhanced lipolysis leading to increased levels of free fatty acids in the oil. Palm oil contains high levels of saturated palmitic acid and a high quantity of unsaturated fats, principally those derived from oleic acid. The high oleic and linoleic acid contents (52.1%) give it a much higher unsaturated acid content than coconut or palm kernel oil. It is a good vegetable cooking oil containing about 50% saturated fatty acids and having characteristics close to lard (Yeong and Mukherjee, 1983). Table 34.1 gives the fatty acid composition of palm oil in comparison with those of palm kernel oil, coconut oil and lard. The most important minor constituents of palm oil are the carotenoids. Among these, the carotenes are so conspicuous that the oil is termed 'red palm oil'. The carotene content of red palm oil is especially high. Oils taken direct from individual palms may occasionally contain more than 2000 ppm of carotene (Hartley, 1977), although general values are in the range 500–700 ppm (Hartley, 1977; Hutagalung *et al.*, 1982). On average, tocopherol content may be as high as 800 ppm, although values of 800–900 ppm (Hutagalung *et al.*, 1982) and 400–600 ppm (Hartley, 1977) have been observed. The general characteristics of palm oil as determined and published by the Nigerian Institute for Palm Research (NIFOR, 1989) are as follows: saponification value 199; melting point range 27–50°C, moisture content 0.1%; impurities 0.01%; iron 10.0 ppm; iodine value 45–56; anisidine value 0–6; peroxide value 0–5 mequiv/kg; oxidation value 0–16.

The peroxide value, anisidine value, oxidation value, iodine value and moisture content are used to assess the quality of palm oil. The moisture content is required to be as low as possible to prevent hydrolysis. Poor quality palm oil may be defined as having any of the following characteristics – high fatty acid content, contamination with water and other impurities, poor bleachability. Palm fruit contains a very active lipase which rapidly affects the breakdown of fats into fatty acids when the cellular structure of the fruits has been disturbed. This occurs during processing because before processing, oil from fresh fruits contains as little as 0.1% free fatty acid (estimated as palmitic acid) but in bruised and crushed fruit the free fatty acids may increase to 50% in a few hours.

34.2.2 Palm kernel meal

Palm kernel meal (PKM) is the by-product obtained from the extraction of oil from the kernel of oil palm nuts. It is a useful source of protein and energy for livestock. It is highly variable in composition especially in its oil and fibre contents. Onwudike (1986a) suggests that such differences in composition may be due to oil palm types which exist in different geographical regions, to the extent of oil extraction from palm kernel and

Table 34.2 Chemical composition of palm kernel meal (% dry matter basis)

Constituent	*Devendra* *(1978)*	*Fetuga* et al. *(1973)*	*Onwudike* *(1986a)*	*Hartley* *(1977)*	*Longe and Adetola* *(1983)*	*Hutagalung* et al. *(1982)*	*Nwokolo* et al. *(1977)*
Crude protein	19.0	18.70	19.2	19.0	19.4	16.0–21.3	21.3
Crude fibre	16.0	10.3	11.2	13.0	21.2	14.0–16.0	17.5
Ether extract	2.0	6.74	7.94	5.0	9.0	0.7–6.1	7.8
Ash	4.2	4.22	5.12	4.0	4.8	3.0–4.0	5.0
N-free extract	58.8	49.61	47.8	48.0	45.6	43.6–55.3	40.4
Calcium	0.34	0.28	–	–	–	0.21–0.34	0.36
Phosphorus	0.69	0.74	–	–	–	0.48–0.71	0.8
Magnesium	0.16	–	–	–	–	0.16–0.24	0.44
Copper (mg/kg)	–	–	–	–	–	22–36	27
Mn (mg/kg)	–	–	–	–	–	163–340	135
Iron (%)	–	–	–	–	–	0.03–0.05	0.33
Zinc (%)	–	–	–	–	–	0.00–0.01	0.004
Gross energy (kcal/g)	4.13	4.42	4.37	–	–	3.9–4.5	4.68
ME (kcal/g)	–	–	2.61	–	–	–	2.79

Table 34.3 Amino acid composition of PKM

	Content (% dry matter)				
	Palm kernel meal (PKM)				Soybean meal
Amino acid	*Nwokolo* et al. *(1976)*	*Hutagalung* et al. *(1982)*	*Yeang (1983)*	*Onwudike (1986a)*	*Nwokolo* et al. *(1976)*
Arginine	2.68	2.40	2.18	2.65	3.45
Histidine	0.41	0.34	0.29	0.42	1.23
Isoleucine	0.60	0.61	0.62	0.62	1.92
Leucine	1.23	1.14	1.11	1.20	3.71
Lysine	0.69	0.61	0.59	0.68	2.95
Methionine	0.47	0.34	0.30	0.32	0.75
Phenylalanine	0.82	0.74	0.73	0.74	2.44
Tyrosine	0.58	0.47	0.38	0.53	1.84
Threonine	0.66	0.60	0.55	0.68	1.88
Valine	0.43	0.80	0.93	0.88	1.02
Aspartic acid	1.69	1.60	1.55	1.72	5.64
Glutamic acid	3.62	3.42	3.15	4.01	9.01
Proline	0.50	0.60	0.63	0.62	1.21
Serine	0.90	0.77	0.69	0.92	2.48
Glycine	0.91	0.84	0.83	0.92	2.16
Tryptophan	–	0.19	0.17	–	–
Alanine	0.81	0.82	0.92	0.76	2.16
Cystine	–	–	0.20	–	–

also to the methods of processing used in producing the PKM. Table 34.2 shows the comparative chemical composition of palm kernel meal from various authors. Palm kernel meal generally contains 19–21% protein, 10–17% crude fibre, 4–5% ash and ether extract values of 0.7–9.0% depending on the efficiency of extraction of oil from the kernel (Hartley, 1977; Nwokolo *et al.*, 1977; Devendra, 1978; Hutagalung *et al.*, 1982; Onwudike, 1986a). PKM is also relatively high in calcium, phosphorus, magnesium, manganese and iron but low in zinc and copper when compared to other protein concentrates (Nwokolo *et al.*, 1977). The gross energy values range from 3.9 to 4.5 kcal/g (Fetuga *et al.*, 1973; Nwokolo *et al.*, 1977; Onwudike, 1986a) whereas the metabolizable energy values (poultry) range between 2.64 and 2.79 kcal/g (Nwokolo *et al.*, 1977; Onwudike, 1986a). Table 34.3 shows a comparison of the amino acid composition of PKM from various researchers with the amino acid composition of soybean meal and Table 34.4 shows a comparison of the amino acid composition of PKM with the amino acid requirements of chicks as given by National Research Council (1977). From Table 34.3 it is clear that PKM is deficient in the essential amino acids especially lysine, histidine, threonine and methonine in comparison with soybean meal. Nwokolo *et al.* (1976) found that the lysine content of PKM is only one quarter that of soybean meal when considered as a source of amino acids for meeting the requirements for chick growth.

Table 34.4 *Amino acid composition of PKM in relation to requirements for growth of chicks*

Amino acid	NRC requirement (%)	PKM (%)
Arginine	1.20	2.77
Lysine	1.10	0.71
Histidine	0.40	0.44
Phenylalanine	0.70	0.77
Tyrosine	0.60	0.60
Methionine	0.40	0.33
Leucine	1.40	1.25
Isoleucine	0.75	0.65
Valine	0.85	0.92
Threonine	0.70	0.71

Source: Onwudike (1986a).

34.2.3 Palm kernel oil

Palm kernel oil (PKO) is the oil obtained from the kernels of the palm nut after extraction. Extraction may be by hydraulic press or by solvent extraction of the crushed kernels, or by a prepress solvent-extraction process. PKO is similar to coconut oil and both are referred to as lauric oils because of their high content of lauric acid. The usual ranges of fatty acid composition for PKO as presented by Hartley (1977) are shown in Table 34.1. PKO has a high proportion of saturated fatty acids making up about 82.5% of the total fatty acid composition (Table 34.1). The oil has an iodine value of 17 and melts at a temperature range of 25–30°C (Hartley, 1977). It is used extensively in the margarine and candy industries and it is also used for making glycerine, shampoos and candles. Traces of saturated caproic acid, of unsaturated palmitoleic, and of highly unsaturated linolenic acid are also found (Bezard, 1971). PKO is required to have a low content of free fatty acids and a light yellow colour. The quality of the PKO depends to a large extent on the quality of the kernel from which it is produced. The moisture content of the kernel and the presence or absence of moulds are likely to affect the eventual free fatty acid content and colour of the oil. An average sample of palm kernels is made up of 42–52% oil, 6–8% moisture, 7.5–9.0% protein, 23–24% nitrogen-free extract, 5% cellulose and 2% ash (Hartley, 1977). If the moisture level is high, autocatalytic hydrolysis and lipolysis by fat-splitting enzymes occurs in the kernels leading to increased free fatty acids.

34.3 USE OF PALM NUT AND PALM OIL AS HUMAN FOOD

34.3.1 Palm oil

Palm oil is consumed in large quantities in many countries of the world where it is used as a major cooking oil for various types of dishes, while improving the acceptability of food. It also provides provitamin A in the diet. With its addition to food, palm oil helps in destroying poisons and toxic materials. An example of this is noted in the study by Olarewaju and Boszormenyi (1975) in which yellow garri (grated fried cassava) prepared with palm oil contains a smaller amount of cyanide compared to white garri (grated fried cassava without palm oil). Therefore in this case palm oil helps to detoxify the cyanide from cassava. In populations where cassava is a major staple, palm oil can thus help in reducing cyanide poisoning. Also in some parts of Nigeria when local people inadvertently consume the poisonous variety of cassava, the patients are given palm oil orally to help neutralize the effect of the cyanide contained in the cassava. Palm oil is also used in the manufacture of margarine, cooking fat and vegetable shortening and in commercial baking and biscuit manufacture.

34.3.2 Palm kernel oil

PKO is used extensively in the margarine and candy industries. Refined palm oil is available as various brands of vegetable oil for human consumption. It is used in the bakery trade, for confectionery, ice cream and mayonnaise.

34.3.3 Palm kernel

Palm kernel is eaten directly after the nut is cracked. In Nigeria as in many other places, it is eaten alone or in combination with such foods as corn, tapioca (sliced cassava) or garri.

34.4 USE OF PALM OIL AND PALM KERNEL MEAL AS ANIMAL FEED

34.4.1 Palm oil

Palm oil has long been used in diets of both ruminants and non-ruminants. It is traditionally used in the diets of pigs and poultry up to about 5% to supply energy and vitamins A and D, and to reduce the dustiness of the feeds. Studies in pigs and poultry on the use of palm oil as an energy source have shown superior responses in rate of grain and feed efficiency compared to fats of animal origin (Babatunde *et al.*, 1974; Hutagalung, 1977, 1979; Yeong, 1980). Studies in Malaysia and Nigeria on calorie–protein ratio requirements of pigs indicates that any increase in the level of palm oil in the diet beyond 5% increases fat content and backfat measurements (Babatunde *et al.*, 1974, 1975; Hutagalung, 1979). However, other studies have showed no adverse effects of palm oil above 5% (Fetuga *et al.*, 1975; Devendra and Hew, 1977). It is clear that palm oil apart from its direct supply of energy in the feed, improves the efficiency of utilization of energy by reducing the heat increment of feed and therefore making nutrients more available to the body. It also contributes some vitamins, especially vitamin A, and other factors which improve feed utilization. Fomunyam *et al.* (1981) have demonstrated that, in cassava, the rate of hydrolysis of cyanogenic glucosides to produce poisonous hydrogen cyanide is greatly reduced in the presence of palm oil. These workers have, therefore, suggested that in animals fed on cassava-based diets supplemented with palm oil, the delay in decomposition of the cyanogenic glucosides may prevent absorption of the cyanogenic glucosides.

In contrast to non-ruminants, there is little information on the utilization of palm oil in ruminants. This is because ruminants use carbohydrates as their source of energy rather than fats. Dietary supplementation of palm oil at the rate of 2, 4 and 8% has been reported to increase milk yield and milk fat content (Storrey *et al.*, 1968). Similar results have been reported by Banks *et al.* (1980) who also showed increases in lactose yield and reduction in non-fat solids when palm oil was fed to dairy cows. Addition of 8% palm oil to the diets of sheep caused an apparent reduction in dry matter, crude protein, crude fibre and ash digestibilities (Devendra, 1978).

34.4.2 Palm kernel meal (PKM)

Palm kernel meal (PKM) is used to supply both energy and protein to animals. It is, however, usually treated as a protein source. The protein is of relatively good quality but its high crude fibre content and grittiness have tended to reduce the extent of its use especially in non-ruminant feeding. A number of researchers (Nwokolo *et al.*, 1976; Yeong, 1983; Onwudike, 1986a) have reported on the availability of the amino acids from PKM in comparison with amino acid availabilities in other plant proteins (Table 34.5). Availability of amino acids in PKM averaged 83.3% according to Onwudike (1986a), 84.5% according to Nwokolo *et al.* (1976) and 64.4% according to Yeong (1983). The values reported by Nwokolo *et al.* (1976) and Onwudike (1986a) are much higher than the values reported by Yeong (1983). Onwudike (1986a) had suggested that these differences as well as those seen in the chemical composition of PKM may be due to

Table 34.5 *Percent Amino acid availability of PKM compared with those of soybean meal (SBM) and groundnut cake (GNC)*

	Availability (%)				
	PKM[a]	*PKM*[b]	*PKM*[c]	*SBM*[a]	*GNC*[c]
Arginine	93.2	87.0	92.7	98.8	96.4
Histidine	90.1	66.8	88.7	98.8	95.8
Isoleucine	86.1	64.9	87.5	97.7	90.1
Leucine	88.5	66.7	90.6	98.4	93.3
Lysine	90.0	58.6	88.9	99.0	91.1
Methionine	91.0	72.1	92.1	98.7	92.7
Phenylalanine	90.5	70.4	91.6	98.6	93.7
Tyrosine	85.0	65.7	89.9	98.0	91.8
Threonine	86.5	60.7	85.3	97.9	88.4
Valine	68.4	62.8	66.7	92.9	88.5
Aspartic acid	87.6	64.4	85.3	98.3	90.2
Glutamic acid	90.1	74.4	88.6	98.9	93.6
Proline	68.0	55.0	64.2	93.0	87.5
Serine	88.7	65.0	85.4	98.1	90.5
Glycine	63.3	25.8	52.1	92.9	84.3
Tryptophan	–	–	–	–	–
Alanine	85.5	67.7	83.0	97.4	89.4
Cystine	–	–	–	–	–
Overall mean	84.5	64.4	83.3	97.3	91.1

Source:[a] Nwokolo *et al.* (1976);
[b]Yeong (1983); [c]Onwndike (1986a).

oil palm types which exist in different geographical regions, the extent of oil extraction from the palm kernel, and to methods of processing employed in producing the PKM. Most other plant protein sources, however, show higher amino acid availability than this meal. PKM has been utilized as a supplement up to 20% in diets of broilers and layers and up to 25% for swine, without any adverse effect on performance (Yeong, 1980; Hutagalung, 1980). Onwudike (1986b) demonstrated that diets containing up to 34% PKM can be fed to starter pullets without any adverse effects on performance. Regression analyses have shown significant relationships for daily gain, feed conversion and feed intake values with the levels of PKM. Birds appear to adjust better to higher levels of PKM the longer they stay on the diets. Devendra (1978) also reports that broilers grow well on diets containing 30% PKM. With laying birds, Onwudike found that 40% PKM can be fed without any adverse effect on laying performance and egg quality. As the proportion of PKM increases beyond the 40% level, there is a significant drop in egg production, egg weight, feed intake and feed efficiency. Beyond the 40% level of PKM, birds produce very watery faeces, for a reason that has not been fully explained. From available information, it is evident that PKM can be successfully utilized as a protein source in non-ruminant diets provided it is adequately supplemented with other protein sources or other amino acids to achieve a better amino acid balance.

Nwokolo *et al.* (1977) have also shown that the availability figures for phosphorus (70.80%), calcium (68.6%) and magnesium (56.4%) are high whereas zinc (13.9%) is low relative to other minerals tested.

34.4.3 Palm oil sludge

By-products of palm oil processing have been claimed to possess significant possibilities for use as concentrate supplements in various ruminant animals. Thus, Dalzell (1978) showed that fresh sludge can be effectively utilized, giving an average daily gain of 0.47 kg in cattle. Palm oil sludge may also be fed to sheep although at levels not exceeding 40% of the concentrate diet (Devendra and Muthurajah, 1977). A product from palm oil sludge derived through a three-phase process – concentration by centrifugal means, fermentation, and extraction – has been used in animal feeding. The centrifugal sediment is subjected to anaerobic fermentation to improve the protein content of the solids (Webb *et al.*, 1977). Results of nutritional and toxicological tests in poultry (Hutagalung *et al.*, 1975; Ong and Hutagalung, 1978) and in pigs (Hutagalung *et al.*, 1977) show that a sludge recovered by this three-stage process can be a valuable animal feed.

A fermentation sludge product 'Prolima' has also been found useful in livestock feeding. The process of Prolima production involves biological, chemical and physical treatments but basically it consists in biodegrading the sludge effluent using selected microorganisms. After fermentation, the solid fraction known as Prolima is easily removed by filtration. This solid fraction has about 35–45% protein. The chemical composition is closely similar to soybean meal. Examination of Prolima in terms of protein content, amino acid analysis, oil and fibre content, toxin content, and rat feeding trials have been conducted at the University of Aston, Birmingham. These suggest that Prolima could be a valuable animal feed raw material equivalent to soybean meal (Devendra, 1978). Yeong (1983) has carried out similar studies to show the availability of amino acids in palm oil sludge and Prolima. The results show that amino acid availability is greatly enhanced when palm oil sludge is fermented to produce Prolima. Based on the high amino acid availability of Prolima, Yeong (1983) recommends that it is suitable to be used as a protein source replacing soybean meal and peanut meal in poultry diets.

Effective use of many of the products from oil palm in animal feeding can help to reduce the cost of livestock production and therefore improve the nutrition of the people especially in the developing countries of the world. Also, the use of palm oil mill effluent in animal feeding will help to reduce environmental pollution especially in those countries of the world where large quantities of palm oil mill effluents are produced.

REFERENCES

Babatunde, G.M., Fetuga, B.L. and Oyenuga, V.A. (1974) Comparative studies on the effects of feeding different types of oils at two levels on the performance and carcass characteristics of growing pigs in tropics. *Anim. Prod.*, **18**, 301–8.

Babatunde, G.M., Fetuga, B.L., Odumosu, O. and Oyenuga, V.A. (1975) Palm kernel meal as a major protein concentrate in the diets of pigs in the tropics. *J. Sci. Food Agric.*, **26**, 1279–91.

Banks, W., Clapperton, J.L. and Kelly, M.E. (1980) Effects of oil-enriched diet on the milk yield and composition and on the composition and physical properties of the milk fat of dairy cows receiving a basal diets of grass silage. *J. Dairy Res.*, **47**, 277–85.

Bezard, J.A. (1971) The component triglycerides of palm kernel oil. *Lipids*, **6**, 630.

Child, R. (1974) *Coconuts*, 2nd edn. Longman Group Ltd., London.

Dalzell, R. (1978) A case study on the utilization of effluent and by-products of oil palm by cattle and buffaloes on an oil palm estate, in *Feedingstuffs for Livestock in South East Asia* (eds C. Devendra and R.I. Hutagalung), Malaysian Society of Animal Production, Kuala Lumpur, pp. 132–41.

Devendra, C. (1978) Utilization of feeding stuffs from oil palm, in *Feedingstuffs for Livestock in South East Asia* (eds. C. Devendra and R.I. Hutagalung), Malaysian Society of Animal Production, Kuala Lumpur, pp. 116–31.

Devendra, C. and Hew, V.F. (1977) The utilization of varying levels of dietary palm oil by growing – finishing pigs. *MARDI Res. Bull.*, **4**, 76–87.

Devendra, C. and Muthurajah, R.N. (1977) The utilization of oil palm by-products by sheep, in *International Developments in Palm Oil* (eds D.A. Earp and W. Newall), Incorporated Society of Planters, Kuala Lumpur, pp. 103–23.

Eckey, E.W. (1954) *Vegetable Fats and Oils*. Reinhold, New York.

Fetuga, B.L., Babatunde, G.M. and Oyenuga, V.A. (1973) Protein quality of some Nigerian feedstuffs. Chemical assay of nutrients and amino acid composition. *J. Sci. Food Agric*, **24**, 1505–14.

Fetuga, B.L., Babatunde, G.M. and Oyenuga, V.A. (1975) The effect of varying the level of palm oil in a constant high protein diet on performance and caracass. *E. Afr. Agric. For. J.*, **40**, 264–70.

Fomunyam, R.T., Adegbola, A.A. and Oke, O.L. (1981) The role of palm oil in cassava based rations, in *Tropical Root Crops. Research Strategies for the 1980s*, (eds T.R. Terry, K.A. Odure and F. Caveness), International Development Research Centre, Ottawa, Canada, pp. 152–3.

Hartley, C.W.S. (1977) *The Oil Palm*, 2nd edn, Longman, London.

Hill, F.W. (1952) *Economic Botany*, 2nd edn, MacGraw-Hill, New York.

Hutagalung, R.I. (1977) Additives other than methionine in cassava diets, in *Cassava as Animal Feed* (eds B. Nestel and M. Graham), International Development Research Centre, Ottawa, Canada, pp. 18–32.

Hutagalung, R.I. (1979) Utilization of feedstuffs from tree crops by poultry. *Proc. 2nd Science and Industry Seminar*. Bogor, Indonesia, 1979, 166.

Hutagalung, R.I. (1980) Availability of feedstuffs for farm animals. *Proceedings Abstract First Asia- Australasia Animal Science Congress*, Abstract no. 40, 15.

Hutagalung, R.I., Chang, C.C., Jalaludin, S. and Webb, B.H. (1975) The value of processed oil palm sludge as a feed for chicks. *Malay. Agric. Res.*, **4**, 53–60.

Hutagalung, R.I., Chang, C.C., Toh, K.M. and Chan, H.C. (1977) Potential of palm oil mill effluent as feed for growing – finishing pigs. *Planter (Kuala Lumpur)*, **53**, 2–9.

Hutagalung, R.I., Mahyuddin, M. and Jalaludin, S.C. (1982) Feeds for farm animals from the oil palm, in *The Oil Palm in Agriculture in the Eighties*, Vol. 11 (eds E. Pushparajah and C. Poh-Soon), Incorporated Society of Planters, Kuala Lumpur, pp. 609–22.

Longe, O.G. and Adetola, J.A. (1983) Metabolizable energy values of some agricultural waste and industrial by-products for layers and the effect of these ingredient on gut dimensions. *J. Anim. Prod. Res.*, 3, 1–13.

Ministry of Commerce and Industry, Malaysia (1970) Quoted by S.W. Yeong and T.K. Mukherjee, (1983).

MOPGC (1979) Malaysian Palm Oil. *Technical Bulletin No. 1: Introduction to the Malaysian Palm Oil Industry.* Published by the Malaysian Oil Palm Growers' Council, Kuala Lumpur.

National Research Council (NRC) (1977) *Nutrient Requirements of Domestic Animals.* no. 1. *Nutrient Requirements of Poultry*, 7th edn. NAS-NRC, Washington DC.

NIFOR (Nigerian Institute for Oil Palm Research) (1989) *History, Activities and Achievements of the Nigerian Institute for Oil Palm Research*, 2nd edn. NIFOR, Benin City, Nigeria.

Nwokolo, E.N., Bragg, D.B. and Kitts, W.D. (1976) The availability of amino acids from palm kernel, soy bean, cotton seed and rapeseed meal for the growing chick. *Poultry Sci.*, **55**, 2300–4.

Nwokolo, E.N., Bragg, D.B. and Saben, H.S. (1977) A nutritive evaluation of palm kernel meal for use in poultry rations. *Trop. Sci.*, **19**, 147–54.

Olarewaju, O.C. and Boszormenyi, Z. (1975) The process of detoxification and residual cyanide content of commercial gari samples. *West Afr. J. Appl. Biol. Chem.*, **18**, 7–14.

Ong, H.K. and Hutagalung, R.I. (1978) Utilisation of CENSOR meals by laying hens. *Mal. Appl. Biol.*, **7**(2), 99–103.

Onwudike, O.C. (1986a) Palm kernel as a feed for poultry 1. Composition of palm kernel meal and availability of its amino acids to chicks. *Anim. Feed Sci. Technol.*, **16**, 179–86.

Onwudike, O.C. (1986b) Palm kernel as a feed for poultry 2. Diets containing palm kernel meal for starter and grower pullets. *Feed Sci. Technol.*, **16**, 187–94.

Storrey, J.E., Hall, A.S. and Johnson, V.W. (1968) The effect of increasing amounts of dietary red palm oil on milk secretion. *Br. J. Nutr.*, **22**, 609–14.

Webb, B.L.L., Hutagalung, R.I. and Cheam, S.T. (1977) Palm oil waste as animal feed – processing and utilisation, in *International Developments in Palm Oil* (eds D.A. Earp and W. Newall), Incorporated Society of Planters, Kuala Lumpur, pp. 125–45.

Yeong, S.W. (1980) The nutritive value of palm oil by-products for poultry. *Proc. Abstr. First Asia-Australasia Animal Science Congress*, Abstr. no. 45, 17.

Yeong, S.W. (1983) Amino acid availability of palm kernel cake, palm oil sludge and sludge fermented product (Prolima) in studies with chickens. *MARDI Res. Bull.*, **11**, 84–8.

Yeong, S.W. and Mukherjee, T.K. (1983) The effect of palm oil supplementation in palm kernel cake-based diets on the performance of broiler chickens. *MARDI Res. Bull.*, **11**, 378–84.

PART FIVE

Miscellaneous Oilseeds

CHAPTER THIRTY-FIVE

Rubber (*Hevea brasiliensis* L.) seed, oil and meal

E. Nwokolo

35.1 INTRODUCTION

The rubber tree is a perennial plantation crop, although in many parts of Africa the tree exists in a semi-wild state, the original rubber plantations having fallen into a state of neglect. The tree grows in hot, humid climates, maturing in about 5 years and can be tapped for rubber latex for up to 20 years. About four seeds are borne in a dehiscent fruit which, on drying, disperses the seeds by an explosive mechanism. The Nigerian Rubber Research Institute estimates that seed yield from rubber plantations varies from 100 to 150 kg/ha, depending on soil fertility and crop density. Most humid tropical countries grow or are capable of growing the rubber tree to some extent. Estimates of rubber seed production from 200 000 ha of rubber plantations in Nigeria are about 20 000 tonnes per annum. Yields of kernel from rubber seeds range from 57% to 63%, with a mean of 60.5% and current estimates (Nwokolo and Akpapunam, 1986) are that Nigeria's rubber estates are capable of yielding 12 000 tonnes of full-fat rubber seed meal annually. This in turn, can yield about 5500 tonnes of rubber seed oil and about 6000 tonnes of defatted protein meal for animal and human nutrition. Most countries of West and Central Africa have considerable land areas under rubber tree plantations. In West Africa, Liberia has an even more extensive hectarage under rubber than Nigeria. Countries of South East Asia and Latin America, which currently produce most of the natural rubber of the world, have tremendous potential to produce rubber seed oil and rubber seed protein for human and animal use. Malaysia, Indonesia and Thailand have about 5.0, 3.0 and 1.5 million hectares

Legumes and Oilseeds in Nutrition
Edited by E. Nwokolo and J. Smartt
Published in 1996 by Chapman & Hall. ISBN 0 412 45930 2

respectively, of rubber plantations (FAO, 1982). It has been calculated that these countries have a capability to produce about 137 000, 82 000 and 48 000 tonnes, respectively, of rubber seed oil. They also have a capability to produce 160 000, 96 000 and 48 000 tonnes, respectively, of defatted rubber meal. These huge quantities of meal and oil are capable of being processed into highly nutritious protein concentrates and dietary oils for human use.

35.2 CURRENT FOOD AND FEED USE

Rubber seed oil and protein meal are not used to any appreciable extent as animal feed, and not at all as human food, despite the potential ease of availability of both meal and oil. The huge quantities of defatted meal mentioned above, with a protein content of about 36%, are capable of meeting a considerabl⁻ percentage of each nation's needs for protein concentrates for monogastric animal nutrition. For example, 6000 tonnes of defatted rubber seed meal (36% protein), can be converted into about 3000 tonnes of rubber seed flour (70% protein) or about 2000 tonnes of protein isolate (96% protein). Both the high protein flour and the extremely high protein isolate, are excellent protein sources for human nutrition. Six thousand tonnes of rubber seed oil, refined for human or industrial use, could contribute significantly to national self-sufficiency in vegetable oil production. Rubber seed oil is rich in polyunsaturated fatty acids, especially linoleic and linolenic acids. It is a semidrying oil like rapeseed oil, but unlike the latter, it has no content of erucic acid and it seems, therefore, to have a potential as a dietary oil, when it is refined and its nutritional qualities exhaustively investigated. Little is known about the nutritional quality of rubber seed oil and considerable research attention is now being focused on it.

The need in Africa, Asia and Latin America, for high protein concentrates at a price which the local people can afford is now more than ever before. Because seed yield from a rubber plantation is a by-product, a nuisance in fact, the production costs of the cultivation of an oil seed for protein and oil extraction are eliminated. Countries which cultivate the rubber tree for latex, can virtually get the seed protein and oil for free.

35.3 PROXIMATE COMPOSITION AND NUTRITIVE VALUE

Proximate composition of rubber seed and products is shown in Table 35.1. Whole rubber seed meal has a moderate content of protein (19–20%) and a very high content of fat (46–48%). With fat extraction, there is an increase in content of various nutrients. The fibre component of rubber

Table 35.1 Proximate composition of rubber seed meal (dry matter basis)

Rubber seed meals	*Dry matter(%)*	*Crude protein(%)*	*Ether extract(%)*	*Crude fibre(%)*	*Ash (%)*
Whole	94.2	19.3	48.3	10.3	7.9
Mechanically pressed	90.4	28.8	18.6	12.2	8.4
Solvent-extracted	91.2	38.7	1.0	13.9	8.4
Soybean meal	90.5	48.2	1.1	6.2	6.5

From Nwokolo *et al.* (1988).

Table 35.2 Mineral content and availability in rubber seed meal and soybean meal

	Content (mg/kg)				*Availability(%)*	
Mineral	*Whole RSM*	*MP RSM*	*SE RSM*	*SBM*	*RSM*	*SBM*
Phosphorus	5600	6700	5200	9400	55.08	54.71
Calcium	700	1900	2700	2800	68.97	76.01
Magnesium	2400	3500	2000	4900	66.00	42.38
Potassium	9300	13800	13400	19300	39.23	32.34
Sodium	3700	5500	7600	7700	61.27	48.95
Copper	25	31	16	44	40.38	52.81
Manganese	16	30	35	42	63.19	59.86
Zinc	73	102	36	143	42.56	59.00
Iron	70	225	86	315	31.91	50.80

From Nwokolo *et al.* (1988).
MP RSM = mechanically pressed rubber seed meal.
SE RSM = solvent extracted rubber seed meal.
SBM = soybean meal.

seed meal changes following oil extraction, increasing from 10.3% in whole rubber seed meal, to 13.9% in solvent-extracted meal. Solvent-extracted rubber seed meal has about 80% of the protein in defatted soybean meal, but with about 130% of the ash and over 200% of the fibre. Defatted rubber seed meal has a higher content of most minerals than soybean meal, having about twice the phosphorus and magnesium content, one and half times the potassium content and about the same content of calcium and sodium as in soybean meal (Table 35.2). Mechanically-pressed meals can also be found in many tropical countries. These meals have a lower protein content (28–30%) and higher residual oil content (18–20%) than solvent-extracted rubber seed meal. Content of zinc and iron is three to four times higher in rubber seed meal than in soybean meal, whereas copper is only twice as high. In general,

Table 35.3 Fatty acids and physicochemical properties of rubber seed oil

Fatty acid	*Rubber seed oil*	*Soybean oil*
14:0	0.08	0.11
16:0	9.27	13.07
16:1	0.14	0.14
18:0	10.58	5.53
18:1	26.64	28.16
18:2	34.92	44.44
18:3	17.27	6.45
20:0	0.57	0.49
20:1	0.18	0.19
22:0	0.15	0.51
24:0	0.12	0.27
Physicochemical properties		
Refractive index (25°C)	1.41	1.47
Iodine value (25°C)	138.0	100.2
Saponification value	192.0	190.0
Unsaponified matter (%)	4.8	0.9
Saponified matter (%)	95.2	99.1
Total saturated fatty acids (%)	20.79	20.01
Total unsaturated fatty acids (%)	79.15	79.38
Total essential fatty acids (%)	52.20	50.89
P/S ratio	2.25	2.54

Fatty acid profile from Nwokolo and Sim (1988).
Physicochemical properties from Nwokolo (unpublished data).
P/S ratio = Ratio of polyunsaturated to saturated fatty acids.

rubber seed meal, like most other oilseed meals, is high in potassium, phosphorus, sodium and magnesium.

Rubber seed lipids are highly unsaturated, the unsaturated fatty acids comprising about 79% of total lipids, whereas the polyunsaturated fatty acids comprise about 52% of total lipids (Table 35.3). The ratio of polyunsaturated to saturated fatty acids (P/S ratio) in rubber seed oil is about 2.25, similar to that in corn oil, soybean oil or sunflower oil. Linolenic acid is very high, much higher than in most edible oils, but significantly lower than in linseed oil. Rubber seed oil, however, has a lower content of linoleic acid than many nutritionally preferred vegetable oils (corn oil, sunflower oil, soybean oil, canola oil). The iodine and saponification values of rubber seed lipids are similar to those of other vegetable oils which contain significant amounts of polyunsaturated fatty acids. Rubber seed oil, however, has a fair amount of unsaponified matter (4.8%), in comparison with such edible oils as corn oil or soybean oil.

Content and availability of amino acids in rubber seed meal is presented in Table 35.4. Amino acid content is lower in mechanically pressed rubber seed meal than in solvent-extracted rubber seed meal. These two are, in turn, lower than in soybean meal. In general, the amino

Table 35.4 Content and availability of amino acids in rubber seed meals

	Content (mg/g of meal)			*Availability (%)*	
Amino acid	*MP RSM*	*SE RSM*	*SBM*	*MP RSM*	*SBM*
Aspartic acid	31.0	41.0	51.5	84.9	92.6
Threonine	9.8	13.3	18.3	77.6	90.1
Serine	13.7	18.2	24.2	82.1	94.4
Glutamic acid	45.7	60.8	81.0	86.3	97.2
Proline	16.2	21.5	25.2	80.0	95.9
Alanine	14.9	19.0	20.8	82.7	95.8
Cystine	2.6	3.4	4.6	68.4	70.1
Valine	19.3	25.7	21.7	90.1	97.1
Methionine	4.8	6.4	6.8	83.3	96.9
Isoleucine	10.4	13.8	21.2	75.3	97.6
Leucine	20.2	26.9	37.1	85.5	97.4
Tyrosine	8.9	11.8	17.7	84.1	96.7
Phenylalanine	11.4	15.2	22.3	89.3	96.9
Histidine	13.4	17.8	15.8	59.5	86.2
Lysine	12.5	16.6	32.5	66.2	95.5
Arginine	32.4	43.1	40.8	88.9	93.6
Average availability				80.3	93.6

From Nwokolo *et al.* (1987).
MP RSM = mechanically pressed rubber seed meal.
SE RSM = solvent extracted rubber seed meal.
SBM = soybean meal.

acids in solvent-extracted rubber seed meal are 70–80% lower than those in soybean meal. Rubber seed meal, like many oilseed meals, has a low content of tyrosine, threonine, isoleucine, phenylalanine and lysine and an extremely low content of cystine and methionine. In spite of its relatively low content of many amino acids, in comparison with soybean meal, rubber seed meal is a comparatively good source of amino acids, when compared to other vegetable protein sources like sunflower seed, cotton seed and rape seed meal (Nwokolo *et al.*, 1987).

There is very little information available on nutrient digestibility in rubber seed meal. The few scientific reports prior to 1980, on utilization of rubber seed meal by animals, have concerned mainly the replacement value of rubber seed meal, for other protein sources. Experimental animals have been cattle, and to a lesser extent, pigs and poultry (Rajaguru and Ravindran, 1979; Viswanathan *et al.*, 1979; Ong and Yeong, 1978).

There is no evidence that rubber seed meal has been or is being consumed as human food in any part of the world, not even by native peoples of South America, where it is believed that the rubber tree originated. There is an almost total absence of overt toxic factors in rubber seed meal, as evidenced by growth experiments using rats and chicks.

Consequently, it is proposed that rubber seed meal and rubber seed protein isolates have a tremendous potential as a protein source in human diets. It is also proposed that refined rubber seed oil, has a tremendous potential as a dietary oil for human use.

Nutrient digestibility and availability data for rubber seed meal are few. Nwokolo *et al.* (1987) have compared true protein digestibility, as well as total and individual amino acid digestibilities in rubber seed meal, against soybean meal as a control. True protein digestibility is similar in undefatted, mechanically pressed and solvent-extracted rubber seed meal (72.3%, 74.4% and 75.0%, respectively). True protein digestibility in defatted soybean meal is, however, significantly higher (88.0%) than values observed with the various rubber seed meals. For most protein sources, true protein digestibility values are slightly lower than average amino acid availability values. Availability of many amino acids in rubber seed meal is between 75% and 86%, with a mean availability of 80.3%. For soybean meal, availability of most amino acids is between 90% and 96%, with a mean availability of 93.6%. In both rubber seed meal and soybean meal, availabilities of even the same amino acid vary tremendously. Threonine, cystine, lysine and isoleucine are only moderately available and histidine poorly available in rubber seed meal. In soybean meal, most amino acids are highly available, only cystine showing poor availability. Amino acid availability in rubber seed meal is comparable to that in pigeon pea, bambara groundnuts, cowpeas and other edible beans, but is lower than values reported in many oilseed meals (usually more than 90%).

There is considerable variation in mineral availability between rubber seed meal and soybean meal (Table 35.2). In rubber seed meal, calcium, magnesium, manganese and sodium are highly available, phosphorus is moderately available, and zinc, copper, potassium and iron are poorly available. Average availability of minerals is 53.8%. In soybean meal, calcium is highly available, manganese, zinc, phosphorus, copper and iron are moderately available, and magnesium is poorly available. In general, because of the low content of minerals in both rubber seed meal and soybean meal, these oil meals cannot be considered as reasonable dietary sources of minerals. As a result, no matter how available the constituent minerals may be, these oilseeds and oilseed meals can only be considered as supplementary, to the major dietary sources of minerals in human and animal nutrition.

In a comparison of the digestibility of total lipids and fatty acids in rubber seed oil and corn oil using a chick assay, Nwokolo and Sim (1988) found that lipid digestibility is similar in rubber seed oil (87.6%) and corn oil (86.7%). For most fatty acids, digestibility is similar in both oils, with the exception of palmitic acid and oleic acid, which have lower digestibility in rubber seed oil.

Data on energy digestibility in undefatted, mechanically pressed and solvent-extracted rubber seed meal have also been reported (Nwokolo and Akpapunam, 1986). Gross energy values are extremely high in undefatted meals, and high in the mechanically pressed and solvent-extracted meals (6.95, 5.02 and 4.42 kcal/g, respectively). This is due to the extremely high lipid content of the undefatted meal, and the high residual oil in the mechanically pressed meal. Metabolizable energy (ME) values are proportional to gross energy values, being very high in undefatted rubber seed meal (4.88 kcal/g), high in the mechanically pressed meal (3.97 kcal/g) and only moderately high in the solvent-extracted meal (2.84 kcal/g). In comparison with the ME value of soybean meal (2.40 kcal/g), these are extremely high values, indicating that the considerable residual oil in these meals is responsible for the high values observed. The ME value of 4.88 kcal/g seen in undefatted rubber seed meal, is very much higher than the ME value of full-fat canola meal (4.01 kcal/g).

35.4 TOXICOLOGICAL SUBSTANCES IN RUBBER SEED MEAL

Ravindran *et al.* (1987) have found evidence to show that an antifertility factor exists in rubber seed meal. Using 8-month-old cockerels, they showed that feeding rubber seed meal depressed sperm count, the degree of depression being related to content of rubber seed meal in the diet. Total sperm count was lowered by 10–18%, 50–63% and 56–69% in cockerels fed 10%, 20% and 30% rubber seed meal, respectively. However, sperm motility, percentage dead sperm and sperm dehydrogenase activity were not influenced by content of rubber seed meal in the diet. In another study, Ravindran and Ravindran (1988) confirmed the presence of potentially toxic concentrations of cyanide (164 mg/100 g dry weight) in fresh rubber seeds, although they noted that most of the cyanide was eliminated by storage or cooking. They also reported the absence of antitrypsin agents or tannins in rubber seed meal.

Other studies (Nwokolo and Kitts, 1988; Nwokolo *et al.*, 1988) have shown that when rubber seed oil is fed at the 5% level in cholesterol-free diets to rats, serum and liver cholesterol are lowered. However, when the oil is fed in diets containing cholesterol, a relative hypercholesterolaemic effect is observed. Results indicate that rats fed rubber seed oil at the 10% level grow significantly less than those fed the oil at the 5% level or corn oil at either the 5% or 10% level. No incidence of fatty liver has been observed. Similarly, no adverse effects have been observed in respect of content of haematocrit, serum glucase and serum alpha-amino nitrogen. These are biochemical indicators of toxicity of ingested material, and the normal values observed among these parameters are interpreted as being indicative of absence of overt toxic components. These workers recommend that rubber seed oil be refined for future nutritional evaluation. They suggest that the processes of refining which include degumming, acid neutralization, bleaching and deodorization will be enough to remove low-level toxicants in the oil.

35.5 GROWTH EXPERIMENTS WITH RUBBER SEED OIL

Growth experiments evaluating rubber seed oil as a dietary oil, have been conducted with broiler chicks and growing rats. Rats were fed standard diets which contained 0% or 1% cholesterol, each diet supplemented with either rubber seed oil or corn oil. Rats consuming the cholesterol diets ingested as much food as those fed the cholesterol-free diets. Supplementation with rubber seed oil or corn oil did not affect feed intake. However, rats consuming cholesterol-free diets containing rubber seed oil actually exhibited significantly greater average daily gains than rats fed cholesterol-free diets containing corn oil. In another experiment in which growing chicks were fed rubber seed oil or corn oil at the 8% level for 28 days, there were no significant differences in various growth parameters among chicks fed both oils. Studies such as these would indicate that the addition of rubber seed oil to animal diets has no toxic or antipalatability effect. In the rat study mentioned above, both the serum and liver total cholesterol levels are significantly reduced, but triglycerides are not affected in animals fed 5% rubber seed oil. This is in comparison with the control animals fed the corn oil, cholesterol-free diets. Both the degree of saturation and chain length of dietary triglycerides have recently been shown to be important in regulating serum and liver cholesterol levels. In some cases this property has been attributed to an increase in faecal sterol excretion (Wilson and Siperstein, 1959).

A high percentage of polyunsaturated fatty acids (PUFAs) in the diet has been reported to lower plasma cholesterol or affect cholesterol distribution between tissues and plasma (Grundy and Ahrens, 1970; Kellogg, 1974). For example, lower liver cholesterol levels have been observed in animals fed diets containing safflower oil, compared to more saturated oils such as coconut oil (Kellogg, 1974). The bulk of the evidence relating these effects to certain vegetable oils, has been attributed to their linoleic acid (C18:2 n=6) content. In the rubber seed oil studies, the oil has been shown to contain a considerably higher proportion of alpha-linolenic acid (c18:3 n=3), but the total essential fatty acid content of rubber seed oil (52.2%) is similar to that of corn oil (58.1%). The finding that rubber seed oil has a greater cholesterol lowering effect than corn oil in a cholesterol-free diet (Table 35.5) suggests a possible hypolipiaemic effect of linolenic acid in addition to linoleic acid. This hypothesis is supported by other studies which have shown linolenic acid to be equivalent to the unsaturated fatty acids present in fish oil in lowering cholesterol levels (Sim *et al.*, 1991). In addition, linolenic acid is known to suppress the

Table 35.5 Effect of rubber seed oil on serum and liver lipids in rats

Serum lipids (mg/100 ml)	Rubber seed oil diet	Corn oil diet
Total cholesterol	92.5 ± 9.0	116.3 ± 6.5
Triglycerides	128.3 ± 11.2	127.7 ± 14.4
HDL cholesterol	63.0	89.6
LDL cholesterol	16.25	20.28
CH-VLDL + LDL	202.0	242.0
Liver cholesterol (mg/g)	4.14 ±0.31	5.06 ±0.39

From Nwokolo *et al.* (1988).
HDL=high density lipoprotein; LDL=low density lipoprotein;
VLDL=very low density lipoprotein.

metabolism of linoleic acid (FAO, 1977), and may therefore have indirectly influenced the cholesterolaemic response of these animals by sparing linoleic acid.

In another set of experiments (Nwokolo and Kitts, 1988) in which rats were fed rubber seed oil or corn oil at two dietary levels (5% or 10%), rats fed rubber seed oil at the 5% level grew at the same rate as those fed corn oil at the 5% or 10% level. Rubber seed oil at the 10% dietary level is, however, too high and is associated with a significant depression in growth rate of the rats. Inclusion of most oils at the 10% dietary level, may be considered high, even excessive, though it may not depress growth. It must be remembered that the rubber seed oil used in these experiments was unrefined and that refining of the oil eliminates the growth-depressing factor in the oil, when it is fed at high levels. Growth rates observed in rats fed rubber seed oil at the 5% level are similar to values previously reported in animals fed other dietary oils. Such dietary oils include olive oil or peanut oil (Craig and Beare, 1967), corn oil (Hung *et al.*, 1977), or soybean oil (Kramer *et al.*, 1973). The severe growth depression which is observed in rats fed rapeseed oil, is not noticed in studies with rubber seed oil fed at either 5 or 10%. This is because rubber seed oil shows no detectable level of erucic acid (C22:1, n=9) and only an extremely low content of gandoic acid (C20:1, n=9).

Table 35.6 presents the body weights and individual organ weights of rats fed corn oil or rubber seed oil, respectively. Absolute liver weights (not shown) as well as relative weights (g/100 g body weight) of this organ in rats fed the 10% rubber seed oil diets, have been found to be disproportionately lower than in other rats fed the 5% rubber seed oil and corn oil diets. It is suggested that oxidation-derived antinutritional factors may be responsible for the growth restriction noted in rats fed 10% rubber seed oil. Blood chemistry analyses involving haemoglobin, haematocrit, serum glucose and alpha-amino nitrogen in rats fed corn oil and rubber seed oil are also presented in Table 35.6. There is no evidence of metabolic

Table 35.6 Tissue weights, tissue lipids and blood chemistry in rats fed rubber seed oil

	5% oil diet		*10% oil diet*	
Parameter	*CO*	*RSO*	*CO*	*RSO*
Final body wt (g)	297.0	311.5	317.5	297.0
Body wt gain (g)	147.0	154.8	150.8	132.0
Average daily gain (g)	6.99	7.30	7.09	6.28
Daily feed eaten (g)	20.2	20.8	20.6	19.2
Feed conversion ratio	2.92	2.85	2.91	3.10
Liver wt (g/100g)	4.65	4.78	4.85	4.17
Heart wt (g/100g)	0.34	0.33	0.34	0.33
Liver lipids (% wet wt)	4.79	4.80	5.06	5.04
Heart lipids (% wet wt)	2.95	2.97	3.15	3.17
Haematocrit (%)	47.9	50.2	46.9	49.3
Haemoglobin (g/dl)	13.8	14.1	14.0	14.1
Serum glucose (mg/dl)	130.0	131.0	124.0	132.0
Serum alpha-amino nitrogen (mg/dl)	7.95	9.38	7.95	9.07

CO = corn oil; RSO = rubber seed oil.
Partly from Nwokolo and Kitts (1988).

or physiological disorders attributed to rubber seed oil, at either the 5 or 10% dietary levels despite the fact that the rubber seed oil was unrefined. Although it has been reported that the type as well as the content of fat in the diet can influence net protein utilization in rats, there is no sign of abnormal levels of serum alpha-amino nitrogen in rats fed rubber seed oil.

Babatunde and Pond (1987a, 1988) evaluated rubber seed meal and oil as sources of nutrients for rat growth. They also investigated rubber seed meal as a nutrient source for growing–fattening pigs (Babatunde and Pond, 1990). These experiments all confirmed the great potential which exists for utilization of rubber seed meal in formulation of diets for non-ruminant animals.

35.6 CONCLUSIONS

The studies described have been attempts to evaluate the nutritional properties of rubber seed meal and rubber seed oil. These results indicate that the use of rubber seed meal and rubber seed oil will not be restricted by toxic or antipalatability factors. In western countries, there is a trend towards increased intake of vegetable fat and, particularly, unsaturated fatty acids in the diet. A majority of the vegetable fats and oils consumed are derived from soybean oil which is not only an excellent source of linoleic acid but also contains a relatively high proportion of linolenic acid

(7–10%). In the studies reviewed, rubber seed oil has been shown to be a rich source of alpha-linolenic acid and to a lesser extent linoleic acid. In its raw state, rubber seed oil has a dark colour, off-flavour and an odour. In laboratory experiments in which the oil has been degummed and refined, a golden brown, tasteless and odourless oil has been produced. The usefulness of the refined oil can be enhanced through addition of antioxidants and other stabilizing agents to prevent auto-oxidation and production of off-flavours. It is feasible that the oil can be utilized as a blend with other vegetable fats, such as palm kernel, which contain predominantly saturated fatty acids. With the increasing evidence pointing to *n*=3 fatty acids as being essential nutrients, with distinctly different and important physiological functions to the *n*=6 family (Newringer and Connor, 1986), recent recommendations have prudently encouraged an adequate amount of *n*=3 fatty acids in the diet, especially during pregnancy, lactation and infancy.

REFERENCES

Babatunde, G.M. and Pond, W.G. (1987a) The nutritive value of rubber seed (*Hevea brasiliensis*) meal and oil. 1. Rubber seed meal versus soybean meal as sources of protein in semi-purified diets for rats. *Nutr. Rep. Int.*, **36**, 617–30.

Babatunde, G.M. and Pond, W.G. (1987b) The nutritive value of rubber seed (*Hevea brasiliensis*) meal and oil. 2. Rubber seed oil versus corn oil in semi-purified diets for rats. *Nutr. Rep. Int.*, **36**, 857–65.

Babatunde, G.M. and Pond, W.G. (1988) The nutritive value of rubber seed (*Hevea brasiliensis*) meal and oil. 3. Performance characteristics, relative organ weights, hematocrit and plasma metabolites of growing female rats fed corn diets containing rubber seed meal, soybean meal or casein. *Anim. Feed Sci. Technol.*, **20**, 125–33.

Babatunde, G.M. and Pond, W.G. (1990) The nutritive value of rubber seed (*Hevea brasiliensis*) meal. Utilization by growing pigs of semipurified diets in which rubber seed meal partially replaced soybean meal. *J. Anim. Sci.*, **68**, 392–7.

Craig, B.M. and Beare, J.L. (1968) Nutritional properties of Canadian cambra oil. *Can. Inst. Food Technol. J.*, **1**, 64–7.

FAO (1977) Food and nutrition series: dietary fat and oils in human nutrition, no. 20. Report of an expert consultation by the Food and Agriculture Organization, Rome, and World Health Organization, Geneva.

FAO (1982) *FAO Production Yearbook*, Vol. 36, Food and Agriculture Organization of the United Nations, Rome.

Grundy, S.M. and Ahrens, E.H. (1970) Measurements of cholesterol turnover, synthesis and absorption in man carried out by isotope kinetic and sterol balance methods. *J. Lipid Res.*, **10**, 91–107.

Hung, S., Umemura, T., Yamashiro, S. *et al.* (1977) The effects of original and randomized rapeseed oils containing high or very low levels of erucic acid or cardiac lipids and myocardial lesions in rats. *Lipids*, **13**, 215–21.

Kellogg, T.F. (1974) Steroid balance and tissue cholesterol accumulation in germ-free and conventional rats fed diets containing saturated and polyunsaturated fats. *J. Lipid Res.*, **15**, 574–9.

Kramer, J.K.G., Mahadevan, S., Hunt, J.R. *et al.* (1973) Growth rate, lipid composition, metabolism and myocardial lesions of rats fed rapeseed oils (*Brassica campestris* var Arlo, Echo and Span and *B. napus* var Oro). *J. Nutr.*, **103**, 1696–708.

Newringer, M. and Connor, W.E. (1986) n-3 fatty acids in the brain and retina. Evidence for their essentiality. *Nutr. Rev.*, **44**, 285–94.

Nwokolo, E. and Akpapunam, M.A. (1986) Content and availability of nutrients in rubber seed meal. *Trop. Sci.*, **26**, 83–8.

Nwokolo, E.N. and Kitts, D.D. (1988) Growth parameters and plasma-tissue fatty acid profiles of rats fed rubber seed oil. *Food Chem.*, **30**, 219–29.

Nwokolo, E. and Sim, J.S. (1988) Dietary utilization of rubber seed oil by growing chicks. *Trop. Sci.*, **28**, 147–52.

Nwokolo, E., Bragg, D.B. and Sim, J.S. (1987) Nutritional assessment of rubber seed meal with broiler chicks. *Trop. Sci.*, **27**, 195–204.

Nwokolo, E., Kitts, D.D. and Kanhai, J. (1988) Serum and liver lipids of rats fed rubber seed oil. *Plant Foods Hum. Nutr.*, **38**, 145–53.

Ong, H.K. and Yeong, S.W. (1978) Prospects for the use of rubber seed meal for feeding pigs and poultry, in *Feedingstuffs for Livestock in South East Asia*, (eds C. Devendra and R.I. Hutagalung), Malaysian Society of Animal Production, Kuala Lumpur, pp. 337–44.

Rajaguru, A.S.B. and Ravindran, V. (1979) Rubber seed meal as a protein supplement in growing swine rations. *J. Natl Sci. Council*, (*Sri Lanka*) **7**, 101–4.

Ravindran, V. and Ravindran, G. (1988) Some nutritional and antinutritional characteristics of para-rubber (*Hevea brasiliensis*) seeds. *Food Chem.*, **30**(2), 93–102.

Ravindran, V., Rajaguru, A.S.B. and de-Silva, C. (1987) Evaluation of rubber (*Hevea brasiliensis*) seed in white leghorn cockerel diets. *J. Agric Sci.* (*UK*), **108**, 505–8.

Sim, J.S., Nwokolo, E.N. and Jiang, Z. (1991) Modulation of plasma and tissue cholesterol and fatty acid composition by feeding flax and canola seeds and oils to rats. *Can. J. Anim. Sci.*, **71**, 1207–14.

Viswanathan, T.V., Ananthasubramaniam, C.R. and Menachery, M. (1979) Evaluation of rubber (*Hevea brasiliensis*) seed cake for promoting growth in calves. *Ind. J. Nutr. Diet.*, **16**, 383–9.

Wilson, J.D. and Siperstein, M.D. (1959) Effect of saturated and unsaturated fats on fecal excretion of end products of cholesterol metabolism in the rat. *Am. J. Physiol.*, **196**, 596–8.

CHAPTER THIRTY-SIX

African breadfruit (*Treculia africana* Decne) and Polynesian breadfruit (*Artocarpus altilis* Fosberg)

E. Nwokolo

36.1 INTRODUCTION

The African breadfruit (*Treculia africana* Decne) is a member of the Moraceae, a family that includes the more popular Polynesian breadfruit, *Artocarpus altilis* Fosberg (also called *Artocarpus communis*). The family contains over 50 genera and over 1000 species of tropical trees and shrubs, many of which produce edible fruits and seeds. The members of this family are much-branched trees, up to 20 m high, inhabiting tropical wet coastal plains and well-drained soils. The trees bear large fleshy fruits which are covered with wart-like growths. The African breadfruit (*Treculia africana*), is filled with seeds, the plant being cultivated primarily for its delicious ellipsoidal seeds. Seed production from the African breadfruit is considerable, a mature tree producing up to 30 fruits annually, each fruit yielding 5–10 kg of seed after processing. For processing, the fruits are stacked in a heap and allowed to ferment. The fermented mass is macerated and washed in running water until all adhering slimy, jelly-like substances have been removed. Cleaned seeds are subsequently air-dried, dehulled and available for sale in local markets. Cooked seeds make very delicious meals. Many cultivars of the Polynesian breadfruit (*Artocarpus altilis*) are seedless, the breadfruit being cultivated for its pleasant-tasting fleshy fruit. Fruit production is also

Legumes and Oilseeds in Nutrition
Edited by E. Nwokolo and J. Smartt
Published in 1996 by Chapman & Hall. ISBN 0 412 45930 2

considerable, a mature tree yielding 20–30 fruits per year, each mature fruit measuring 15–20 cm in diameter and weighing 10–20 kg.

The African breadfruit is native to tropical Africa, whereas the Polynesian breadfruit is native to the Malay archipelago, from where it is believed to have spread to the South Pacific and the whole of Polynesia. From Polynesia, the breadfruit spread to the Caribbean, Central and South America and Africa.

36.2 CURRENT AND POTENTIAL USAGE AS FOOD

Seeds of the African breadfruit are edible, but not the fleshy pulp in which the seeds are embedded, which is bitter tasting. The seeds are small and ellipsoidal, about 15 cm long, and covered with a brittle brown shell. Seeds may be roasted, dehulled and eaten as a snack or they may be parboiled and dehulled prior to cooking, a process which lasts 2–3 h. Dehulled seeds may be cooked alone or in combination with shelled corn or other vegetables.

African breadfruit meals may be garnished with dried fish or meat, resulting in a highly nutritious diet. Roasted seeds may also be used to thicken soups and gravies, or milled into a flour for use in the preparation of cookies and cakes. The oil may be processed for use in cooking. The Polynesian breadfruit which is essentially a starchy food, may be cooked, baked, roasted or fried into a wide variety of dishes. The breadfruit may also be sliced, dried and milled into a flour which is used in baking snacks, cakes and cookies or as a base for various sauce mixes.

36.3 NUTRIENT COMPOSITION AND NUTRITIVE VALUE

In the past 20 years, there has been a large increase in the number of scientific studies and evaluations of the breadfruit (*Artocarpus* spp.) and the African breadfruit (*Treculia africana*). A chemical characterization of the carbohydrate and other fractions in breadfruit was conducted by Arango-Acosta and Quijano-Toban (1977) and by Sunarto (1978). Compositional analyses of African breadfruit (Nwokolo, 1985; Makinde *et al.*, 1985; Ekpenyong, 1985; Edet *et al.*, 1985) indicate that there is considerable variation in nutrient content of *Treculia africana*. Thus protein varies from 13.4% to 23.2%, fat from 10.4 to 18.9% and carbohydrates from 53.7% to 62.2%. Data in Table 36.1 on the nutrient composition of flours prepared from three varieties of the Polynesian breadfruit indicate that this breadfruit, more than the African breadfruit, is a major source of dietary carbohydrates. In the Polynesian breadfruit, starch content ranges from 66.2 to 73.1%, the stage of maturity not being particularly important. Average starch content in Polynesian breadfruit (69.3%) far exceeds the

Table 36.1 Composition of flours from the Polynesian breadfruit

	Content (% of dry matter)						
Varieties	*Moisture*	*Protein*	*Fat*	*Fibre*	*Ash*	*Sugars*	*Starch*
Ulu puou							
Immature	11.65	4.76	0.80	3.83	2.94	14.8	72.1
Mature	13.73	4.46	0.96	3.75	2.84	11.5	73.1
Ulu maopo							
Immature	8.60	4.60	1.33	6.56	2.99	15.1	66.7
Mature	6.17	3.34	1.93	5.01	2.55	15.6	71.3
Ulu maafala							
Immature	3.61	3.57	1.11	3.20	1.69	21.7	66.2
Mature	2.84	3.70	1.35	2.87	1.90	22.1	66.7

From Wotton and Tumaalii (1984).

starch content of such common pulses as peas, lentils, chickpeas, common beans, cowpeas, lima beans and pigeon peas. Average total sugar content is variable, being inversely related to the starch content. Varieties containing higher starch levels have lower sugar content and vice versa. Total sugar content in Polynesian breadfruit far exceeds values reported for various pulses. Fructose is the major sugar (9–10.5%), glucose is fairly high in some varieties and sucrose in others (Table 36.2). Just as in pulses in which high carbohydrate content is associated with a relatively small content of lipids, fat content in Polynesian breadfruit is very low, mainly in the form of phospholipids. The proximate composition and the lipid content of the African breadfruit, is shown in Table 36.3. The African breadfruit has a relatively high lipid content (11–19%), as much fat as in lupins and winged beans, less than in sunflower seed and soybean and very much less than in groundnuts. Onwuka (1983) described the oil as slightly sweet, with an acid value of 5.4 and an iodine value of 66.75. The major lipid fraction is primarily as triacylglycerol, with much lesser amounts of phospholipids. The fatty acid profile is distinctly dissimilar to that of soybean oil and corn which are considered standard dietary oils. There is 233% more stearic acid and 54% more palmitic acid in the African breadfruit than in soybean oil. Content of oleic acid is similar in both the African breadfruit and soybean oil, but linoleic acid is 36% lower in the African breadfruit than in soybean oil. Soybean has a much higher content of linolenic acid than the African breadfruit which has only trace amounts of this fatty acid (Table 36.4). Protein content is fairly high in African breadfruit, with an average of 19%, and a range 14–23%. In the Polynesian breadfruit, protein content is low, with an average of 4% and a range of 3.5–4.8%. The protein level in the African breadfruit is similar to protein levels in garden pea, lentil, chickpea, the common bean, cowpea, pigeon pea and lima bean. It is, however, lower than the protein of groundnut and soybean. That seeds of the African breadfruit have a good amino acid

Table 36.2 Fructose, glucose and sucrose in Polynesian breadfruit flour

	Content (% dry matter)		
Varieties	*Fructose*	*Glucose*	*Sucrose*
Ulu puou			
Immature	9.90	4.90	–
Mature	4.80	6.40	0.40
Ulu maopo			
Immature	10.40	2.90	1.80
Mature	9.00	3.60	3.00
Ulu maafala			
Immature	10.20	4.60	6.90
Mature	9.50	5.00	7.50

From Wotton and Tumaalii (1984).

Table 36.3 Proximate composition of seeds of the African breadfruit

	African breadfruit samples			
Nutrients	*1*	2	3	4
Dry matter (%)	91.00	90.50	–	92.2
Moisture (%)	9.00	9.50	–	7.8
Crude protein (%)	19.83	19.01	23.2	13.4
Ether extract (%)	10.82	10.41	11.4	18.9
Crude fibre (%)	2.45	2.40	1.34	1.4
Ash (%)	4.17	4.23	1.98	2.1
Carbohydrate (%)	53.70	54.45	62.20	58.1
Food energy (kcal/g)	4.45	4.40	4.18	4.56

Samples 1 and 2 from Nwokolo, unpublished data; sample 3 from Ekpenyong (1985); sample 4 from Edet *et al.* (1985)

Table 36.4 Fatty acid profile of oil from the African breadfruit and soybean (% of total fatty acids).

	African breadfruit samples		
Fatty acids	*1*	*2*	*Soybean*
14:0	–	–	0.11
16:0	20.20	19.20	13.07
18:0	18.44	9.80	5.53
18.1	25.30	13.10	28.16
18:2	32.77	44.00	44.44
20:0	0.90	–	0.50
18:3	0.94	–	6.45

Sample 1 and soybean oil from Nwokolo (1985).
Sample 2 from Ikediobi (1981).

Table 36.5 Amino acid composition of seeds of the African breadfruit

	Content (g/kg dry matter)		
Amino acids	*1*	2	3
Aspartic acid	17.4	12.5	9.9
Threonine	9.3	7.1	5.4
Serine	12.0	8.5	6.3
Glutamic acid	23.9	15.7	10.2
Proline	8.4	5.2	3.9
Glycine	12.7	9.2	7.3
Alanine	7.2	5.8	4.0
Cystine	0.8	1.2	–
Valine	12.3	8.5	6.7
Methionine	1.6	1.6	2.0
Isoleucine	10.4	7.0	6.5
Leucine	14.3	8.7	7.0
Tyrosine	10.6	6.2	6.0
Phenylalanine	12.2	7.2	6.2
Histidine	6.8	3.9	2.7
Lysine	13.5	7.9	5.5
Arginine	12.3	6.0	7.2

Sample 1 from Nwokolo (1985);
sample 2 from Ekpenyong (1985);
sample 3 from Ikediobi (1981).

profile and chemical score was confirmed by various workers (Ikediobi, 1981; Achinewhu, 1982; Makinde *et al.*, 1985; Nwokolo, 1985). Makinde *et al.* (1985) found that the meal is capable of furnishing most essential amino acids in the human diet, and noted that the protein is particularly rich in the aromatic amino acids although it is deficient in the sulphur amino acids and tryptophan. The amino acid profile of the African breadfruit (Table 36.5) is similar to the profile of many edible beans and pulses and, in common with all of them, has a very low content of sulphur amino acids. The very low level of sulphur amino acids in various pulses as well as in the African breadfruit would impede the optimal utilization of these proteins when they are consumed as the sole food by babies, children and young adults. However, consumption of large amounts of legume proteins and probably of the African breadfruit protein, may meet adult human amino acid requirements. Norton *et al.* (1985) note that the sulphur amino acid limitation to protein quality in legume seeds is very important when food legumes are fed alone. However, when these legumes are consumed with considerable quantities of such cereals as corn, millet or guinea corn, the significance of the inherent sulphur amino acid deficiency may be diminished.

The African breadfruit is a good source of potassium, phosphorus and magnesium, and a poor source of calcium (Table 36.6). Trace mineral

Table 36.6 Mineral content of African breadfruit seed

Minerals	*Content* 1	2	3
Phosphorus, (%)	278	364	381.8
Calcium (%)	68	18	17.5
Sodium (%)	–	5	7
Magnesium (%)	169	–	183.5
Potassium (%)	372	601	585
Iron (mg/kg)	2	1.5	1.6
Zinc (mg/kg)	1.6	–	7.5
Manganese (mg/kg)	0.7	–	–
Copper (mg/kg)	1.0	–	3.9

Sample 1 from Nwokolo (1985);
sample 2 from Edet *et al.* (1985);
sample 3 from Ekpenyong (1985).

Table 36.7 Nutrient content of processed and unprocessed *Treculia africana*

Nutrients	*Whole fruit*	*Unprocessed seed*	*Parboiled seeds*	*Cooked diet*
Dry matter (%)	11	60	40	31
Crude protein (%)	12	14.4	17.7	17.6
Crude fat (%)	7.3	10.6	7.1	4.4
Ash (%)	4.9	6.7	4.1	1.1
Crude fibre (%)	5.8	5.2	4.1	3.5
Minerals				
Sodium (mg/100 g)	241	261	235	204
Potassium (mg/100 g)	260	283	255	243
Calcium (mg/100 g)	10	15	12	12
Phosphorus (mg/100 g)	129	139	121	90
Manganese (mg/100 g)	0.16	0.17	0.11	0.10
Zinc (mg/100 g)	0.85	0.89	0.85	0.81
Iron (mg/100 g)	0.62	0.86	0.60	0.46
Magnesium (mg/100 g)	75	90	70	61
Copper (mg/100 g)	0.11	0.11	0.10	0.09

content is very low. As in legumes, it is likely that the high phosphorus content is primarily phytate phosphorus. Edet *et al.* (1985) have reported on vitamin content of *Treculia africana*. In general, thiamine (0.5 mg % DM) and riboflavin (0.3 mg % DM) content are low, beta-carotene content (6.0 mg % DM) is moderate, whereas ascorbic acid content (45.1 mg %) is high. In general, content of vitamins in the African breadfruit follows a pattern similar to the content of similar vitamins in pulses. Lawal (1986) observed that roasting, a very common form of processing of *Treculia africana* seeds for consumption, had no influence on protein content of the seed although there was considerable reduction in vitamin content. Thiamine (152 mg), riboflavin (235 mg), ascorbic acid (13.5 mg), pyridoxal (0.4 mg), beta-carotene (17.68 IU) and vitamin A (5972 IU/100 g dry matter) decreased by 31, 26, 40, 13, 25 and 24%, respectively. There was no effect on niacin content. Bassir and Lawal (1985) in an investigation of the changes in major nutrients in seeds of *Treculia africana* during cooking, found that nutrient content in whole fruit, unprocessed seed, parboiled seed and cooked meal were as shown in Table 36.7. The influence of alkaline cooking of *Treculia africana* with trona ('akanwu'), and of supplementation of the cooked meal with corn or crayfish was studied by Nzomiwu (1990). Whereas alkaline cooking was detrimental to nutritive value, supplementation with corn or crayfish greatly enhanced the protein quality of the cooked meal.

36.3.1 Processing

Olatunji and Akinrele (1978) studied the potential for industrial utilization of *Treculia africana* to investigate the possibility of extending wheat flours with various tropical tuber flours or *Treculia africana* flour. Product quality was poor. The possibility of processing seeds of *Treculia africana* to flours for use as beverages as well as in food, feed, and other industrial applications was discussed by Ejiofor *et al.* (1988).

36.4 ANTINUTRITIONAL FACTORS IN BREADFRUIT SEEDS

Although the presence of antinutritional factors in raw African breadfruit seed has been suspected for a long time, Uzogara *et al.* (1987) were the first to demonstrate the existence of trypsin inhibitors in raw *Treculia africana* seeds. They compared trypsin inhibitor activity in castor and breadfruit seeds. Trypsin inhibitor activity in breadfruit meal was 620 x 10^{-4} per mg protein, whereas it was 1319 x 10^{-4} in castor (*Ricinus communis*). The presence of antinutritional factors in the African breadfruit may partially explain the depression in many biological parameters observed when unautoclaved seeds are fed to chickens. In a preliminary experiment, Ayodele and Nwokolo (1981) observed that raw breadfruit seeds produced a significant increase in relative liver and pancreas weight of chicks compared with those fed on non-breadfruit seed diets. This increase was not observed in chicks fed on diets containing steamed breadfruit seeds. Chicks fed on raw breadfruit seeds lost weight so rapidly, and mortality was so high that the experiment had to be abandoned after 10 days. Several heat-labile antinutritional factors have been identified in almost all pulses. These include carbohydrase and protease inhibitors, and haemagglutinins (lectins). Other antinutritional

Table 36.8 Nutrient intake from cooked African breadfruit and cowpeas (100 g of diet)

	Contents (per 100 g of diet)	
Nutrients	*African breadfruit*	*Cowpeas*
Water (g)	78.80	88.90
Protein (g)	4.80	3.27
Total lipids (g)	2.73	0.23
Fibre (g)	0.62	1.04
Ash (g)	1.05	0.54
Minerals		
Calcium (mg)	17.16	42
Potassium (mg)	93.89	240
Magnesium (mg)	42.65	26
Phosphorus (mg)	70.20	55
Iron (mg)	0.53	1.97
Zinc (mg)	0.40	0.37
Copper (mg)	0.25	0.08
Manganese (mg)	0.18	0.17
Essential amino acids		
Threonine (mg)	235	115
Isoleucine (mg)	262	188
Leucine (mg)	361	267
Lysine (mg)	341	235
Methionine (mg)	40	13
Phenylalanine (mg)	308	105
Arginine (mg)	310	157
Histidine (mg)	172	20
Valine (mg)	310	319

From Nwokolo (1985).

factors include saponins, phytins, tannins, and flatulence factors (Norton *et al.*, 1985). The protease inhibitors include trypsin inhibitors as well as others which inhibit chymotrypsin and other serine proteases. Haemagglutinins are widely distributed in legume seeds and are primarily glycoproteins. These protease inhibitors and lectins are easily destroyed by heat, and therefore pose no threats when these legumes and possibly African breadfruit, is thoroughly cooked prior to consumption. Some antiphysiological factors specifically identified in raw African breadfruit include oxalate, phytate, tannin and hydrocyanic acid (Edet *et al.*, 1985). Levels of these are as follows: total oxalates 3.0%, soluble oxalates 2.5%, phytic acid, phytate 20.0 mg%, tannin 15.0 mg% and HCN 1.1 mg%. Oxalates and phytates interfere with utilization of minerals contained in food and may therefore inhibit mineral utilization from the African breadfruit. Tannins are polyphenolic compounds which interfere with protein digestibility and consequently affect the protein quality of various foods. The presence of tannins in raw African breadfruit is likely to inhibit utilization of the protein of this food, although the extent of this interference will be expected to be little, given the low content of tannin in the African breadfruit. There are small levels of HCN in raw African breadfruit. However, HCN is generally heat-labile and is lost during the process of cooking. Since cooking of the African breadfruit is usually a prolonged process, it is likely that no appreciable HCN will survive this cooking.

36.5 FEEDING EXPERIMENTS

Some of the earliest animal feeding experiments with *Treculia africana* was by Omole *et al.* (1978) who estimated the nutritional and growth-promoting qualities of *Treculia africana* and other tropical foods fed to rats at 30% dietary level. A significant contribution to information on the nutritional quality of *Treculia africana* was made by Achinewhu (1982, 1983) who reported lower apparent dry matter digestibility (69%) and poorer feed conversion efficiency (0.129) than was observed in melon seed or casein. The African and Polynesian breadfruit have been used as food for hundreds of years, and interviews conducted with individuals who have consumed the African breadfruit for years indicate that when the product is properly cooked, there are no symptoms of gastrointestinal hypersensitivity manifested as diarrhoea and vomiting. There are also no symptoms of flatulence, characterized by large amounts of gas which may contain significant amounts of methane and hydrogen. It is possible to conclude that in all likeliness, the three oligosaccharides (raffinose, stachyose and verbascose) which are known to be responsible for flatulence are not present to any relevant degree. Table 36.8 shows the nutrient intake of human subjects consuming 100 g of cooked, drained seeds of the African breadfruit and cowpea. Such human subjects will have a higher intake of total protein and essential amino acids from the African breadfruit than from cowpea.

REFERENCES

Achinewhu, S.C. (1982) The nutritive qualities of plant foods: chemical and nutritional composition of breadfruit (*Artocarpus altilis*) and climbing melon seed (*Colocynthis vulgaris*). *Nutr. Rep. Int.*, **25**, 643–7.

Achinewhu, S.C. (1983) The nutritive qualities of plant foods: 2. Protein quality of breadfruit (*Artocarpus altilis*), climbing melon seed (*Colocynthis vulgaris*) and creeping melon (*Citrullus vulgaris*) *Nutr. Rep. Int.*, **27**, 541–5.

Arango-Acosta, G.J. and Quijano-Toban, J. (1977) Chemical analytical study of the fruit of the breadfruit tree [*Artocarpus altilis* (Parkinson) Fosberg]. *Actual. Biol. Colombia*, **6**, 8–13.

Ayodele, A. and Nwokolo, E. (1981) Biological evaluation of nutritive value of African breadfruit (*Treculia africana*, Decne). BSc student thesis, Rivers State University of Science and Technology, Port Harcourt, Nigeria.

Bassir, O. and Lawal, R.O. (1985) Changes in the major nutrients of the seeds of *Treculia africana* during dietary preparation. *J. Plant Foods*, **6**, 247–51.

Edet, E., Eka, E. and Ifon, E.T. (1985) Chemical evaluation of nutritive value of seeds of African breadfruit. *Food Chem.*, **17**, 41–7.

Ejiofor, M.A.N., Obiajulu, O.R. and Okafor, J.C. (1988) Diversifying utilities of African breadfruit as food and feed. *Int. Tree Crops J.*, **5**(3), 125–34.

Ekpenyong, T.E. (1985) Chemical composition and amino acid content of African breadfruit (*Treculia africana*, Decne). *Food Chem.*, **17**, 59–64.

Ikediobi, C.O. (1981) Amino and fatty acid composition of *Pentraclethra macrophylla* and *Treculia africana* seeds. *J. Am. Oil Chem. Soc.*, **58**, 30–1.

Lawal, R.O. (1986) Effect of roasting on the chemical composition of the seeds of *Treculia africana*. *Food Chem.*, **22**, 305–14.

Makinde, M.A., Elemo, B.O., Arukwe, U. and Pellett, P. (1985) Ukwa seed (*Treculia africana*) protein. I. Chemical evaluation of the protein quality. *J. Agric. Food Chem.*, **33**, 70–2.

Norton, G., Bliss, F.A. and Bressani, R. (1985) Biochemical and nutritional attributes of grain legumes, in *Grain Legume Crops* (eds R.J. Summerfield and E.H. Roberts), Collins Professional and Technical Books, London.

Nwokolo, E. (1985) Nutritional quality of the seeds of the African breadfruit (*Treculia africana*, Decne). *Trop. Sci.*, **27**, 39–47.

Nzomiwu, N.R. (1990) The effect of alkaline treatment (akanwu) and supplementary value of corn or crayfish on the protein quality of breadfruit (*Treculia africana*). *Plant Foods Hum. Nutr.*, **40**, 99–105.

Olatunji, O. and Akinrele, I.A. (1978) Comparative rheological properties and bread qualities of wheat flour diluted with tropical tuber and breadfruit flours. *Cereal Chem.*, **55**, 1–6.

Omole, A., Adewusi, S.A., Adeyemo, A. and Oke, O.L. (1978) The nutritive value of tropical fruits and root crops. *Nutr. Rep. Int.* **17**, 575–80.

Onwuka, N.D. (1983) Solvent extraction and characterization of oils from the african breadfruit (AB), the african mango (AM) and the oil bean seed (OB). *Progr. Food Eng.*, pp. 175–9.

Sunarto, A.T. (1978) The breadfruit as a carbohydrate resource. *Bul. Kebun. Raya.*, **3**, 129–32.

Uzogara, S.G., Udoh, C.G. and Eka, O.U. (1987) Trypsin inhibitors in some lesser known seeds in Nigeria. *Food Chem.*, **24**, 227–32.

Wotton, M. and Tumaalii, F. (1984) Composition of flours from Samoan breadfruit. *J. Food Sci.*, **49**, 1396–7, 1400.

CHAPTER THIRTY-SEVEN

Castor (*Ricinus communis* L.) seed, oil and meal

S.S.Negi

37.1 INTRODUCTION

Castor (*Ricinus communis* L.), a heat loving plant, is grown widely in tropical and subtropical countries of Asia, Africa and America between latitudes 40° north and south latitudes. However, its species are cultivated in temperate climatic zones of China, Iran, Syria and Armenia. It requires good soil fertility conditions and responds well to the application of fertilizers and manures (Moshkin, 1986). Castor is a major oilseed and has been known since ancient times. It contains approximately 50% oil and 18% protein. About 500 000 tonnes of seed are processed annually on a global basis, India ranking second in production after Brazil. Castor oil is best known for its medicinal properties as a laxative but numerous aspects of its industrial applications include use in surface coatings, soaps and other cosmetic products. It is a source of sebacic acid, some esters of which retain their viscosity over a wide temperature range and are therefore used in the lubrication of jet engines. Castor seed meal or pomace contains a high percentage of protein, undecorticated pomace 20% and decorticated pomace 45%. However, it is unsuitable as a feed constituent, because of the presence of toxic substances. In the absence of large-scale detoxification procedures, castor seed cake is used almost entirely as a fertilizer (Vaughan, 1970).

Legumes and Oilseeds in Nutrition
Edited by E. Nwokolo and J. Smartt
Published in 1996 by Chapman & Hall. ISBN 0 412 45930 2

37.2 POTENTIAL USE AS HUMAN FOOD

Watt (1986) mentions a variety of castor that contains no poisonous principle, the oil of which is edible. In many areas of Nigeria, this variety of castor seed is used as food after fermentation (Coursey, 1964). The Nigerian castor variety is a tall slender plant which is grown commercially in large family holdings and in plantations. Mature, dry castor seeds are harvested, decorticated, and cooked for several hours. The cooking water which contains some toxic components is thrown out, the cooked seeds are washed thoroughly, and ground in mortars. The milled castor seed is wrapped in clean plantain leaves and left to ferment over a period of 15–30 days. Fermented castor meal is used extensively in a large variety of dishes as a spice and a food condiment.

37.3 POTENTIAL USE AS ANIMAL FEED

Castor leaves are used to feed a species of indigenous silkworm in Assam. The green leaves are reported to be eaten by cattle, sheep and goats with relish and are believed to increase the flow of milk in cows and buffaloes. The oil cake is generally thought to contain a variety of poisonous materials and hence is not used as cattle feed (Watt, 1986).

37.4 NUTRIENT COMPOSITION AND NUTRITIVE VALUE OF CASTOR OIL CAKE

There is wide variation in the nutrient content of castor seed meal, depending on whether it is decorticated or not. Protein content varies from 20.5% to 46% depending on the method of processing. Thus decorticated pomace contains more than double the protein in undecorticated pomace. Carbohydrates and fibre may range from 26% to 49%, and mineral content from 10.5% to 15%. The oil content of pomace can be reduced by efficient crushing and solvent extraction to below 2%. The ash of the cake is rich in minerals and contains CaO 17%, P_2O_5 25%, MgO 6%, K_2O 10% and Fe_2O_3 6% (Kulkarni, 1959). Castor seed meal, detoxified by steam treatment at 5 kg/cm^2 for 15–30 min could be included in concentrate mixtures for sheep at the 10% level (Punj, 1988). The typical nutrient composition of castor bean meal (Table 37.1), (AICRP, 1967–85) shows that the meal is rich in digestible protein and total digestible nutrients. The amino acid composition of castor pomace is presented in Table 37.2.

Table 37.1 Nutrient composition of castor meal

Constituent	*% in dry matter*
Crude protein	32.96
Ether extract	26.59
Crude fibre	12.10
Nitrogen-free extract	19.15
Total ash	9.20
Calcium	0.70
Phosphorus	0.80
Digestible protein	22.50
Total digestible nutrients	63.99

From AICRP (1967–85).

Table 37.2 Amino acid composition of castor pomace

Amino acid	*g/100 g of protein*
Valine	6.6
Leucine	7.2
Isoleucine	5.3
Phenylalanine	4.2
Tryptophan	0.6
Threonine	3.6
Methionine	1.5
Arginine	11.0
Histidine	2.5
Lysine	3.1
Aspartic acid	4.6
Glutamic acid	18.0

Kodras *et al.* (1949).

37.5 TOXINS IN CASTOR SEED MEAL AND THEIR DETOXIFICATION

Castor seed and castor seed meal contain at least three toxic principles: (1) a protein ricin, (2) a powerful allergen which is a protein polysaccharide of lower molecular weight, active even at concentrations of 1 in 10^6 and (3) a mildly toxic alkaloid ricinine (Kodras *et al.*, 1949). Detoxification of castor seed meal is not carried out on a large scale even though a cake of excellent value would result. It is, therefore, suggested that in those areas of the world where there is potential for large-scale/commercial production of castor seed, serious consideration should be given to the increased production of this high protein seed for processing into animal feed. It would be necessary to decorticate castor seed prior to use since the jagged shell pieces may damage the intestines of livestock (Kulkarni, 1959). Solvent extraction followed by dilute hydrochloric acid and hot water extraction, precipitation with salt and final extraction with alcohol may completely detoxify castor seed meal. A scheme for simultaneous separate isolation of the three toxic fractions for detoxification of the pomace has been described (Waller and Negi, 1958).

37.6 CASTOR POMACE IN ANIMAL DIETS

Commercially and laboratory-prepared samples of castor pomace with or without the toxic principles removed, have been fed to white rats. Mortality occurred in rats fed castor pomace containing the toxic principles. In rats fed pomace from which the toxic principles have been extracted, there was no mortality. Rats fed the detoxified pomace at two levels (1:1 and 2:1 mixture of pomace and a complete rat diet), showed no weight loss over a period of 8 weeks. The main problem encountered with extracted castor pomace was one of palatability, although the rats overcame their objections within 2–4 days and subsequently ate the blended pomace as readily as the control diet (Waller and Negi, 1958). Soybean meal in the diet of Large White pigs has been replaced by detoxified castor meal at the levels of one-third, two-thirds or all of the soybean in the diet. In two other test diets, the highest castor diet was either reautoclaved at 120°C for 30 min or supplemented with lysine and tryptophan (Martins *et al.*, 1982). Serum levels of P, Ca, Mg, Fe, Zn and Cu did not differ significantly among test groups and controls. Feed intake and weight gains decreased with increasing castor meal content. Rats fed the highest castor meal diets which were supplemented with amino acids performed as well as those fed the lowest level of castor seed meal. All indices of performance were extremely poor in rats fed the highest level of castor pomace, whether reautoclaved or not.

Purushotham *et al.* (1986) fed groups of weaned lambs for 150 days on isocaloric and isonitrogenous diets containing autoclaved castor meal at 0, 10, 20 and 30% levels replacing a standard concentrate mixture. No significant differences in feed intake, nutrient digestibility and nitrogen balance were reported. However, weight gain declined progressively with increasing levels of castor meal in the concentrate mixture. In simultaneous histopathological studies, Rao *et al.*, (1986) observed deleterious changes in vital organs of sheep fed castor pomace at levels of 20 and 30%. They concluded that the pomace could be used without any ill effects at the 10% level in concentrate mixtures.

The utilization of castor pomace in animal diets is limited by toxicity problems. However, the material has the apparent potential of being upgraded into an excellent source of protein, if commercial extraction and detoxification processes are put in place.

REFERENCES

AICRP (1967–85) Terminal Report. All India Coordinated Research Project on Utilization of Agricultural By-products and Industrial Wastes for Evolving Economic Rations for Livestock. Andhra Pradesh Agricultural University, Hyderabad, India.

Coursey, D.G. (1964) Nigerian Federal Institute of Industrial Research, *Technical Monograph 22*.

Kodras, R., Whitehair, C.K. and MacVicar, R. (1949) Studies on the detoxification of castor seed pomace. *J. Am. Oil Soc.*, **26**, 641–4.

Kulkarni, L.G. (1959) *Castor*, Monograph of the Indian Central Oilseeds Committee, Hyderabad, India.

Martins, M.F., Ferreira-Neto, J.M. and Silva, R. (1982) *Nutr. Abstr. Rev.*, **52B**, 2771.

Moshkin, V.A. (1986) *Castor*. Translated from Russian. Amerind Publishing Co. Pvt Ltd. New Delhi.

Punj, M.L. (1988) *Bulletin of the All India Coordinated Research Project*. Indian Council on Agricultural Research, New Delhi, India.

Purushotham, N.P., Rao, M.S. and Raghavan, G.V. (1986) Utilization of castor bean meal in the concentrate mixture of sheep. *Ind. J. Anim. Sci.*, **56**, 1090–3.

Rao, M.S., Purushotham, N.P., Raghvan, G.V. *et al.* (1986) Studies in the pathology of experimental feeding of detoxified castor bean meal in sheep. *Ind. Vet. J.*, **63**, 944–6.

Vaughan, J.G. (1970) *The Structure and Utilization of Oilseeds*. Chapman & Hall, London.

Waller, G. and Negi, S.S. (1958) Isolation of ricin, ricinine and the allergic reaction from castor seed pomace from two different sources *J. Am. Oil Soc.*, **35**, 409–12.

Watt, G. (1986) *The Commercial Product of India*. Today and Tomorrow Printers and Publishers, New Delhi, India.

CHAPTER THIRTY-EIGHT

Salseed (*Shorea robusta* Gaertn.) oil and meal

S.S. Negi

38.1 INTRODUCTION

Sal (*Shorea robusta* Gaertn.) is a subdeciduous tree which grows extensively in Northern and Central India and as far north as the outer Himalayan region. Tree height may range from 18 to 45 m and girth from 1.8 to 3.6 m or more. Leaves, twigs, bark and seeds of the sal tree are well endowed with tannins, twigs and leaves containing 20–22% on a dry basis. Salseed contains 19–20% of oil and resins which are valuable in the manufacture of shoe polish, carbon paper and typewriter ribbon (CSIR, 1972). Availability of salseed is estimated at over 5.5 million tonnes per annum (Negi, 1982).

38.2 POTENTIAL USE AS HUMAN FOOD AND ANIMAL FEED

Salseeds are sometimes eaten after roasting, especially in times of food scarcity. However, these are said not to be palatable (Nayak *et al.*, 1967). Salseed fat has a buttery texture, with a fatty acid composition of palmitic acid (8.3%), stearic acid (34.7%), arachidic acid (12.3%), oleic acid (41.9%) and linoleic acid (2.8%). The oil is used for cooking and lighting and is suitable as a substitute for cocoa butter in the manufacture of chocolate (CSIR, 1972).

By virtue of its availability, salseed and salseed meal have a tremendous potential for utilization as animal feed. Recommendations have been made for its incorporation in poultry mashes at the 5–7% level

Legumes and Oilseeds in Nutrition
Edited by E. Nwokolo and J. Smartt
Published in 1996 by Chapman & Hall. ISBN 0 412 45930 2

Table 38.1 Typical composition of salseed meal

Nutritional constituent	*% in dry matter*
Crude protein	7.8–9.7
Ether extract	0.8–9.4
Crude fibre	1.6–3.0
NFE	74.6–84.2
Ash	3.1–3.6
Tannins	6.2–13.7
Acid detergent fibre	21.6
Cellulose	5.7
Lignin	11.9
Soluble carbohydrates	57.8
Starch	30.1

From Negi (1990).

(IVRI, 1978) and in concentrate mixtures for growing and lactating cattle at the 10–20% level (ICAR, 1972).

However, a critical review of the published work on the utilization of salseed and salseed meal as feed for cattle, poultry and pigs shows that tannins limit their utilization (Negi, 1982, 1990).

38.3 NUTRIENT COMPOSITION AND NUTRITIVE VALUE OF SALSEED MEAL

The typical chemical composition of salseed meal is given in Table 38.1. The high content of nitrogen-free extract (NFE) erroneously suggests its use as a possible source of cheap energy in livestock diets in place of cereals like maize and barley. In reality, the meal has a low proportion of starch associated with high proportions of lignin and tannins. The metabolizable energy (ME) value of the meal is rather low, estimated at 1.86 Mcal/kg in goats (Khan *et al.*, 1986). Because of its low protein content, salseed meal has little value as a protein supplement in spite of its reasonable amino acid composition (Table 38.2). On a dry matter basis, salseed meal has a low content of calcium (0.18–0.29%) and phosphorus (0.16–0.18) (ICAR, 1967). Estimates of total digestible nutrients (TDN) in salseed meal in feeding trials in cattle, obtained by three sets of researchers, have been 58, 41 and 19%, respectively (Negi, 1982). The considerable variations in TDN apparently resulted from the absence of adequate experimental safeguards in the handling and utilization of agricultural wastes as livestock feed. Pal *et al.* (1973) who obtained the lowest value of 19% TDN, observed that inclusion of salseed meal in equal proportions with groundnut cake had a depressing effect on overall digestibility of dry matter, organic matter and crude protein. Robb (1975) reported a negative digestibility of protein in salseed meal fed to dairy

Table 38.2 Amino acid composition of salseed meal protein

Amino acid	*% of protein*
Arginine	6.8
Histidine	4.8
Threonine	7.8
Valine/methionine	7.6
Norleucine/leucine/isoleucine	12.0
Tyrosine	4.3
Cystine	2.2
Lysine	6.7
Phenylalanine/tryptophan	4.3

From Mitra and Mishra (1967).

cows and ascribed it to tannins in *Shorea robusta*, which formed complexes with proteins in the diet.

38.4 TANNINS IN SALSEED MEAL AND THEIR DETOXIFICATION

The tannins in salseed meal are of the pyragallol type (CSIR, 1972). Panda *et al.* (1969) observed that 60% of tannins in salseed meal could be removed by simple water washing, in keeping with the findings (Vijjan and Katiyar, 1973) that 64% of tannins in salseed are in the hydrolysable form. Gandhi *et al.* (1975) recommended soaking in 30% acetone or 0.1M sodium hydroxide for removal of tannins. Washing with water after sodium hydroxide treatment was fairly effective but removed 20–70% of the dry matter, resulting in leaching of soluble nutrients. However, tannins in the meal could be inactivated by calcium hydroxide spray but the postulated complex was broken down in the rumen to release tannins (Sharma *et al.*, 1977; Wah *et al.*, 1977). Better palatability and higher energy value of alkali-treated salseed meal has been claimed (Singh and Arora, 1978), but the residual tannins were identified as condensed tannins (Singh and Arora, 1980), which, according to McLeod (1974), bind feed proteins in irreversible complexes, and not likely to be hydrolysed in the post-ruminant gut.

38.5 SALSEED MEAL IN DIETS OF PIGS AND POULTRY

38.5.1 Pigs

Murty *et al.* (1969), following replacement of 30% of maize with salseed meal (20%) and molasses (10%) in isonitrogenous swine diets, observed adverse effects of salseed inclusion manifested as depressed live weight gains, reduced nutrient digestibility, lower intake and retention of

Table 38.3 Performance of grower pigs fed salseed meal

	Level of salseed meal in diets (%)			
	0	*5*	*10*	*15*
Daily gain (kg)	0.40	0.37	0.31	0.33
Daily intake (kg)	1.99	1.89	1.90	1.88
Feed efficiency (feed/gain)	4.97	5.10	6.13	5.71

From Reddy (1972).

nitrogen. In a similar study, Das and Mohanty (1970–71) fed young pigs with isonitrogenous diets containing 0, 25, 37.5 and 50% of salseed meal for 8 months. There was a corresponding decrease in nutrient digestibility and live weight with increasing content of salseed meal in diets. Reddy (1972) studied the weight gains and feed efficiency of grower pigs (15–45 kg), between 8 and 12 weeks of age, fed on diets incorporating salseed meal at 0.5, 10 and 15% replacing corresponding quantities of maize. Feed efficiency of pigs receiving salseed meal was lower, with smaller, but not significantly different, live weight gains. A decrease in digestibility of protein with increasing content of salseed meal was also observed (Table 38.3). Pathak and Ranjhan (1978) studied the performance of three groups of Large White Yorkshire female pigs, weaned at 5 weeks of age, and fed for 112 days on control diets containing 40% maize, or experimental diets in which half or all of the maize was replaced by salseed meal. Although there was a slight depression in digestibility and retention of protein and other nutrients, live weight gains in the groups of pigs were similar. There was a detrimental effect on the kidneys of pigs on salseed meal diets although no mention is made of any degenerative changes in vital organs. This can not be ruled out in view of the degeneration observed in kidney, liver, spleen and intestines of cattle (Gupta *et al.*, 1977) and poultry (Panda *et al.*, 1975; Mohapatra and Panda, 1978) fed on salseed meal.

38.5.2 Poultry

Growth of White Leghorn starter chicks has been studied from 0 to 8 weeks of age on diets incorporating unextracted salseeds at 2.5 and 5% levels replacing maize or groundnut cake (Nayak *et al.*, 1967). The conclusion that salseed can be used as an economic replacement for maize at 5% level is not supported by the data presented by these researchers (Table 38.4). Another study, (Saxena, 1967) in which maize was replaced with salseed meal (at 5, 10, 15, 20% level) in mash for starter chicks, concluded that salseed meal could replace maize at up to 5% of the diet. In this study, there was a slight depression in live weight at 5% inclusion rate when Hyline chicks were used. However, with Arbor Acres chicks, there was no growth depression up to 20% level of inclusion. Rather, the feed efficiency at 5 and 10% level of inclusion was better than in the control

Table 38.4 Weight gains and feed conversion ratios of starter chicks fed salseed diets during 0–8 weeks of age

	Normal starter diet	*Maize replaced by salseed*		*GNC replaced by salseed*	
		2.5%	5.0%	2.5%	5.0%
Gain in weight (g)	340	312	263	261	248
Feed/gain ratio	2.4	2.6	3.4	3.8	4.2

From Nayak *et al.* (1967).
GNC = groundnut cake.

diet. In a later study, Saxena (1971) observed a depression in growth of chicks fed on diets containing salseed meal beyond the 3% level, and reported a reduction in egg production and an increase in abnormal coloration of eggs on layer diets containing salseed meal beyond the 10% level of inclusion.

The recommendation of Panda *et al.* (1969) that salseed meal processed to partially remove tannins could be used as a source of energy up to a level of 5–7% in diets of growing chicks, and a level of 7–10% in layer diets, was based merely on proximate composition rather than on feeding trials. Salseed meal has been compared with autoclaved poultry litter and dried cow dung at the 3% level of inclusion in poultry starter diets (Mohapatra and Panda, 1978). There were no significant differences in productivity or mortality. Paul *et al.* (1972) studied the dressing percentage of Rhode Island Red chicks fed diets in which maize was replaced by salseed meal at levels of 0, 10 or 15%. Basing their conclusions on the percentage of dressed meat at 8 or 24 weeks of age, they concluded that replacement of maize at the 10% level had no adverse effects on chick growth. Negi (1982), however, criticized these conclusions. Panda *et al.* (1975) replaced maize in starter diets with salseed or salseed meal at levels of 2.5–45%. Even at the lowest level of inclusion, body weights of chicks on salseed or salseed meal were 80 and 93% of those on the control diets. Use of higher levels of salseed or salseed meal was associated with increased depression in live weight, and was only partly counteracted by supplementation with extra protein from groundnut cake. Mortality on higher levels of salseed or salseed meal was rather heavy. There were histological lesions in the liver, kidney and spleen of dead chicks, accompanied by anaemia characterized by paleness of musculature. Such adverse effects were ascribed to tannins in salseed or salseed meal (11–13.7%), which even at 2.25% level of incorporation were not completely safe. These pathological findings have been substantiated by Mohapatra and Panda (1978). It would appear that the use of salseed meal in livestock diets may be motivated more by cost than by merit as a nutritionally rich feed ingredient.

REFERENCES

CSIR (1972) *The Wealth of India, Raw Materials. Dictionary of Indian Raw Materials and Industrial Products*. Publication and Information Directorate, Council of Scientific and Industrial Research, New Delhi, India.

Das, B.K. and Mohanty, G.P. (1970–71) Annual progress report of investigations on agricultural waste material for evolving economic rations for livestock. A Coordinated project of the Indian Council of Agricultural Research.

Gandhi, V.M., Cherian, M.K., Mulky, M.J. and Menon, K.K.G. (1975) Utilization of non-traditional indigenous oilseed meals. 2. Studies with deoiled salseed meal. *J. Oil Technol. Assoc.*, **7**, 44–7.

Gupta, B.S., Gupta, B.N., Saha, R.C. *et al.* (1977) Studies on long term feeding of sal (*Shorea robusta*) seed meal to adult cattle. *Ind. Vet. J.*, **54**, 463–9.

ICAR (1967) *Report of the study group for exploring the possibility of using food by-products as poultry feed*. Indian Council of Agricultural Research, New Delhi.

ICAR (1972) *Report of the Subcommittee on 'Studies on Salseed Meal'*. Indian Council of Agricultural Research, New Delhi.

IVRI (1978) *Brochure on Research, Education and Extension, Division of Poultry Research*, Indian Veterinary Research Institute, Izatnager, U.P.

Khan, M.Y., Lal, M. and Krishan, J. (1986). Nutritional evaluation of deoiled oilseed (*Shorea robusta*) meal for goats. *Ind. J. Anim. Nutr.*, **3**, 29–32.

McLeod, M.N. (1974) Plant tannins–their role in forage quality: a review. *Nutr. Abstr. Rev.*, **44**, 803–15.

Mitra, C.R. and Mishra, P.S. (1967) Amino acid content of processed oilseed meals. *J. Agric. Food Chem.*, **15**, 697–9.

Mohapatra, H.C. and Panda, N.C. (1978) Effect of feeding sal (*Shorea robusta*) seed and sal oil meal to chicks. *Ind. Vet. J.*, **55**, 559–66.

Murty, V.N., Joshi, B.C. and Agarwal, D.K. (1969) Evaluation of salseed meal in the ration of pigs. Annual Report of the Division of Animal Nutrition, Indian Veterinary Research Institute, Izatnagar, U.P.

Nayak, N.C., Tripathi, S.B. and Mohanty, M.K. (1967) Use of decorticated salseed as a poultry feed–A preliminary trial. *Ind. Vet. J.*, **44**, 706–11.

Negi, S.S. (1982) Tannins in salseed (*Shorea robusta*) and salseed meal limit their utilization as livestock feeds. *Anim. Feed Sci. Technol.*, **7**, 161–83.

Negi, S.S. (1990) Salseed meal, in *Non Traditional Feed Resources For Use in Swine Production*, (eds P.A. Thacker and R.N. Kirkwood), Butterworths, Boston, MA, pp. 383–9.

Pal, R.N., Daniel, S.J. and Murty, V.N. (1973) Studies on salseed cake as cattle feed. *Ind. J. Anim. Sci.*, **43**, 267–71.

Panda, B., Jayraman, M., Ramamurty, N.C. and Nair, R.B. (1969) Processing and utilization of salseed as a source of energy in poultry feed. *Ind. Vet. J.* **46**, 1073–7.

Panda, N.C., Mitra, A., Kedary, S. and Parichha, S.N. (1975) Salseed and salseed oil meal as cereal substitutes in chick mashes. *Ind. Vet. J.*, **52**, 186–92.

Pathak, N.N. and Ranjhan, S.K. (1978) Nutritional studies with salseed meal as a component of finisher rations in Large White-Yorkshire pigs. *Ind. J. Anim. Sci.*, **43**, 424–7.

Paul, A.K., Sen Gupta, S.N., Mondal, G.B. and Sarkar, S.K. (1972) Effects of feeding salseed cake on chicks. *Ind. J. Anim. Sci.*, **42**, 1061–5.

Reddy, K.J. (1972) Studies on supplementation of deoiled salseed meal in growing swine. A.P. Agricultural University, Hyderabad vide *Thesis Abstracts* (1978) **4**(2), 92.

Robb, J. (1975) Alternatives to conventional cereals, in *Feed Energy Sources for Livestock*, (eds H. Swan and D. Lewis), Butterworths, London.

Saxena, H.C. (1967) Use of salseeds, silkworm pupae and coca shells in poultry rations. Indian Poultry Science Association Seminar, Dec.

Saxena, H.C. (1971) Effect of feeding sal (*Shorea robusta*) and sal oil meal to chicks. *Ind. Poult. Rev.*, **3**, 5.

Sharma, K., Wah, C.S. and Jackson, M.C. (1977) A comparison of calcium hydroxide spray-treated and untreated salseed meal as ingredients in goat and chick diets. *Ind. J. Anim. Sci.*, **47**, 473–8.

Singh, K. and Arora, S.P. (1978) Studies on detanning and determining the nutritive value of alkali-treated salseed meal. *Ind. J. Anim. Sci.*, **48**, 84–9.

Singh, K. and Arora, S.P. (1980) Studies on tannins of seed meal of sal (*Shorea robusta*). *Ind. J. Anim. Sci.*, **50**, 1043–51.

Vijjan, V.K. and Katiyar, R.C. (1973) Types of tannins in deoiled salseed meal and oak kernels and their metabolic behaviour in sheep. *Ind. J. Anim. Sci.*, **43**, 398–401.

Wah, C.S., Sharma, K. and Jackson, M.C. (1977) Studies of various chemical treatments of salseed meal to remove or inactivate tannins. *Ind. J. Anim. Sci.*, **47**, 9–12.

PART SIX

Future Developments in Improvement

CHAPTER THIRTY-NINE

Molecular approaches to improving the nutritional and functional properties of oilseeds, grains and legumes

B.O. de Lumen

The purpose of this review is to discuss the rationale for improving the nutritional and functional properties of plant seeds through genetic engineering, and to point out recent developments in molecular biology of seed development and plant biotechnology that are relevant to this goal. Since this is not meant to be a comprehensive review, the reader is referred to more complete reviews on biochemistry and molecular biology of seed proteins written in the last ten years (Shewry and Miflin, 1984; Higgins, 1984; Croy and Gatehouse, 1985; Casey *et al.*, 1986; Gatehouse *et al.*, 1986; Goldberg, 1986; Goldberg *et al.*, 1989; Shewry *et al.*, 1989).

39.1 OILSEEDS, GRAINS AND LEGUMES CONTRIBUTE SIGNIFICANTLY TO THE HUMAN DIET

Seeds of cereals and legumes and their by-products make up a major part of the human diet. Rice is the staple food of half of the world's population and is the single largest source of protein in the world (Juliano, 1985). Cereals contribute 50% of the per capita energy intake worldwide and 65% in developing countries and Asian centrally planned economies (FAO, 1980). Cereals account for about 45% of the daily per capita protein supply in the world and approximately 63% in developing countries. Although the worldwide production of grain legumes is small relative to cereals, its relative contribution to human nutrition is greater than its

Legumes and Oilseeds in Nutrition
Edited by E. Nwokolo and J. Smartt
Published in 1996 by Chapman & Hall. ISBN 0 412 45930 2

relative production. Approximately 700 million people use legumes as an essential part of their diet (Abbott, 1966). Legumes make up 9% of the total plant food production in developing countries and serve as the major source of proteins in the diet of a number of countries (Toenniessen, 1985).

Animal products which have replaced plant foods in human diets in the developed market economies of the world are derived from farm animals. Non-ruminants and intensively reared ruminants consume seeds and their by-products supplemented with protein concentrates processed from oil seeds. In addition, the consumption of vegetable oils and fats, mostly derived from oilseeds account for 4–10% of the daily per capita energy supply in the world (Duffus and Slaughter, 1980). Soybean is the largest source of vegetable oil in the world (Soya Bluebook, 1982). Thus, the total contribution of seeds to the human diet, as energy and protein sources, can be as high as 85% of total calories.

The protein content of cereals varies from 7 to 14% whereas that of legumes is from 20 to 40%. Lysine is the first nutritionally limiting essential amino acid in most cereals, tryptophan is the second limiting amino acid in maize and threonine in other cereals (Eggum and Beames, 1983). The high content of prolamin (alcohol-soluble fraction) in cereals accounts for the low content of essential amino acids. Not only are prolamins very low in lysine, but they are also low in threonine, valine and isoleucine. Rice and oats have a better balance of essential amino acids than other cereals due to a lower content of prolamins (Doll, 1984; Shewry and Miflin, 1984). Legumes are uniformly limiting in methionine and cysteine because their major storage proteins, the globulins (salt-soluble fraction) are low in these amino acids (Duffus and Slaughter, 1980; Gupta, 1983).

39.2 BIOLOGY OF PLANT GENES

The application of recombinant DNA techniques and related methodologies to plants has opened up the potential to improve agronomic characters, food processing traits and food quality properties of plants as food sources. The rapid accumulation of knowledge on gene expression processes in plants in the last six years has given us an understanding of structure, function and organization of plant genes that has brought about some applications.

In the last six years, there has been a rapid accumulation of knowledge on gene expression processes in higher plants (Kuhlemeir *et al.*, 1987; Schell, 1987; Goldberg, 1988; Goldberg *et al.*, 1989). Much of this information has been obtained from genes that are intensively studied because they code for abundant gene products, are inducible, and are involved in unique plant processes that are economically important.

During seed development, proteins, carbohydrates, lipids, vitamins,

Table 39.1 Utilization problems of legumes and cereals and possible causes

	Problems	*Causes*
Legumes	Incomplete protein	Limiting amounts of methionine and cysteine
	Incomplete availability of amino acids	Unknown
	Low protein digestibility	Nature of protein (?) protease inhibitors (?)
	Antinutritional factors	Protease inhibitors, haemagglutinins, goitrogens, etc.
	Flatulence	Raffinose oligosaccharides, dietary fibre (?), undigestible starch (?)
	Undesirable 'beany flavour'	Lipoxygenases
Cereals	Low protein content	
	Incomplete protein	Limiting amounts of lysine, and in some cereals tryptophan, threonine

minerals and non-nutrients are laid down, leading to the unique chemical composition of plant seeds that determine their nutritional and functional properties. Understanding the molecular genetic mechanisms by which photosynthetic assimilates are partitioned into these components will allow the design of genetic engineering strategies to improve the usefulness of seeds as food sources. Utilization problems of legumes and cereals and their possible causes are summarized in Table 39.1.

39.3 IS THERE A NEED TO GENETICALLY ENGINEER LEGUMES AND CEREALS FOR IMPROVED NUTRITIONAL PROPERTIES?

In practice, cereals and legumes are eaten with other foods. The overall protein quality of cereal–legume mixtures is better than that of either protein source alone due to the complementary nature of their amino acid profiles. Cereals are low in lysine and relatively high in methionine and cysteine whereas legumes are high in lysine and low in methionine and cysteine. Supplementation of cereals and legumes with crystalline amino acids improves their value to support growth of animals and increases the efficiency of utilization of dietary protein (Bressani and Elias, 1968). Methionine is used as a supplement in the formulations of poultry and swine rations of the highly competitive feed industry where soybean is the main protein source (Abbott, 1966). There has been little practical use made of supplementation for planning human diets, although clinical studies with human beings have demonstrated their effectiveness (Bressani and Elias, 1968).

Since cereals are the major sources of protein worldwide, increasing the protein content alone will improve the protein intake of a large number of people, especially those in developing countries, where protein-calorie malnutrition is prevalent. For instance, an increase of 3% in rice protein without a decrease in grain yield will add an additional 12 million tons of protein in the diet of rice-eating people, based on 411 million tons of total rice production in 1981/82 (Juliano, 1985). Improving the protein quality of the rice grain will bring additional significant benefits because the nitrogen-digestibility and nitrogen-retention of rice diets are low (Roxas *et al.*, 1979).

Because the protein contents of legumes are already high, improving the protein quality is the more useful goal. Why is there a need for a nutritionally balanced protein in legumes when supplementation and complementation work? Several reasons can be given. First, in supplemented foods and feeds, leaching of free methionine during processing leads to losses and subsequent bacterial fermentation resulting in objectionable odours and flavours. Acceptability is perhaps the biggest problem in supplementation of human diets with crystalline methionine. Addition of minute amounts of methionine to the diets of non-technical, village societies may also be a problem. Secondly, a large number of people who are dependent on root crops as their main sources of energy will benefit from 'high protein quality legumes' because the protein content of root crops is extremely low and their amino acid profile is not complementary to that of legumes. For instance, cassava, a major root crop, is the basic food for 200–400 million inhabitants of the tropics (Odigboh, 1983). Finally, the nutritionally complete protein of a genetically engineered bean could be added to the list of advantages of legumes as food such as 'no cholesterol', dietary fibre source, possible beneficial effects of plant proteins on blood lipid status (Bodwell, 1987), and the capability of legumes to produce the greatest amount of protein per unit of land (Butz, 1973). These characteristics could be used as marketing attributes to encourage more people to eat legumes as sources of protein.

The economic implications of modifying traits of soybean as the major source of protein in animal feed have been analysed (Iowa State University, 1990). Increasing by 1% the protein content, lysine content and methionine content would lead to improved animal nutrition and additional values of 12, 49 and 21 cents/bushel, respectively. Based on an annual soybean production of 1750 M bushels, these are calculated to bring about additional total values of $204 M, $851 M and $375 M, respectively for a 1% increase in protein, lysine and methionine.

There are reports emphasizing that protein from plant sources meets the amino acid requirements of adults and that protein from some plant sources meets those of young children (Bodwell, 1987). However, recent reassessment of the requirements for essential amino acids in healthy adults led to the conclusion that the requirements for lysine, leucine, valine and threonine are likely to be 2–3 times higher than those reported in the 1985 WHO/FAO/UNU report (FAO/WHO/UNU, 1985; National Research Council, 1988).

39.4 FUNCTIONAL PROPERTIES OF PROTEINS

Functional properties of proteins are broadly defined as those properties other than nutritional attributes that affect its utilization. These include sensory properties (flavour, odour, colour, texture), water interacting properties (hydration, dispersibility, solubility), surface active properties (emulsification, foaming, adsorption) and rheological properties (gelation, texturization) (Nakai and Powrie, 1981). The overall functional property of a food system is the result of composite properties of individual protein components as they interact with one another and with non-protein components. Clearly, the functionality of proteins is very important in foods, yet very little is known about its molecular basis.

The correlation between the proportion of high-molecular-weight (HMW) glutenin polymers and breadmaking quality in bread wheat, is perhaps the only example there is of elucidating the molecular basis of a complex functional property (Payne *et al.*, 1981; Shewry *et al.*, 1989). The initial evidence came from plant breeding work that produced allelic variation in polypeptide composition of the HMW subunits and correlation with breadmaking quality. Confirmation of this observation in many laboratories has led to the routine use of HMW glutenin subunit screening in many plant-breeding programmes. Subsequent cloning of the HMW subunit genes and the derived amino acid sequences led to more detailed physicochemical analyses and the proposal of conformational models to explain gluten structure and functionality (Shewry *et al.*, 1989).

Computer modelling of structure-function relationships in proteins, mainly enzymes, is increasingly being used to design proteins rationally for specific functions (Blundell and Sternberg, 1985; Lin, 1986; Hol, 1987). Such an approach could be used to determine how site-specific mutations would affect functional properties of proteins. Information on the three-dimensional structure of seed storage proteins will also be useful (Lawrence *et al.*, 1990). Another strategy is to express modified plant seed proteins in microorganisms to determine the effects of modification on level of expression, processing and assembly, and on functional properties before transformation of plants (Kim *et al*, 1990a, b). Since only minute amounts of proteins are currently available from molecular techniques, micro-systems are needed to test the effects of structural changes on protein functionalities. Such information, combined with a knowledge of the critical sequences in seed proteins needed for post-translational processing and packaging, would allow the rational design of proteins with specific functional attributes without adversely affecting seed biology.

39.5 MOLECULAR BIOLOGY OF SEED PROTEINS

The seed proteins are classified into two major classes, namely storage proteins and the 'housekeeping proteins'. The storage proteins which make up a considerable part of the total protein, provide the germinating seed with carbon, nitrogen, sulphur and other nutrients whereas the 'housekeeping proteins' are necessary for the normal metabolism of the seed.

Because of their economic importance, the storage proteins of legumes have been intensively studied. Logically, the seed proteins became the focus of studies on plant gene expression because they are abundant gene products that accumulate rapidly at mid-maturation stage of seed development and because much is known about their biochemical and chemical properties as food proteins. The seed proteins also offer to basic plant biologists a model for temporal and tissue specific regulation during seed development. Consequently, the first plant gene cloned and sequenced completely was that of the French bean phaseolin and established the presence of introns in plant genes (Sun *et al.*, 1981; Slightom *et al.*, 1983).

39.5.1 Seed development begins after a double fertilization process unique to angiosperms (Raven *et al.*, 1987)

One of the two male nuclei in the pollen tube unites with the egg cell in the ovule to form the diploid zygote that develops into the embryo. The other unites with the two polar nuclei to form a triploid endosperm which provides nourishment to the developing seed as well as the germinating seedling. In legumes, the endosperm is absorbed by the developing embryo while the fleshy food-storing cotyledons develop. Thus, in legumes, the seed differentiates into two organ systems, the axis and the cotyledons which have different developmental pathways. The axis develops into the roots, the stem and the leaves whereas the cotyledon is a terminally differentiated organ that provides the nutrients to the germinating seedling. In most legumes, the cotyledon makes up at least 80% of the mature seed weight. In cereals, the starchy endosperm constitutes the major part of the seed.

39.5.2 The seed undergoes major and distinct physiological processes during development (Spencer and Higgins, 1982; Goldberg *et al.*, 1989)

The early stages are characterized by cell division followed by a phase of cell expansion in which reserve synthesis and deposition occur. Starch, proteins and lipids are synthesized at the cell expansion stage in which DNA content also increases. The last stage is characterized by a decrease in RNA and protein synthesis, loss of water and finally dormancy.

39.5.3 The synthesis of a number of seed storage proteins is characterized by post-translational processing including glycosylation, endoproteolytic processing and intracellular transport (Higgins, 1984; Muntz, 1989)

The primary translation products have 'signal sequences' that are cleaved as the proteins enter the lumen of the endoplasmic reticulum. Once inside the endoplasmic reticulum, the polypeptides may undergo proteolytic cleavage and, in most cases, are transferred via the Golgi apparatus into the membrane-bound protein bodies. Inside the protein bodies, the proteins aggregate, evidently determined by the primary structure of the protein. A structural model for the 19 and 22 kDa zein proteins has been proposed (Argos *et al.*, 1982), implying that certain critical sequences need to be preserved for proper packaging in addition to targeting signals for intracellular sorting.

39.5.4 The major storage proteins in legumes and cereals are coded by several homologous multigene families that vary in size, organization and chromosomal location (Higgins, 1984; Kreis *et al.*, 1985; Casey *et al.*, 1986; Gatehouse *et al.*, 1986; Elliston *et al.*, 1988; Goldberg *et al.*, 1989;)

In peas, the 50 kDa subunit of vicilin (7S) is coded by at least three families with up to six members within each family, indicating that as many as 18 genes code for this subunit. At least four homologous genes code for the 11S proteins (legumin) of pea (Croy *et al.*, 1982). The legumin (glycinin) subunits of soybean are coded by five genes belonging to two subfamilies (Nielsen *et al.*, 1989). The three genes belonging to one subfamily (Gy1, Gy2, Gy3) occupy two distinct domains in the soybean genome. Two subfamilies of legumin-like proteins are also found in pea (Domoney and Casey, 1985) and the bean *Vicia faba* (Baumlein *et al.*, 1986). The 7S soybean proteins (β – conglycinin) are coded by two distinct genes with at least 15 members clustered in six regions in the soybean genome (Harada *et al.*, 1989). Similarly, the 7S proteins (phaseolin) which are the major storage proteins in *Phaseolus vulgaris* are encoded by two unique gene families made up of 7–9 members (Talbot *et al.*, 1984; Slightom *et al.*, 1985).

The prolamin (zein) family in maize is coded by three to ten genes for each family and since there may be ten families, there could be as many as 30–100 zein genes in maize (Wienand and Feix, 1980; Hagen and Rubinstein, 1981; Burr *et al.*, 1982; Pedersen *et al.*, 1982; Viotti *et al.*, 1982). The prolamin genes characterized to date do not contain introns, unlike other plant genes (Kreis *et al.*, 1985). Glutelin, the major rice protein is encoded by a small multigene family (Takaiwa *et al.*, 1987a). In contrast with the maize prolamin genes, the rice glutelin gene has three introns (Takaiwa *et al.*, 1987b).

The genes for the 7S proteins from different species of legumes show remarkable homologies, suggesting a common ancestor (Casey *et al.*, 1986). Likewise, the genes coding for the 11S proteins from different species have strong homologies (Casey *et al.*, 1986; Nielsen *et al.*, 1989). The rice glutelin precursor sequence also shows homologies with leguminous 11S proteins (Takaiwa *et al.*, 1986). The 5'-flanking region of rice glutelin gene contains a sequence similar to the legumin box (Gatehouse *et al.*, 1986; Takaiwa *et al.*, 1987b). Comparison of gene sequences not only suggests common evolutionary origins but reveals sequences that may be useful in genetic engineering such as putative seed-specific promoters, quantitative-control sequences and variable protein sequences where gene modifications can conceivably be made without adversely affecting seed biology.

The number and complexity of seed storage protein genes present a problem if mutations are to be made to modify the amino acid composition of the proteins they code for. Ideally, the most strongly expressed gene would be modified to bring about a significant change in the overall amino acid composition of the seed. However, it is possible to enhance seed specific expression significantly by inserting a seed specific DNA sequence into a strong constitutive promoter such as cauliflower mosaic virus (CaMV) 35S promoter (Chen *et al.*, 1988).

39.5.5 The differential expression of seed protein genes is controlled temporally during seed development and spatially within the plant and within the seed

Using RNA blotting and hybridization with labelled cDNA, detailed analysis of seed mRNAs has been carried out during legume seed development. In general, each mRNA shows a characteristic period during which it increases and then declines although all the seed mRNAs overlap and are maximal at some stage within the period of rapid protein accumulation (Goldberg *et al.*, 1982; Higgins, 1984; Gatehouse *et al.*, 1986). In addition, the seed mRNAs are either non-detectable or present at levels several orders of magnitude lower in mature plant organs (Goldberg *et al.*, 1981; Walling *et al.*, 1986). Within the organ system of the seed, differential expression of seed protein genes is further observed. For instance, lectin mRNA is present at lower concentration in the embryonic axis than in the cotyledon (Walling *et al.*, 1986). *In situ* hybridization with labelled antisense mRNA reveals a finer degree of differential expression in that seed mRNAs are localized within specific cells in the cotyledon and axis but not in the seed coat (Barker *et al.*, 1988).

That seed protein gene expression is controlled by transcriptional and post-transcriptional processes is concluded from comparing accumulation of mRNA and rates of transcription. Transcription rates are obtained by *in vitro* synthesis of RNA from isolated nuclei and mRNA accumulation by RNA dot blots (Evans *et al.*, 1984; Beach *et al.*, 1985; Chappell and Chrispeels, 1986; Gatehouse *et al.*, 1986; Walling *et al.*, 1986). Features of transcription rates that support transcriptional control include observations that fluctuations in transcription rates parallel lower or absent levels in plants where mRNA levels are very low or absent.

The evidence that post-transcriptional processes play a role in controlling gene expression is based on the general observation that relative rates of transcription do not reflect the relative accumulation of mRNA (Beach *et al.*, 1985; Boston *et al.*, 1986; Chappell *et al.*, 1986; Walling *et al.*, 1986). For example, 22 kDa zein genes are transcribed at about twice the rate of 19 kDa zein genes yet the relative amount of 19 kDa mRNA is almost three times the 22 kDa transcript (Boston *et al.*, 1986). Mechanisms of post-transcriptional control are not known but efficiency of transport of mRNA through the nuclear membrane and stability of mRNA in the cytoplasms have been suggested (Goldberg *et al.*, 1989).

39.5.6 DNA sequences that control seed-specific and quantitative expression have been identified in seed protein genes

These DNA regions are being determined experimentally by analysing the expression of seed protein genes in transgenic plants and conceptually by comparing seed protein gene sequences and identifying common, conserved sequences. A conserved region of 28 bases (approximately 50% G+C) in legumin genes is present in pea, soybean and *Vicia faba* and is now called the 'legumin box' (Gatehouse *et al.*, 1986). A similar 'vicilin box' of conserved sequences in the vicilin gene family is found in pea, *Vicia faba* and *Phaseolus vulgaris* (Gatehouse *et al.*, 1986). The 'vicilin box' is not homologous to the 'legumin box' and is probably an example of a regulatory sequence specific to a gene family. A consensus sequence of 5′ T(A, C)AACACA(A, C)T(A, C) 3′ is proposed as a candidate seed-specific *cis*-control element because it is found in the 5′ regions of five soybean seed proteins (Goldberg, 1986). A 'universal seed-specific DNA sequence' remains to be identified and tested by mutations and functional analysis in transgenic plants.

Expression of seed protein genes in transgenic tobacco and petunia plants has revealed DNA sequences that control seed specific expression (Chen *et al.*, 1986, 1988; Bray *et al.*, 1987; Okamuro *et al.*, 1986; Barker *et al.*, 1988; Goldberg *et al.*, 1989). Only 159 nucleotides contiguous to the transcription initiation site are needed for a low level (5% of maximum) expression of the α-subunit of β-conglycinin gene in transgenic petunia plants (Chen *et al.*, 1986). An additional 98 nucleotides increased the expression to maximum, equal to that of the complete gene. Analysis of the enhancing sequence of 127 nucleotides (-131 to -257) 5′ to the gene revealed four repeats of a 6-base-pair (G+C)-rich sequence (AG[A, C]CCCA), suggesting the critical role of the (G+C)-rich repeats in determining the level of expression of the β-conglycinin gene (Chen *et al.*, 1986). Interestingly, insertion of a longer version of this sequence (-78 to -257) into the 5′ and 3′ ends of a constitutive promoter (CaMV 35S promoter) conferred seed-specific enhancement of 25–40-fold (Chen *et al.*, 1988). This demonstrates that it is possible to enhance the expression of a weakly expressed seed protein gene significantly by inserting key DNA sequences in its 5′ flanking region into a CaMV 35S promoter. Significant increases in the expression of foreign genes in plants in a non-tissue specific manner has also been achieved by duplicating the CaMV 35S promoter in tandem (Kay *et al.*, 1987).

39.5.7 DNA binding proteins that interact with seed protein gene regulatory regions have been identified in plants

A model proposed to explain differential expression of seed protein genes during seed development is that each gene has *cis*-control elements that interact with cellular factors (*trans*-elements) to activate transcription (Goldberg, 1986). These transacting factors may be coded by regulatory genes. In higher organisms, sequence-specific DNA binding proteins are directly involved in the regulation of mRNA transcription initiation (Dynan and Tjian, 1985; Kadonaga *et al.*, 1986). A DNA-binding protein that interacts with the 5′-flanking region of the soybean lectin gene and that of two other seed protein genes has been identified (Jofuku *et al.*, 1987; Goldberg *et al.*, 1989). The DNA binding protein activity increases and decreases during seed development, parallels the transcription rate and is associated with a 60 kDa protein (Jofuku *et al.*, 1987). A similar DNA binding protein has been identified in corn endosperm and interacts with the 15 bp DNA sequence found in the 5′ region of all zein genes sequenced so far (Maier *et al.*, 1987). The physiological significance of these DNA binding proteins remains to be established. Regulatory regions in other plant genes have been shown to interact with nuclear protein (Allen *et al.*, 1989; Bustos *et al.*, 1989; Jordano *et al.*, 1989; Riggs *et al.*, 1989).

In maize, the Opaque-2 gene that regulates zein deposition has been cloned (Schmidt *et al.*, 1987; Motto *et al.*, 1988). The Opaque-2 gene encodes a protein containing domains similar to the 'leucine zipper' motif identified in DNA binding proteins of animal proto-oncogenes and transcriptional activators (Hartings *et al.*, 1989; Schmidt *et al.*, 1990), (see section 39.7).

The CaMV 35S promoter is the most widely used promoter in plant biotechnology because it confers high expression on heterologous genes in plants. Recent studies on molecular dissection of the 35S promoter indicate that transcriptional regulation can not be explained by a simple model in which regulation is controlled by the mere absence or presence of transacting factors. Instead, what is emerging is that regulation of transcription leading to different patterns of expression is mediated by combinations of a limited number of *cis* regulatory elements and transacting factors (Benfey and Chua, 1990). Thus, the CaMV stretch of approximately 35S promoter is now known to be made up of six subdomains within a stretch of approximately 300 base pairs (bp) upstream of the minimal TATA promoter (-343 to -46 from transcription initiation site). Specific combinations of these subdomains confer tissue- and cell-specific expression that are not detected from isolated subdomains. Furthermore, these combinatorial codes of *cis* regulatory elements could be interpreted differently in different plant species, as evidenced by different expression patterns in petunia and tobacco, conferred by specific combinations of 35S subdomains. In terms of genetic engineering applications, these findings suggest the use of specific combinations of promoter subdomains to confer highly specific and localized expression of desirable genes (or antigenes) in target cells or tissues in the plant.

39.6 GENETIC ENGINEERING ENHANCEMENT OF METHIONINE CONTENTS IN LEGUMES

Modification of seed proteins, if it is to be considered a success, must not have any adverse effect on seed development and germination. Several approaches are possible to increase the level of methionine, the limiting essential amino acid in legumes.

One strategy is to introduce methionine residues or methionine-rich peptides into non-conserved, and presumably non-critical regions of storage proteins. A methionine-enriched β-phaseolin gene driven by its own promoter, when transformed into tobacco did not accumulate in the transgenic seeds, due to degradation (Hofman *et al.*, 1988). Availability of a three-dimensional structure of phaseolin at 3 Å resolution revealed that the insertion of the synthetic DNA rich in methionine codons was internal to a helix structure which could have disrupted the protein structure and led to degradation (Lawrence *et al.*, 1990). Transformation of the *Arabidopsis* 2S albumin 1 (AT2S1) gene enriched with methionine codons into *Arabidopsis thaliana* resulted in accumulation of modified protein at a level of 1–2% of total protein (Table 39.2) (de Clerq *et al.*, 1990). This level of expression was not high enough to bring about an increase in seed methionine for accurate determination.

Another strategy is to transfer genes coding for methionine-rich proteins (MRP) from other species, i.e. heterologous genes. Complementary DNA (cDNA) clones for MRPs from Brazil nut (Altenbach *et al.*, 1987)

Table 39.2 Genetic engineering experiments to increase methionine contents in seed proteins

Promoter	*Methionine-rich protein domain*	*Target plant/organism*	*Results*	*Reference*
β-phaseolin gene	Insertion of methionine-rich synthetic oligonucleotide into β-phaseolin gene	Tobacco	Tissue- and temporal-specific expression in seed but protein was degraded.	Hofman *et al.* (1988)
β-phaseolin gene	Brazil nut methionine-rich protein gene.	Tobacco	Tissue- and temporal-specific expression, 8% of total protein in seed, increase methionine content of seed by 30%.	Altenbach *et al.* (1989)
Soybean lectin gene	Brazil nut methionine-rich protein gene.	*Brassica napus*	Tissue- and temporal-specific expression, 0.02–0.06% of total protein in seed.	Guerche *et al.* (1990)
Arabidopsis thaliana 2S albumin (AT2S1) gene.	Methionine enrichment in non-conserved region of AT 251 gene.	*Brassica napus*, *Arabidopsis thaliana*, tobacco	1–2% of total protein in *Arabidopsis* seed.	de Clerq *et al.* (1990)
	Brazil nut methionine-rich protein gene.	*Brassica napus*, *Arabidopsis thaliana*, tobacco	0.05%, 0.1% and 0.3% of total protein in seed of tobacco, *Arabidopsis* and *Brassica*, respectively	de Clerq *et al.* (1990)
trc promoter in expression vector PKK 233–2	$A_{1a}B_{1b}$ proglycinin subunit with methionine-rich insert in variable domain	*Escherichia coli* JM 105	Zero to high level expression depending on location of insert; gelation and emulsifying properties of expressed protein better than native glycinin.	Kim *et al.* (1990b)

and sunflower seed (Lilley *et al.*, 1989) have been isolated and sequenced. The Brazil nut MRP gene fused to a β-phaseolin promoter, when transformed into tobacco, resulted in the MRP making up to 8% of the total protein of the seed and increases in methionine of up to 30% in the seeds of transgenic plants (Altenbach *et al.*, 1989). Ligation of the Brazil nut MRP gene to a soybean lectin gene promoter and transformation into *Brassica napus* led to very low accumulation of the MRP in the seed (0.02–0.06% of total protein, Guerche *et al.*, 1990). Similarly small accumulations of Brazil nut MRP in the seeds of tobacco, *Arabidopsis thaliana* and *Brassica napus* occurred when the MRP gene was driven by an *Arabidopsis thaliana* 2S albumin gene (AT2S1) promoter (de Clerq *et al.*, 1990). These levels of expression are too low to cause overall increases in methionine contents of the seed (Table 39.2). These results, however, demonstrate the feasibility of altering the methionine content of plant seeds through the transformation of heterologous genes coding for methionine-rich proteins driven by seed specific plant promoters. A variation on this theme is to introduce synthetic genes coding for a protein with a high level of essential amino acids (Jaynes *et al.*, 1986; Yang *et al.*, 1989).

For the reason that increasing the level of endogenous proteins would be least likely to be deleterious to the seed, our approach is to increase the level of endogenous MRP. Relatively methionine-rich protein was identified in soybean (Kho and de Lumen, 1988) with a method developed for identifying methionine-containing proteins and quantitating their methionine contents (de Lumen and Kho, 1987). Further resolution using two-dimensional electrophoresis identified a 10.8 kDa protein that has 12.10% methionine for which cDNA is now being cloned (George and de Lumen, 1991). The level of this MRP has to be increased about 17-fold from its level of 0.6% of the total protein to bring the overall methionine content of soybean seed to that of the FAO reference protein (FAO, 1957, 1973).

To determine the effects of amino acid modification on the assembly and the functional properties of glycinin, Kim *et al.* (1990a, b) used *E. coli* expression vectors. The $A_{1a}B_{1b}$ proglycinin subunit was enriched with methionine by inserting a synthetic DNA containing four continuous methionine residues into the variable domains of the coding regions (Wright, 1988) and ligating into *E. coli* expression vectors. Expression in *E. coli* yields modified soluble glycinin proteins (analysed by SDS-PAGE) in the cell that self-assemble at moderate to high levels depending on the position of the synthetic DNA insert. The functional properties (heat induced gelation and emulsifying property) of the modified glycinins are better than the native proteins. The amino acid compositions of the modified glycinins, however, were not determined. This approach typifies a strategy of testing the effects of protein modifications on processing and assembly in microorganisms before transformation into plants.

39.7 HIGH-LYSINE CEREAL MUTANTS

The discovery of opaque-2 (O2) high-lysine mutants in corn (Mertz *et al.*, 1964) laid down the foundation for the continuing molecular analysis of the O2 mutation (Soave *et al.*, 1981; Burr and Burr, 1982; Marks *et al.*, 1985; Kodrzycki *et al.*, 1989) and the recent cloning of the O2 gene (Schmidt *et al.*, 1987; Motto *et al.*, 1988) which encodes a DNA-binding protein (Hartings *et al.*, 1989; Schmidt *et al.*, 1990).

The nutritional superiority of O2 corn which was clearly demonstrated after its discovery (Mertz *et al.*, 1965; Bressani *et al.*, 1969) and confirmed by many other workers, is explained by the higher levels of lysine and tryptophan and more available niacin compared with normal corn. The high-lysine phenotype is due to a significant reduction in the lysine-poor zein proteins, especially the 22 kDa polypeptides which are virtually absent (Jones *et al.*, 1977), leading to a more nutritionally balanced amino acid profile compared with normal corn. It is noteworthy that there are other high-lysine mutants in corn, namely the opaque-7 (di Fonzo *et al.*, 1980) and floury-2 (Jones, 1978) which affect the levels of the zein proteins differently from the O2 mutation. Furthermore, high-lysine phenotypes are also found in barley (Doll *et al.*, 1974) and sorghum (Singh and Axtell, 1973).

The reduced levels of the 22 kDa zein protein in the O2 mutant, is associated with decreased levels of corresponding transcripts, suggesting that the mutation affects transcription (Burr and Burr, 1982; Marks *et al.*, 1985; Kodrzycki *et al.*, 1989). Genetic linkage analysis has located the O2 locus on chromosome 7, whereas several of the 22 kDa genes are found on chromosome 4 and 10 (Soave and Salamini, 1984). These results suggest that the O2 gene is a *trans*-acting regulator of zein gene expression. The O2 gene was subsequently cloned by transposon tagging techniques using the mobile elements Spm (Schmidt *et al.*, 1987) and Ac (Motto *et al.*, 1988). Sequence analysis of the O2 genomic and cDNA clones revealed that the encoded protein contains domains similar to the 'leucine zipper' motif found in DNA binding proteins of animal proto-oncogenes and transcriptional activators (Hartings *et al.*, 1989; Schmidt *et al.*, 1990). The O2 gene presents a model to study the mechanisms by which a gene differentially promotes the transcription of a specific subset of genes belonging to a family of genes that are coordinately expressed.

Whereas the O2 mutants have provided an exciting intellectual puzzle to molecular biologists, it is worth noting that the continuing efforts of plant breeders to improve the O2 corn has led recently to the development of quality protein maize (QPM), an improved version without the undesirable properties of the original O2. The desirable nutritional quality of O2 corn was accompanied by a number of undesirable properties. The

O2 kernels were soft, opaque and chalky, instead of the preferred shiny kernels in most regions, yielded lower than traditional corn varieties, and dried more slowly and were therefore more susceptible to fungi and insects than normal corn (National Research Council, 1988). Plant breeders at Centro Internacional de Mejoramiento de Maiz y Trigo (CIMMYT) have continued to correct the problems associated with the original O2 corn by combining the O2 gene with genetic modifiers, genes that influence the expression of a non-allelic gene(s) (CIMMYT, 1985).

A key step in the early development of improved varieties was the development of sensitive analytical methods that required very small amounts of samples. Tissue samples could be obtained from a single kernel without damaging its ability to germinate, thus allowing identification and subsequent propagation of individual kernels with desirable agronomic and nutritional traits (Villegas *et al.*, 1984).

The basic QPM germplasm at CIMMYT, made up of 30 experimental populations, is being tested in international trials in 40 countries. In 1987, four QPM varieties were released commercially in Guatemala, Paraguay, China, Vietnam and Eastern Europe (National Research Council, 1988).

39.8 MOLECULAR BIOLOGY OF PROTEASE INHIBITORS

The structure, chemistry and mechanism of action of serine protease inhibitors are well understood (Ryan, 1981). The inability of raw soybean to support the growth of rats (Liener and Kakade, 1969) and the lethal effect of raw winged bean (de Lumen and Chan, 1986) are attributed to antinutritional factors, primarily protease inhibitors and haemagglutinins (Higuchi *et al.*, 1983). Since their adverse effects can be removed by proper processing, the practical significance of protease inhibitors in human diets in terms of their native function is questionable. In fact, the presence of denatured protease inhibitors in human diets may be beneficial because they are sources of sulphur amino acids and may have anticarcinogenic effects in the lower intestine (Ryan, 1989).

Thus, the molecular biology of protease inhibitors has focused on their function as defence compounds against predators and as models of seed proteins that are differentially expressed during development. The genes and/or cDNAs for several protease inhibitors have been isolated (Hammond *et al.*, 1984; Graham *et al.*, 1986; Lee *et al.*, 1986; Walling *et al.*, 1986; Cleveland *et al.*, 1987; Hilder *et al.*, 1987; Joudrier *et al.*, 1987; Sanchez-Serrano *et al.* 1987; Thornburg *et al.*, 1987; Williamson *et al.*, 1987; Jofuku and Goldberg, 1989; Perez-Grau and Goldberg, 1989; Ryan, 1989). The defence capability of foreign protease inhibitors against predators in transgenic plants was demonstrated by Hilder *et al.* (1987). Transformation of a cowpea trypsin inhibitor DNA construct into tobacco resulted in a high constitutive expression in the leaves and a significant resistance to tobacco bud worm. Other protease inhibitors isolated from potato and tomato are similarly being evaluated (Ryan, 1989). Investigations on the Kunitz trypsin inhibitor gene family demonstrate their usefulness as models of seed protein genes that are differentially expressed during embryogenesis and are also represented in leaf, stem and root mRNA populations. Studies on Kunitz trypsin inhibitor genes suggest that each gene contains *cis*-elements that control its expression in the developing seed and in the mature plant (Jofuku and Goldberg, 1989) and that the cell-specific expression programme is determined very early during embryogenesis (Perez-Grau and Goldberg, 1989).

39.9 MOLECULAR BIOLOGY OF CARBOHYDRATE SYNTHESIS

39.9.1 Starch biosynthesis

Starch is the major component in cereals and legumes except in certain oilseeds. Cloning the genes for key enzymes in the biosynthetic pathways and studying the regulation of their expression during seed development may provide information useful for molecular genetic strategies to improve functional properties of carbohydrates.

A comprehensive discussion of starch biosynthesis in plants is covered in reviews (Preiss, 1982; Juliano, 1985; Manners, 1985). The formation of new glucosidic bonds in a growing amylose or amylopectin molecule is catalysed by starch synthetase via the transfer of a glucosyl group from ADP- or UDP-glucose. The nucleotide sugars are synthesized either through the pyrophosphorylase reaction or the reversal of the sucrose synthetase reaction. The current thinking is that ADP-glucose pyrophosphorylase is a key regulatory enzyme in the starch biosynthetic pathway in the leaf (Preiss, 1982). However, the sucrose synthesis reaction may play a very important role in sucrose–starch conversion in the developing maize and rice seed (Chourey and Nelson, 1976; Lee and Su, 1982).

The availability of mutants that are defective in starch biosynthesis has allowed extensive genetic and biochemical studies of starch biosynthesis in maize seed. These mutants lead to altered levels of starch or its quality (Creech, 1968). Four mutants in maize have been identified with their enzyme products: (a) shrunken (sh) associated with decreased sucrose synthetase and a collapsed phenotype due to reduced level of starch in the endosperm (Chourey and Nelson, 1976); (b) shrunken-2 (sh-2) and brittle-2 (bt-2) associated with reduction in endosperm starch content and a complete loss of ADP-glucose pyrophosphorylase (Tsai and Nelson, 1966); and (c) waxy (wx) associated with an altered starch. The wx mutation produces starch with 100% amylopectin compared with 25% amylose and 75% amylopectin in wild type (Chourey, 1982) and starch granules that have much reduced UDP-glucose starch synthetase compared with wild type (Nelson and Rines, 1962).

Similar endosperm mutations like amylose extender (ae), shrunken (sh), sugary (su), dull (du), waxy (wx), white-core and floury have been induced in rices using chemical mutagens and irradiations (Okuno, 1976; Satoh and Omura, 1981; Yano *et al.*, 1985). Since many of the cooking and eating qualities of rice are determined by the ratio of amylose to amylopectin, molecular analysis of these mutations might provide useful information for manipulation of amylose: amylopectin ratios through genetic engineering.

The cDNA for rice ADP-pyrophosphorylase has been cloned (Krishnan *et al.*, 1986). Studies by this group showed that there are at least two tissue specific forms of ADP-pyrophosphorylase encoded by distinct mRNA in the leaf and seed of wheat and rice.

The genetic expression of sucrose synthetase in maize has been investigated intensively after it was shown that mutants deficient in sucrose synthetase had significantly reduced levels of starch compared with wild type (Chourey and Nelson, 1976; Chourey *et al.*, 1986; Taliercio and Chourey, 1989; Gupta *et al.*, 1988; McCarty *et al.*, 1986). In maize, there are two tissue-specific isozymes of sucrose synthetase encoded by two sets of genes which are differentially expressed in endosperm and embryo tissues (McCarty *et al.*, 1986; Chourey, 1981). Both sets of enzymes have been cloned and characterized (Sheldon *et al.*, 1983; McCarty *et al.*, 1986; Gupta *et al.*, 1988).

Strategies to increase the protein content of cereals, especially rice, will have to take into consideration the inverse relationship between protein content and grain yield (Doll, 1984; Juliano, 1985). In a given rice variety, grain yield and protein content may be increased up to a certain threshold, beyond which further increases in protein result in a sacrifice in grain yield. Since the decrease in grain yield is due completely to reduction in starch (Doll, 1984), understanding the genetic relationships between starch biosynthesis and storage protein biosynthesis in the developing cereal grain may provide useful information to increase protein content without affecting yield adversely. Competition for photoassimilates between the carbohydrate and protein biosynthetic pathways may be a simple explanation. Nitrogen translocation may also be a factor, since high-protein rice varieties seem to have a more efficient nitrogen transport from the leaves to the developing grain, as reflected in increased free amino acids in the culm sap and developing grain, rather than increased absorption of nitrogen from the soil (Cagampang *et al.*, 1971; Perez *et al.*, 1973). Rice mutants defective in starch biosynthesis that have reduced starch and increased protein and lipid contents might be useful models to study the starch–protein relationships (Juliano, personal communication).

39.9.2 Biosynthesis of raffinose oligosaccharides

The raffinose family of oligosaccharides which are α-D-galactosides of sucrose and include raffinose, stachyose and verbascose are second only to sucrose in abundance in legume seeds (Dey, 1985). They accumulate at the late stage of seed maturation as the seed desiccates towards dormancy and have been thought to play roles in response of plants to cold, seed viability and to produce flatulence in humans upon consumption of beans (Calloway and Murphy, 1968; Ovcharov and Koshelev, 1974; Kandler and Hopf 1980; Rackis, 1981; Saravitz *et al.*, 1987). The raffinose oligosaccharides are synthesized by the sequential transfer of galactosyl units to sucrose via galactinol catalysed by specific transferases (Dey, 1985). Since the biosynthetic pathway of these oligosaccharides is known, understanding the regulation of genes for key enzymes in the pathway would be useful in manipulating the levels of raffinose oligosaccharides in legume seeds to clarify their roles in plants and in flatulence formation (Castillo *et al.*, 1990).

39.10 MOLECULAR BIOLOGY OF LIPID METABOLISM

Legumes are the world's major source of vegetable oil, soybean being the single largest source (Hymowitz, 1987). The nutritional properties, food applications and industrial uses of vegetable oils are largely dependent on their fatty acid composition. Thus, a major objective of oilseed plant breeding programmes is to obtain cultivars with fatty acid compositions that are suited to certain market needs.

The application of techniques of molecular biology to plant lipid metabolism is in its infancy, and mainly limited by a poor understanding of biosynthetic pathways (Battey *et al.*, 1989). Mutants of *Arabidopsis thaliana* that are deficient in fatty acid desaturation are being used to study the regulation and functional significance of lipid unsaturation (Somerville *et al.*, 1987; Hugly *et al.*, 1989; Kunst *et al.*, 1989). Since acyl carrier protein (ACP) plays a central role in plant lipid metabolism and is well characterized, one approach is to study its protein and gene structure which may reveal control mechanisms that regulate fatty acid biosynthesis in plants (Elhussein *et al.*, 1987; Guerra *et al.*, 1987; Ohlrogge *et al.*, 1987; Slabas *et al.*, 1987). Another approach is suggested by the differences in fatty acid composition in seed storage lipids and those found in membrane phospholipids in other organs of plants. Studies are being carried out to determine if oilseeds have mechanisms to target unusual fatty acids to storage triglycerides where they will have less deleterious effects and to exclude them from phospholipids in the membranes where their presence can disrupt membrane structure (Battey and Ohlrogge,

1989). Such information will give insights into designing genetic engineering techniques to modify fatty acid composition without disrupting normal plant metabolism.

Lipoxygenases are a group of polyunsaturated fatty acid oxidases that are thought to play important roles in the generation of undesirable flavours and aroma in legume seed protein and oil products (Kinsella, 1979). Mutants lacking various isoenzymes of lipoxygenases have been identified in soybean (Hildebrand and Hymowitz, 1981; Kitamura *et al.*, 1983; Kitamura, 1984). cDNA clones for soybean lipoxygenases have been isolated and used as probes to demonstrate that lipoxygenase null mutants accumulate reduced levels of lipoxygenase transcripts (Start *et al.*, 1986). Soybean genomic clones have been isolated and characterized and are being transformed into soybean and tobacco to obtain transformants with up- and down-regulated lipoxygenase activity. These are being correlated with generation of volatile compounds and with pest resistance to better understand the role of lipoxygenases in plant metabolism (Hildebrand *et al.*, 1987).

39.11 TRANSFORMATION TECHNIQUES

A key to any genetic engineering strategy requires introducing the desirable gene or set of genes into the genome of the target plant. Very rapid progress has been achieved in this area in the last four years and the reader is referred to several reviews on plant transformation systems for detailed discussions (Fraley *et al.*, 1986; Cocking and Davey, 1987; Klee *et al.*, 1987; Schell, 1987; Gobel and Lorz, 1988; Weising *et al.*, 1988; Gasser and Fraley 1989; Uchimiya *et al.*, 1989; Potrykus, 1990).

Currently used gene transfer techniques in plants can be broadly classified into one group that uses a delivery vehicle for DNA transfer and another that involves the direct transfer of DNA. The most commonly used vector is *Agrobacterium tumefaciens* which causes tumorous crown galls on infected species. The virulence of *Agrobacterium* is due to tumour-inducing (Ti) plasmids which contain two regions necessary for the transfer and integration of the bacterial genes: a transfer DNA (T-DNA) domain that is transferred to the plant and a virulence (vir) domain that catalyses the transfer but which is not itself transferred. Molecular analysis of the Ti plasmid revealed that the T-DNA and the vir region do not have to be in the same piece of DNA (Fraley *et al.*, 1986), leading to the design of intermediate vectors that can be multiplied in *E. coli* and manipulated before introduction into *Agrobacterium* cells and infection of plants. A typical vector contains the border sequences that define the limits of the DNA transferred to the plant, a plant selectable marker such as kanamycin resistance, an origin of replication that allows multiplication in *Agrobacterium*, another origin of replication for *E. coli* and an *E. coli* selectable marker such as spectinomycin resistance (Gasser and Fraley, 1989). The piece of DNA transferred to the plants bounded by the border sequences contains the foreign gene and the selectable marker. The intermediate vector is introduced into *Agrobacterium* cells containing the plasmid with the vir region and inactivated pathogenesis genes (disarmed Ti). Upon infection of the host, the vir DNA acts in *trans* to mobilize the transfer of the DNA within the border sequences and the phenotype conferred by the kanamycin resistance gene is used to screen transformed cells during plant regeneration. The most commonly used infection procedure is by exposure of leaf discs to *Agrobacterium* cells (Horsch *et al.*, 1985).

An important limitation of *Agrobacterium*-mediated gene transfer is that the major cereals and legume crops have not been successfully transformed because they are poor hosts and because of the inability to regenerate whole plants from leaf discs. Recent efforts to deliver DNA directly into the genome of these plants have been reported. Such techniques include facilitated uptake of DNA by protoplasts (Fromm *et al.*, 1986; Uchimiya *et al.*, 1986), DNA application into reproductive organs (de la Pena *et al.*, 1987, Luo and Wu, 1988), microinjection into cells of immature embryos (Neuhaus *et al.*, 1987) and rehydration of desiccated embryos (Topfer *et al.*, 1989). Of practical value is the stable transformation of soybean by DNA-coated gold particles introduced into meristems by electric discharge acceleration (Christou *et al.*, 1988; McCabe *et al.*, 1988). The introduced genes are inherited in a Mendelian manner and stable for up to six generations (Christou *et al.*, 1989, 1990). Also noteworthy, is the regeneration of fertile rice plants from transformed protoplasts (Shimamoto *et al.*, 1989) since easy and reproducible production of transgenic cereals has not been reported previously.

39.12 SUMMARY

In summary, rapid developments in our knowledge of plant gene structure and organization and their regulation bring us closer to actual applications of genetic engineering to improve the nutritional properties of legumes and cereals. Recent experiments demonstrate the feasibility of altering methionine contents of plant seeds through the use of methionine-rich or methionine-enriched protein genes driven by seed-specific promoters. Such techniques can be used to modify lysine, tryptophan and threonine contents in cereals and legumes. Improving the digestibility of proteins and starch, reducing flatulence, altering fatty acid composition, manipulating the amylose: amylopectin ratios in rice starch and minimizing formation of undesirable flavours and aroma will have to be long-term objectives because of the very limited knowledge we have of the molecular biology of carbohydrate and lipid metabolism in the developing seed. To design seed proteins with specific functional attributes without adversely affecting seed biology requires knowledge of the molecular basis of functional properties and of the critical features of seed protein structures that are essential in post-translational processing and packaging.

With the rapid progress being made in gene cloning and transformation methodologies in plants, our knowledge of basic plant metabolic processes can be the limiting factor in applying genetic engineering to improve the usefulness of plant seeds as food sources. Any genetic changes must not have any adverse effects on normal seed biology.

REFERENCES

Abbott, J.C. (1966) Protein supplies and prospects: the problem, in *World Protein Resources*, (ed. R.F. Gould), ACS Washington, DC, pp. 1–15.

Allen, R.D., Bernier, F., Lessard, P.A. and Beachy, R.N. (1989) Nuclear factors that interact with a soybean β-conglycinin enhancer. *Plant Cell*, **1**, 623–31.

Altenbach, S.B., Pearson, K.W., Leung, F.W. and Sun, S.S.M. (1987) Cloning and sequence analysis of a cDNA encoding a Brazil nut protein exceptionally rich in methionine. *Plant Mol. Biol.*, **8**, 239–50.

Altenbach, S.B., Pearson, K.W., Meeker, G. *et al.* (1989) Enhancement of the methionine content of seed proteins by the expression of a chimeric gene encoding a methionine-rich protein in transgenic plants. *Plant Mol. Biol.*, **13**, 513–22.

Argos, P., Pedersen, K., Marks, M.D. and Larkins, B.A. (1982) A structural model for maize zein proteins. *J. Biol. Chem.*, **257**, 9984–90.

Barker, S.J., Harada, J.J. and Goldberg, R.B. (1988) Cellular localization of soybean storage protein mRNA in transformed tobacco seeds. *Proc. Nat Acad. Sci. USA*, **85**, 458–62.

Battey, J.F. and Ohlrogge, J.B. (1989) A comparison of the metabolic fate of fatty acids of different chain lengths in developing oilseeds. *Plant Physiol.*, **90**, 835–40.

Baumlein, H., Wobus, U., Pustell, J. and Kafatos, F.C. (1986) The legumin gene family: structure of a B type gene of *Vicia faba* and a possible legumin gene specific regulatory element. *Nucl. Acids Res.*, **14**, 2707–20.

Beach, L.R., Spencer, D., Randall, P.J. and Higgins, T.J.V. (1985) Transcriptional and posttranscriptional regulation of storage protein gene expression in sulfur deficient pea seeds. *Nucl. Acids Res.*, **13**, 999–1013.

Benfey, P.N. and Chua, N.H. (1990) The cauliflower mosaic virus 35S promoter: combinatorial regulation of transcription in plants. *Science*, **250**, 959–66.

Blundell, T. and Sternberg, M.J. (1985) Computer aided design in protein engineering. *Trends Biotechnol.*, **3**, 229–31.

Bodwell, C.E. (1987) Nutritional implications of cereals, legumes and their products, in *Cereals and Legumes in the Food Supply*, (eds J. Dupont and E.M. Osman), Iowa State University Press, Ames, IA, pp. 259–75.

Boston, R.S., Kodrycki, R. and Larkins, B.A. (1986) Transcriptional and posttranscriptional regulation of maize zein genes, in *Molecular Biology of Seed Storage Proteins and Lectins*, (eds L.M. Shannon and M.J. Chrispeels), American Society of Plant Physiologists.

Bray, E.A., Naito, S., Pan, N.S. *et al.* (1987) Expression of the β-subunit of β-conglycinin in seeds of transgenic plants. *Planta*, **172**, 364–70.

Bressani, R. and Elias, L.G. (1968) Processed vegetable protein mixtures for human consumption in developing countries. *Adv. Food Res.*, **16**, 1–103.

Bressani, R., Elias, L.G. and Gomez–Brenes, R.A. (1969) Protein quality of opaque-2 corn evaluation in rats. *J. Nutr.*, **97**, 173–80.

Burr, F.A. and Burr, B. (1982) Three mutations in *Zea mays* affecting zein accumulation: a comparison of zein polypeptides, *in vitro* synthesis and processing, mRNA levels and genomic organization, *J. Cell Biol.*, **94**, 201–6.

Burr, B., Burr, F.A., St John, T.P. *et al.* (1982) Zein storage gene family of maize: an assessment of heterogeneity with cloned mRNA sequences. *J. Mol. Biol.*, **154**, 33–49.

Bustos, M.M., Guiltinan, M.J., Jordano, J. *et al.* (1989) Regulation of B-glucuronidase expression in transgenic tobacco plants by an A/T-rich, cis-acting sequence found upstream of a French bean B-phaseolin gene. *Plant Cell*, **1**, 839–53.

Butz, E.L. (1973) World protein markets – a supplier's view. *Plenary Session of World Protein Markets, World Soybean Conference* Nov. 11–14, Munich, Germany.

Cagampang, G.B., Cruz, L.J. and Juliano, B.O. (1971) Free amino acids in the bleeding sap and developing grain of the rice plant. *Cereal Chem.*, **48**, 533–9.

Calloway, D.H. and Murphy, E.L. (1968) The use of expired air to measure intestinal gas formation. *Ann. NY Acad. Sci.*, **150**, 82–5.

Casey, R., Downey, C. and Ellis, N. (1986) Legume storage proteins and their genes. *Oxford Surv. Plant Mol. Cell Biol.* **3**, 1–95.

Castillo, E.M., de Lumen, B.O., Reyes, P.S. and de Lumen, H.Z. (1990) Raffinose synthase and galactinol synthase in developing seeds and leaves of legumes. *J. Agric. Food Chem.*, **38**, 351–5.

Chappell, J. and Chrispeels, M.J. (1986) Transcriptional and posttranscriptional control of phaseolin and phytchemagglutinin gene expression in developing cotyledons of *Phaseolus vulgaris*. *Plant Physiol.*, **81**, 50–4.

Chen, Z.L., Schuller, M.A. and Beachy, R.N. (1986) Functional analysis of regulatory elements in a plant embryo-specific gene. *Proc. Natl Acad. Sci. USA*, **83**, 8560–4.

Chen, Z.L., Pan, N.S. and Beachy, R.N. (1988) A DNA sequence element that confers seed-specific enhancement to a constitutive promoter. *EMBO J.*, **7**, 297–302.

Chourey, P.S. (1981) Genetic control of sucrose synthetase in maize endosperm. *Mol. Gen. Genet.*, **184**, 372–6.

Chourey, P.S. (1982) Starch mutants and their protein products, in *Maize for Biological Research* (ed. W.F. Sheridan), Plant Molecular Biology Association Charlottesville, VA, pp. 129–33.

Chourey, P.S. and Nelson, O.E. (1976) The enzymatic deficiency conditioned by the shrunken-1 mutations in maize. *Biochem. Genet.*, **14**, 1041–55.

Chourey, P.S., Latham, M.D. and Still, P.E. (1986) Expression of two sucrose synthetase genes in endosperm and seedlings of maize: evidence of tissue specific polymerization of protomers. *Mol. Gen. Genet.*, **203**, 251–5.

Christou, P., McCabe, D.E. and Swain, W.F. (1988) Stable transformation of soybean callus by DNA-coated gold particles. *Plant Physiol.*, **87**, 671–4.

Christou, P., Swain, W.F., Yang, N.S. and McCabe, D.E. (1989) Inheritance and expression of foreign genes in transgenic soybean plants. *Proc. Natl Acad. Sci USA*, **86**, 7500–4.

Christou, P., McCabe, D.E., Martinell, B.J. and Swain, W.F. (1990) Soybean genetic engineering – commercial production of transgenic plants. *TIBTECH*, **8**, 145–51.

CIMMYT (Centro Internacional de Mejoramiento de Maiz y Trigo). (1985)

Boosting protein quality in maize. *CIMMYT Research Highlights 1984*, CIMMYT Mexico.

Cleveland, T.E., Thornburg, R.W. and Ryan, C.A. (1987) Molecular characterization of a wound-inducible inhibitor I gene from potato and processing of its mRNA and protein. *Plant Mol. Biol.*, **8**, 119–207.

Cocking, E.C. and Davey, M.R. (1987) Gene transfer in cereals. *Science*, **236**, 1259–62.

Creech, R.G. (1968) Carbohydrate synthesis in maize. *Adv. Agron.*, **20**, 275–322.

Croy, R.R.D. and Gatehouse, J.A. (1985) Genetic engineering of seed proteins: current and potential applications, in *Plant Genetic Engineering*, (ed. J.H. Dodds), Cambridge University Press, Cambridge, pp. 143–268.

Croy, R.R.D., Lycett, G.W., Gatehouse, J.A. *et al.* (1982) Cloning, characterization and sequence analysis of cDNAs encoding pea storage protein precursors–legumin. *Nature*, **295**, 76–9.

de Clerq, A., Vandewiele, M., Van Damme, J. *et al.* (1990) Stable accumulation of modified 2S albumin seed storage proteins with higher methionine contents in transgenic plants. *Plant Physiol.*, **94**, 970–9.

de la Pena, A., Lorz, H. and Schell, J. (1987) Transgenic rye plants obtained by injecting DNA into young floral tillers. *Nature*, **325**, 274–5.

de Lumen, B.O. and Chan, J. (1986) Antinutritional and biochemical properties of winged bean trypsin inhibitors, in *Nutritional and Toxicological Significance of Enzyme Inhibitors in Foods*, (ed. M. Friedman), pp. 413–27.

de Lumen, B.O. and Kho, C.J. (1987) Identification of methionine containing proteins and quantitation of their methionine contents. *J. Agric Food Chem.*, **35**, 688–91.

Dey, P.M. (1985) D-Galactoside containing oligosaccharides, in *Biochemistry of Storage Carbohydrates in Green Plants*, (eds P.M. Dey, and R.A. Dixon) Academic Press, New York, pp. 53–129.

di Fonzo, N., Gentinetta, E., Salamini, F. and Soave, C. (1980) Action of the opaque-7 mutation on the accumulation of storage products in maize endosperm, *Plant Sci.*, **14**, 345–54.

Doll, H. (1984) Nutritional aspects of cereal proteins and approaches to overcome their deficiencies. *Phil. Trans. R. Soc. Lond.*, **B304**, 373–80.

Doll, H., Kole, B. and Eggum, B.O. (1974) Induced high-lysine mutants in barley, *Radiat. Bot.*, **14**, 73–80.

Domoney, C. and Casey, R. (1985) Measurement of gene number for seed storage proteins in *Pisum*. *Nucl. Acids Res.*, **13**, 687–99.

Duffus, C.M. and Slaughter, J.C. (1980) *Seeds and Their Uses*. Wiley, New York.

Dynan, W.S. and Tjian, R. (1985) Control of eukaryotic mRNA synthesis by sequence specific DNA binding proteins. *Nature*, **316**, 774–8.

Eggum, B.O. and Beames, R.M. (1983) The nutritive value of seed proteins, in *Seed Proteins: Biochemistry, Genetics, Nutritive Value*, (eds W. Gottschalk and H.P. Muller), Martinus Nijhoff, The Hague, pp. 499–531.

Elhussein, S.A., Miernyk, J.A. and Ohlrogge, J.B. (1987) A preliminary characterization of plant holo-acyl carrier protein synthase, in *The Metabolism, Structure, and Function of Plant Lipids*, (eds P.K. Stumpf, J.B. Mudd and W.D. Ness), Plenum Press, New York, pp. 709–13.

Elliston, K.O, Imran, S. and Messing, J. (1988) Cloned genes of storage proteins. *Plant Mol. Biol. Rep.*, **6**, 22–6.

Evans, I.M., Gatehouse, J.A., Croy, R.R.D. and Boulter, D. (1984) Regulation of transcription of storage protein mRNA in nuclei isolated from developing pea cotyledons. *Planta*, **160**, 559–68.

FAO (1957) *Protein Requirements FAO Nutr. Studies* No.16, FAO, Rome.

FAO (1980) *Food Balance Sheets, 1975–1977*. Food and Agriculture Organization of the UN.

FAO/WHO (1973) Energy requirements and protein requirements. WHO Tech Rep Ser No. 522.

FAO/WHO/UNU (1985) Energy and Protein Requirements. *WHO Technical Report Series* 724. WHO Geneva, Switzerland.

Fraley, R.T., Rogers, S.G. and Horsch, R.B. (1986) Genetic transformation in higher plants. *CRC Crit. Rev. Plant Sci.*, **4**, 1–46.

Fromm, M., Taylor, L.P. and Talbot, V. (1986) Stable transformation of maize after gene transfer by electroporation. *Nature*, **319**, 791–3.

Gasser, C.S. and Fraley, R.T. (1989) Genetically engineering plants for crop improvement. *Science*, **236**, 1293–9.

Gatehouse, J.A., Evans, I.M., Croy, R.R.D. and Boulter, D. (1986) Differential expression of genes during seed development. *Phil. Trans. R. Soc. Lond.*, **B314**, 367–84.

George, A. and de Lumen, B.O. (1991) A novel methionine rich protein in soybean seed: identification, amino acid composition and N-terminus sequence. *J. Agric. Food Chem.*, **39**, 224–70

Gobel, E. and Lorz, H. (1988) Genetic manipulation of cereals. *Oxford Surv. Plant Mol. Cell Biol.*, **5**, 1–22.

Goldberg, R.B. (1988) Plants: novel developmental processes. *Science*, **240**, 1460–6.

Goldberg, R.B. (1986) Regulation of plant gene expression. *Phil. Trans. R. Soc. Lond.*, **B314**, 343–53.

Goldberg, R.B., Hoschek, G., Tam, S.H. *et al.* (1981) Abundance, diversity and regulation of mRNA sequence sets in soybean embryogenesis. *Dev. Biol.*, **83**, 201–17.

Goldberg, R.B., Barker, S. and Perez-Grau, L. (1989) Regulation of gene expression during plant embryogenesis. *Cell*, **56**, 149–60.

Graham, J., Hall, G., Pearce, G. and Ryan, C.A. (1986) Regulation of synthesis of proteinase inhibitors I and II mRNAs in leaves of wounded tomato plants. *Planta*, **169**, 399–405.

Guerche, P., De Almeida, E.R.P., Schwartzstein, M.A. *et al.* (1990) Expression of the 2S albumin from *Berthellotia excelsa* in *Brassica napus*. *Mol. Gen. Genet.*, **221**, 306–14.

Guerra, D.J., Ohlrogge, J.B. and Frentzen, M. (1987) A possible differential role for plant acyl carrier protein isoforms in higher plants, in *The Metabolism, Structure and Function of Plant Lipids* (eds P.K. Stumpf, J.B. Mudd and W.D. Ness), Plenum Press, New York, pp. 705–8.

Gupta, M., Chourey, P.S., Burr, P. and Still, P. (1988) cDNAs of two non-allelic sucrose synthase genes in maize: cloning expression, characterization and molecular mapping of the sucrose synthase-2 gene. *Plant Mol. Biol.*, **10**, 215–24.

Gupta, Y.P. (1983) Nutritive value of food legumes, in *Chemistry and Biochemistry of Legumes*. (ed. S.K. Arora), Oxford and IBH Publishing Co., New Delhi, India.

Hagen, G. and Rubinstein, I. (1981) Complex organization of zein genes in maize. *Gene*, **13**, 239–49.

Hammond, R.W., Foard, D.E. and Larkins, B.A. (1984) Molecular cloning and analysis of a gene coding for the Bowman–Birk protease inhibitor in soybean. *J. Biol. Chem.*, **259**, 9883–90.

Harada, J.J., Barker, S.J. and Goldberg, R.B. (1989) Soybean β-conglycinin genes are clustered in several DNA regions and are regulated by transcriptional and posttranscriptional processes. *Plant Cell*, **1**, 415–25.

Hartings, H., Maddaloni, M., Lazzaroni, N. *et al.* (1989) The 02 gene which regulates zein protein deposition in maize endosperm encodes a protein with structural homologies to transcriptional activators. *EMBO J.*, **8**, 2795–801.

Higgins, T.J.V. (1984) Synthesis and regulation of major seed proteins in seeds.

Annu. Rev. Plant Physiol., **35**, 191–221.

Higuchi, M., Suga, M. and Iwai, K. (1983) Participation of lectin in biological effects of raw winged bean seeds on rats. *Agric. Biol. Chem.*, **47**, 1879–86.

Hildebrand, D.F. and Hymowitz, T. (1981) Two soybean genotypes lacking lipoxygenase. *J. Am. Oil Chem. Soc.*, **58**, 583–6.

Hildebrand, D.F., Altschule, M. *et al.* (1987) Physiological and transformational analyses of lipoxygenase, in *The Metabolism, Structure and Function of Plant Lipids*, (eds P.K. Stumpf, J.B. Mudd and W.D. Ness), Plenum Press, New York, pp. 715–17.

Hilder, V.A., Gatehouse, A.M.R., Sheerman, S.E. *et al.* (1987) A novel mechanism of insect resistance engineered into tobacco. *Nature*, **330**, 160–3.

Hofman, L.M., Donaldson, D.D. and Herman, E.M. (1988) A modified storage protein is synthesized, processed and degraded in the seeds of transgenic plants. *Plant Mol. Biol.*, **11**, 717–29.

Hol, W.G.J. (1987) Applying knowledge of protein structure and function. *Trends Biotechnol.*, **5**, 137–43.

Horsch, R.B., Fry, J.E. *et al.* (1985) A simple and general method for transferring genes into plants. *Science*, **227**, 1229–31.

Hugly, S., Kunst, L., Browse, J. and Somerville, C.R. (1989) Enhanced thermal tolerance of photosynthesis and altered choloplast ultrastructure in a mutant of *Arabidopsis* deficient in lipid desaturation. *Plant Physiol.*, **90**, 1134–42.

Hymowitz, T. (1987), increasing yields, food and nutritional quality through breeding: grain legumes, in *Cereals and Legumes in the Food Supply*, (eds J. Dupont and E.M. Osman) Iowa State Agriculture and Home Economics Experiment Station, Iowa State University, Ames, IA, (1990) *Economic Implications of Modified Soybean Traits*. Special Report 92, pp. 15–24. 88pp.

Jaynes, J., Yang, M., Espinoza, M. and Dodds, J. (1986) Plant protein improvement by genetic engineering: use of synthetic genes. *Trends Biotechnol.*, **4**, 314–20.

Jofuku, K.D. and Goldberg, R.B. (1989) Kunitz trypsin inhibitor genes are differentially expressed during the soybean life cycle and in transformed tobacco plants. *Plant Cell*, **1**, 1079–93.

Jofuku, K.D., Okamuro, J.K. and Goldberg, R.B. (1987) Interaction of an embryo DNA binding protein with a soybean lectin gene upstream region. *Nature*, **328**, 734–7.

Jones, R.A., (1978) Effects of floury-2 locus on accumulation and RNA metabolism during endosperm development. *Biochem. Genet.*, **16**, 27–38.

Jones, R.A., Larkins, B.A. and Tsai, C.Y. (1977) Storage protein synthesis in maize II. Reduced synthesis of a major zein component by the opaque-2 mutant of maize. *Plant Physiol.*, **59**, 525–9.

Jordano, J., Almoguera, C. and Thomas, T.L. (1989) A sunflower helianthin gene upstream sequence ensemble contains an enhancer and sites of nuclear protein interaction. *Plant Cell*, **1**, 855–66.

Joudrier, P.E., Foard, D.E., Floener, L.A. and Larkings, B.A. (1987) Isolation and sequence of cDNA encoding the soybean protease inhibitors PI IV and C-II. *Plant Mol. Biol.*, **10**, 35–42.

Juliano, B.O. (ed.) (1985) *Rice: Chemistry and Technology*. American Association of Cereal Chemists, St Paul, MN, pp. 1–16.

Kadonaga, J.T, Jones, K.A. and Tjian, R. (1986) Promoter-specific activation of RNA polymerase II transcription by Sp1. *Trends Biochem.*, **11**, 20–3.

Kandler, O. and Hopf, H. (1980) Metabolism and function of oligosaccharides, in *Biochemistry of Plants*, Vol. 3 Carbohydrates. (ed. J. Preiss), Academic Press, New York, pp. 231–70.

Kay, R., Chan, A., Daly, M. and McPherson, J. (1987) Duplication of CaMV 35S promoter sequences creates a strong enhancer for plant genes. *Science*, **236**, 1299–302.

Kho, C.J. and de Lumen, B.O. (1988) Identification and isolation of methionine–cysteine rich proteins in soybean seed. *Plant Foods Hum. Nutr.*, **38**, 287–96.

Kim, C.S., Kamiya, S., Kanamori, J. *et al.* (1990a) High level expression, purification and functional properties of soybean proglycinin from *Escherichia coli*. *Agric. Biol. Chem.* **54**, 1543–50.

Kim, C.S., Kamiya, S., Sato, T. *et al.* (1990b) Improvement of nutritional and functional properties of soybean glycinin through protein engineering. *Protein Eng.*, **3**, 725–31.

Kinsella, J. (1979) Functional properties of soy proteins. *J. Am. Oil Chem. Soc.*, **56**, 242–58.

Kitamura, K. (1984) Biochemical characterization of lipoxygenase lacking mutants, L-1 Less, L-2 Less, L-3 Less soybeans. *Agric. Biol. Chem.*, **48**, 2339–46.

Kitamura, K., Davies, C.S., Kaizuma, K. and Nielsen, N.C. (1983) Genetic analysis of a null allele for lipoxygenase-3 in soybean seeds. *Crop Sci.*, **23**, 924–7.

Klee, H., Horsch, R. and Rogers, S. (1987) *Agrobacterium*-mediated plant transformation and its further application to plant biology. *Annu. Rev. Plant Physiol.*, **38**, 467–86.

Kodrzycki, R., Boston, R.S. and Larkins, B.A. (1989) The opaque-2 mutation of maize differentially reduces zein gene transcription. *Plant Cell*, **1**, 105–14.

Kreis, M., Shewry, P.R., Forde, B.G. *et al.* (1985) Structure and evolution of seed storage proteins and their genes with particular reference to those of wheat, barley and rye. *Oxford Surv. Plant Mol. Cell Biol.*, **2**, 253–317.

Krishnan, H.B., Reeves, C.D. and Okita, T.W. (1986) ADP-Pyrophosphorylase is encoded by different mRNA transcripts in leaf and endosperm of cereals. *Plant Physiol.*, **81**, 642–5.

Kuhlemeir, C., Green, P.J., Chua, N.H. (1987) Regulation of gene expression in higher plants. *Annu. Rev. Plant Physiol.*, **38**, 221–466.

Kunst, L. and Somerville, C.R. (1989) A mutant of *Arabidopsis* deficient in desaturation of palmitic acid in leaf lipids. *Plant Physiol.*, **90**, 943–7.

Lawrence, M.C., Susuki, E., Varghese, J.N. *et al.* (1990) The three-dimensional structure of the seed storage protein phaseolin at 3 Å resolution. *EMBO J.*, **9**, 9–15.

Lee, J.L., Brown, W.E., Pearce, G. *et al.* (1986) Molecular characterization and phylogenic studies of a wound-inducible proteinase inhibitor I gene in *Lycopersicon* species. *Proc. Natl Acad. Sci. USA*, **83**, 7277–81.

Lee, P.D. and Su, J.C. (1982) Sucrose–starch transforming system in rice grains – a tracer feeding study. *Proc. Natl Sci. Counc. Rep. China*, **B6**, 189–96.

Liener, I.E. and Kakade, M.L. (1969) Protease inhibitors, in *Toxic Constituents of Plant Foodstuffs*, (ed. I.E. Liener), Academic Press, New York.

Lilley, G.G., Caldwell, J.B. and Kortt, A.A. (1989) Isolation and primary structure for a novel, methionine-rich protein from sunflowerseeds (*Helianthus annuus* L.) *Proceedings of the World Congress on Vegetable Protein Utilization in Human Foods and Animal Feedstuffs*, (ed. T.H. Applewhite), pp. 497–502.

Lin, Y.L. (1986) Genetic engineering and process development for production of food processing enzymes and additives. *Food Technol.*, **40**, 104–6.

Luo, Z. and Wu, R. (1988) A simple method for the transformation of rice via the pollen tube pathway. *Plant Mol. Biol. Rep.* **6**, 165–74.

Maier, U.G., Brown, J.W.S. and Faix, G. (1987) Binding of a nuclear factor to a consensus sequence in the 5′ -flanking regions of zein genes from maize. *EMBO J.*, **6**, 17–22.

Manners, D.J. (1988) Starch, in *Biochemistry of Storage Carbohydrates in Plants*, (eds P.M. Dey and R.A. Dixon), Academic Press, London, pp. 149–203.

Marks, M.D., Lindell, J.S. and Larkins, B.A. (1985) Quantitative analysis of the accumulation of zein mRNA during maize endosperm development. *J. Biol. Chem.*, **260**, 16445–50.

McCabe, D.E., Swain, W.F. *et al.* (1988) Stable transformation of soybean (*Glycine max*) by particle acceleration. *Bio/Technology*, **6**, 923–6.

McCarty, D., Shaw, J. and Hannah, L. (1986) The cloning, genetic mapping, and expression of the constitutive sucrose synthase locus of maize. *Proc. Natl Acad. Sci. USA*, **83**, 9099–103.

Mertz, E.T., Bates, L.S. and Nelson, O.E. (1964) Mutant gene that changes protein composition and increases lysine content of maize endosperm. *Science*, **145**, 279–80.

Mertz, E.T., Veron, O.A., Bates, L.S. and Nelson, O.E. (1965) Growth of rats fed on opaque-2 maize. *Science*, **148**, 1741–2.

Motto, M., Maddaloni, M., Onziani, G, *et al.* (1988) Molecular cloning of the o2-m5 allele of *Zea mays* using transposon marking. *Mol. Gen. Genet.*, **212**, 488–94.

Muntz, K. (1989) Intracellular protein sorting and the formation of protein reserves in storage tissue cells of plant seeds. *Biochem. Physiol. Pflanzen.*, **185**, 315–35.

Nakai, S. and Powrie, W.D. (1981) Modifications of proteins for functional and nutritional improvements, in *Cereals – A Renewable Resource, Theory and Practice*, (eds. Y. Pomeranz and L. Munck), American Association of Cereal Chemists, St Paul, MN, pp. 217–42.

National Research Council (1988) *Quality Protein Maize.* National Academy Press, Washington, DC, p.6.

Nelson, O.E. and Rines, H.W. (1962) The enzymatic deficiency in waxy mutants of maize. *Biochem. Biophys. Res. Commun.*, **9**, 297–300.

Neuhaus, G., Spangenberg, G. *et al.* (1987) Transgenic rapeseed plants obtained by the microinjection of DNA into microspore-derived embryoids. *Theor. Appl. Genet.*, **74**, 30–6.

Nielsen, N.C., Dickinson, C.D., Cho, T.J. *et al.* (1989) Characterization of the glycinin gene family in soybean. *Plant Cell.*, **1**, 313–28.

Odigboh, E.U. (1983) Cassava: production, processing and utilization, in *Handbook of Tropical Foods.* (ed. H.T. Chan), Marcel Dekker, New York, pp. 145–202.

Ohlrogge, J.B., Beremand, P.D., Hannapel, D.J. *et al.* (1987) Acyl carrier protein as a probe of the molecular biology of plant fatty acid synthesis, in *The Metabolism, Structure, and Function of Plant Lipids*, (eds. P.K. Stumpf, J.B. Mudd and W.D. Ness), Plenum Press, New York, pp. 689–95.

Okamuro, J.K., Jofuko, K.D. and Goldberg, R.B. (1986) Soybean lectin gene and flanking non-seed protein genes are developmentally regulated in transformed tobacco plants. *Proc. Natl Acad. Sci. USA*, **83**, 8240–4.

Okuno, K. (1976) A low-amylose mutant of rice. *Div. Genet. Natl. Inst. Agric. Sci. Japan. Ann. Rep.*, **1**, 28–9.

Ovcharov, K.G. and Koshelev, Y.P. (1974) Sugar content in corn seeds with different viability. *Fiziol. Rast.*, **21**, 969–4.

Payne, P.I., Corfield, K.G., Holt, L.M. and Blackman, J.A. (1981) Correlations between the inheritance of certain high-molecular-weight subunits of glutenin and breadmaking quality in progenies of six crosses of bread wheat. *J. Sci. Food Agric.*, **32**, 51–60.

Pedersen, K., Deverux, J, Wilson, D. *et al.* (1982) Cloning and sequence analysis reveal structural variation among related zein genes in maize. *Cell*, **29**, 1015–26.

Perez, C.M., Cagampang, G.B., Esmana, B.V. *et al.* (1973) Protein metabolism in leaves and developing grains of rices differing in grain protein content. *Plant Physiol.*, **51**, 537–42.

Perez-Grau, L. and Goldberg, R.B. (1989) Soybean seed protein genes are regulated spatially during embryogenesis. *Plant Cell*, **1**, 1095–109.

Potrykus, I. (1990) Gene transfer to plants: assessment and perspectives. *Physiol. Plant.*, **79**, 125–34.

Preiss, J. (1982) Regulation of the biosynthesis and degradation of starch. *Annu. Rev. Plant Physiol.*, **33**, 431–54.

Rackis, J. (1981) Flatulence caused by soya and its control through processing. *J. Am. Oil Chem. Soc.*, **58**, 503–9.

Raven, P.H., Evert, R.F. and Eichhorn, S.E. (1987) *Biology of Plants*, 4th edn, Worth Publishers, New York, pp. 373–83.

Riggs, C.D., Voelker, T.A. and Chrispeels, M.J. (1989) Cotyledon nuclear proteins bind to DNA fragments harboring regulatory elements of phytohemagglutinin genes. *Plant Cell*, **1**, 609–21.

Roxas, B.V., Intengan, C. Ll. and Juliano, B.O. (1979) Protein quality of high protein and low protein milled rices in preschool children. *J. Nutr.*, **109**, 832–9.

Ryan, C.A. (1981) Proteinase inhibitors, in *The Biochemistry of Plants. A Comprehensive Treatise*, (eds P.K. Stumpf and E.E. Conn), Vol.6, Academic Press, New York, pp. 351–71.

Ryan, C.A. 1981 Proteinase inhibitor gene families: strategies for transformation to improve plant defences against herbivores. *Bioessays*, **10**, 20–4.

Sanchez-Serrano, J.J., Keil, M., O'Connor, A. *et al.* (1987) Wound-induced expression of a potato proteinase inhibitor II gene in transgenic plants. *EMBO J.*, **6**, 303–6.

Saravitz, D.M., Pharr, D.M. and Carter, T.E. (1987) Galactrinol synthase activity and soluble sugars in developing seeds of four soybean genotypes. *Plant Physiol.*, **83**, 185–9.

Satoh, H. and Omura, T. (1981) New endosperm mutations induced by chemical mutagens in rice, *Oryza sativa. Jpn J. Breeding*, **31**, 316–26.

Schell, J. (1987) Transgenic plants as tools to study the molecular organization of plant genes. *Science*, **237**, 1176–83.

Schmidt, R.J., Burr, F.A. and Burr, B. (1987) Transposon tagging and molecular analysis of the maize regulatory locus opaque-2. *Science*, **238**, 960–3.

Schmidt, R.J., Burr, F.A., Aukerman, M.J. and Burr, B. (1990) Maize regulatory gene opaque-2 encodes a protein with 'leucine-zipper' motif that binds to zein DNA. *Proc. Natl Acad. Sci. USA*, **87**, 46–50.

Sheldon, E., Ferl, R., Fedoroff, N. and Hannah, L.C. (1983) Isolation and analysis of a genomic clone encoding sucrose synthetase in maize: evidence for two introns in Sh. *Mol. Gen. Genet.*, **190**, 421–6.

Shewry, P.R. and Miflin, B.J. (1984) Seed storage proteins of economically important cereals. *Adv. Cereal Sci. Technol.*, **7**, 1–83.

Shewry, P.R., Halford, N.G. and Tatham, A. (1989) The high molecular weight subunits of wheat, barley, and rye: genetics, molecular biology, chemistry and role in wheat gluten structure and functionality. *Oxford Surv. Plant Mol. Cell. Biol.*, **6**, 163–219.

Shimamoto, K., Terada, R., Izawa, T. and Fujimoto, H. (1989) Fertile transgenic rice plants regenerated from transformed protoplasts. *Nature*, **337**, 274–6.

Singh, R. and Axtell, J.D. (1973) High lysine mutant gene (hi) that improves protein quality and biological value of grain sorghum, *Crop Sci.*, **13**, 535–9.

Slabas, A.R., Harding, P. *et al.* (1987) Oil seed acyl carrier protein (ACP) protein and gene structure, in *The Metabolism Structure, and Function of Plant Lipids*, (eds P.K. Stumpf, J.B. Mudd and W.D. Ness), Plenum Press, New York, pp. 697–700.

Slightom, J.L., Drong, R.F., Klassy, R.C. and Hoffmann, L.M. (1985) Nucleotide sequences from phaseolin cDNA clones: the major storage protein from *Phaseolus vulgaris* are encoded by two unique gene families. *Nucl. Acids Res.*, **13**, 6483–98.

Slightom, J.S., Sun, S.M. and Hall, T.C. (1983) Complete nucleotide sequence of a French bean storage protein gene: phaseolin. *Proc. Natl Acad. Sci.*, **80**, 1897–901.

Soave, C. and Salamini, F. (1984) Organization and regulation of zein genes in maize endosperm. *Phil. Trans. R. Soc. Lond.*, (B) **304**, 341–7.

Soave, C., Tardani, L., di Fonzo, N. and Salamini, F. (1981) Zein level in maize endosperm depends on a protein under control of the opaque-2 and opaque-6 loci. *Cell*, **27**, 403–10.

Somerville, C.R., McCourt, P., Kunst, L. and Browse, J. (1987) Mutants of *Arabidopsis* deficient in fatty acid desaturation, in *The Metabolism, Structure and Function of Plant Lipids* (eds P.K. Stumpf, J.B. Mudd and W.D. Ness), Plenum Press, New York, pp.669–76.

Soya Bluebook (1982) American Soybean Association, St Louis, MO.

Spencer, D. and Higgins, T.J.V. (1982) Seed maturation and deposition of storage proteins, in *Molecular Biology of Plant Cells*, (ed. H. Smith), Botanical Monographs, Vol.14, 306–36.

Start, W.G., Ma, Y., Polacco, J.C. *et al.* (1986) Two soybean seed lipoxygenase nulls accumulate reduced levels of lipoxygenase transcripts. *Plant Mol. Biol.*, **7**, 11–23.

Sun, S.M., Slightom, J.L. and Hall, T.C. (1981) Intervening sequences in a plant gene: comparison of the partial sequence of cDNA and genomic DNA of French bean phaseolin. *Nature*, **289**, 37–41.

Takaiwa, F., Kikuchi, S. and Oono, K. (1986) The structure of rice storage glutelin precursor deduced from cDNA. *FEBS Lett.*, **206**, 33–5.

Takaiwa, F., Kikuchi, S. and Oono, K. (1987a). A rice glutelin gene family – a major type of glutelin mRNAs can be divided into two classes. *Mol. Gen. Genet.*, **208**, 15–22.

Takaiwa, F., Ebinuma, H., Kikuchi, S. and Oono, K. (1987b) Nuleotide sequence of a rice glutelin gene. *FEBS Lett.*, **221**, 43–7.

Talbot, D.R., Adang, M.J., Slightom, J.L. and Hall, T.C. (1984) Size and organization of a multigene family encoding phaseolin, the major seed storage protein of *Phaseolus vulgaris*. *Mol. Gen. Genet.*, **198**, 42–9.

Taliercio, E.W. and Chourey, P.S. (1989) Post-transcriptional control of sucrose synthase expression in anaerobic seedlings of maize. *Plant Physiol.*, **90**, 1359–64.

Thornburg, R.W., An, G., Cleveland, T.E. *et al.* (1987) Wound inducible expression of a potato inhibitor I-chloramphenical acetyl transferase gene fusion in transgenic tobacco plants. *Proc. Natl Acad. Sci. USA*, **84**, 744–8.

Toenniessen, G.H. (1985) Plant molecular biology and the international agricultural research system. *Plant Mol. Biol. Rep.*, **3**, 1–11.

Topfer, R., Gronenborn, B. *et al.* (1989) Uptake and transient expression of chimeric genes in seed derived embryos. *Plant Cell*, **1**, 133–9.

Tsai, C.Y. and Nelson, O.E. (1966) Starch deficient maize mutants lacking ADP-glucose pyrophosphorylase activity. *Science*, **151**, 341–3.

Uchimiya, H., Fushimi, T. *et al.* (1986) Expression of a foreign gene in callus derived from DNA-treated protoplasts of rice (*Oryza sativa*). *Mol. Gen. Genet.*, **204**, 204–7.

Uchimiya, H., Handa, T. and Brar, D.S. (1989) Transgenic plants. *J. Biotechnol.*, **12**, 1–20.

Villegas, E., Ortega, E. and Bauer, R. (1984) *Chemical Methods Used at CIMMYT for Determining Protein Quality*, CIMMYT, Mexico.

Viotti, A., Abildsten, D., Pogna, N. *et al.* (1982) Multiplicity and diversity of cloned zein cDNA sequences and their chromosomal localization. *EMBO J.*, **1**, 53–8.

Walling, L., Drews, G.N. and Goldberg, R.B. (1986) Transcriptional and posttranscriptional regulation of soybean seed protein mRNA levels. *Proc. Natl Acad. Sci. USA*, **83**, 2123–7.

Weising, K., Schell, J. and Kahl, G. (1988) Foreign genes in plants: transfer, structure, expression and application. *Annu. Rev. Genet.*, **22**, 421–77.

Wienand, U. and Feix, G. (1980) Zein specific restriction enzyme fragments of maize DNA. *FEBS Lett.*, **116**, 14–16.

Williamson, M.S., Forde, J., Buxton, B. and Kreis, M. (1987) Nucleotide sequence of barley chymotrypsin inhibitor-2 (CL-2) and its expression in normal and high-lysine barley. *Eur. J. Biochem.*, **165**, 99–106.

Wright, D.J. (1988) The seed globulins – Part II, in *Development in Food Proteins*, (ed. B.J.F. Hudson), Elsevier Applied Science Publishers, London, Vol. 6, 119–77.

Yang, M.S., Espinoza, N.O., Nagpala, P.G. *et al.* (1989) Expression of a synthetic gene for improved protein quality in transformed potato plants. *Plant Sci.*, **64**, 99–111.

Yano, M., Okuno, K., Kawakami, J. *et al.* (1985) High amylose mutants of rice, *Oryza sativa*. *Theor. Appl. Genet.*, **69**, 253–7.

Author index

Subject index

MIX
Papier aus verantwortungsvollen Quellen
Paper from responsible sources
FSC® C105338

If you have any concerns about our products,
you can contact us on
ProductSafety@springernature.com

In case Publisher is established outside the EU,
the EU authorized representative is:
Springer Nature Customer Service Center GmbH
Europaplatz 3, 69115 Heidelberg, Germany

Printed by Libri Plureos GmbH
in Hamburg, Germany